Manual of Enviromental Management

'I have been successfully using Adrian's materials for many years to train people from a variety of industries and I am pleased to see that he has now made this information available in this useful book. I highly recommend this book for practitioners who are tasked with implementing and operating environmental management systems within their place of work. It will be particularly useful to help assess the appropriate actions needed to avoid the risk of pollution to the environment and damage to the company reputation.'

Tony Cullen, Full Member and Environmental Auditor,
Institute of Environmental Management and Assessment

Manual of Environmental Management is a practical guide for those involved in the control and reduction of environmental impacts in organisations. This comprehensive and practical guide takes you through the main environmental challenges organisations face and the improvement strategies used to manage them.

Chapter by chapter, *Manual of Environmental Management* discusses the fundamental issues and principles surrounding environmental policy, law and management and provides crucial information on how to respond and implement environmental programmes.

This book is the perfect reference tool for the environmental professional and an invaluable study text for those preparing for professional examinations such as the NEBOSH Environmental Diploma and IEMA Associate Membership.

▶ Complete with images and diagrams throughout to support understanding
▶ Written by an expert in the field of environmental management
▶ Suitable for study and professional use.

Adrian Belcham is a Chartered Environmentalist, full member of the Institute of Environmental Management and Assessment, and an EARA accredited Environmental Auditor. Since 1998 he has been lead tutor of the UK-based training organisation Cambio Environmental Ltd, which specialises in the delivery of IOSH, IEMA and NEBOSH environmental qualifications. This book is based in large part on the author's 15 years' experience of communicating this broad and multi-faceted subject to a wide variety of students in an accessible and practical manner.

Manual of Environmental Management

Adrian Belcham

Routledge
Taylor & Francis Group

LONDON AND NEW YORK

First published 2015
by Routledge
2 Park Square, Milton Park, Abingdon, Oxon OX14 4RN

and by Routledge
711 Third Avenue, New York, NY 10017

Routledge is an imprint of the Taylor & Francis Group, an informa business

British Library Cataloguing-in-Publication Data
A catalogue record for this book is available from the British Library

Library of Congress Cataloging-in-Publication Data
Belcham, Adrian.
Manual of environmental management / Adrian Belcham.
pages cm
Includes bibliographical references and index.
1. Environmental management. 2. Management–Environmental aspects. 3. Environmental sciences–Information resources. 4. Environmental policy. 5. Sustainable development. I. Title.
GE300.B45 2014
333.7–dc23
2014009633

ISBN: 978–1–138–01466–4 (hbk)
ISBN: 978–1–315–77946–1 (ebk)

Typeset in Univers LT by
Servis Filmsetting Ltd, Stockport, Cheshire

Printed and bound in Great Britain by
TJInternational Ltd, Padstow, Cornwall

Dedication

This book is dedicated to all the students who have attended courses with Cambio Environmental Ltd over the years.

Contents

Contents

Contents

Contents

Contents

Contents

Contents

Figures

Figures

Tables

Tables

Preface

This Manual is the result of more than 15 years of teaching environmental management principles and practices to students from a variety of industries and professional backgrounds. It is an attempt to offer the practitioner a practical yet comprehensive reference that can be used as a reliable first point of inquiry for questions relating to the environmental management in organisations. Most other volumes available in this subject area are either basic introductory texts or specialised, academic or highly technical volumes. Our goal is to provide a text that is detailed enough to give a good overview of environmental and sustainability issues and is written in accessible language and presented from a practitioner's viewpoint. Each chapter begins with a contents summary and chapter index for ease of navigation. At the end of each chapter is a set of links and references for those readers wanting to delve deeper into the topic area covered.

While primarily intended as a practitioner reference, we have organised the contents of this Manual in such a way as to facilitate its use as a study text for the IEMA Associate Membership examination. Readers intending to use it for this purpose are recommended to also view the companion study guide volume. It will also prove to be a valuable resource to those studying for the NEBOSH Diploma in Environmental Management.

Overview of the book

Chapter 1 provides an overview of the key environmental and sustainability issues and principles that are being considered at international, national and local levels.

Chapter 2 focuses on environmental policy at an international, regional and national level and considers examples of policy instruments and key principles.

Chapter 3 is an introduction to environmental law operational at an international, European and UK national level.

Chapter 4 sets out environmental management and sustainable development in a business context, identifying the role of and implications for organisations.

Chapter 5 describes the key approaches used to collect, analyse and report on environmental information and data. Technical assessment methods and sampling and analysis methods are described in an approachable and non-specialised manner.

Chapter 6 is an important chapter providing a comprehensive overview of key environmental management and assessment tools used to help assess, prioritise and manage an organisation's impact.

Chapter 7 is about practical responses to consider and identify sustainable solutions to problems and opportunities in relation to both pollution and resource consumption impacts.

Chapter 8 sets out the general principles used in developing and implementing programmes to deliver environmental performance improvement.

Chapter 9 focuses on communicating strategies to ensure effective engagement with internal and external stakeholders.

Chapter 10 considers some ideas on embedding change and influencing behaviour to improve sustainability in an organisation over both the short and long term.

At the back of the book is a Glossary of terms.

Acknowledgements

I'd like to express my thanks to those who have generously contributed time and advice in the development and writing of this book. In particular, Helen Mann, Julian Hill, Tony Cullen and most importantly, Julia Hill, without whom this project and so many others, would never have made it to completion.

CHAPTER 1

Understanding environmental and sustainability principles

Chapter summary

This chapter begins by exploring key definitions and concepts from the natural sciences that relate to environmental management. The activity–aspect–impact model is introduced as a standard way to relate to the interactions between an organisation and the wider environment.

An overview is then provided of key environmental problems considered at a variety of levels and scales, from global to local, and covering both pollution and resource consumption issues. This section may be considered as both a 'state of the environment' overview and as a summary of impact categories that an organisation may be linked to, directly or indirectly, as a result of its activities.

Finally, we explore the concept of sustainability and highlight its place both in terms of the international agreements that gave rise to it and in relation to the longer-term thinking that is emerging at the forefront of government policy and business planning.

I ENVIRONMENTAL PROCESSES AND PRINCIPLES

Environmental issues are intrinsically important to everyone. Even talking about 'the environment' as if it were something separate from us makes no sense, in fact. With every breath, movement, meal and action, we interact with other elements of what we refer to as the environment. Whether considering each of us as individuals, business organisations, industries or public bodies, we are all reliant upon the environment as the basis of our being. In the sixth century BCE, the classic Chinese philosophical text, the *Tao Te Ching*, captured this truth in a typically eloquent manner that can only be more relevant today than it was then:

> Many people with influence and wealth
> treat the earth as something to be owned.
> To be used and abused to suit their own ends.
>
> But the earth is a living being,
> a great spiritual source.
> To disregard this source
> is to call forth catastrophe,
> since all creatures great and small
> are an inherent and interdependent part
> of this very being.
> (Inspired by Verse 29, We are the World,
> trans. R. A. Dale, 2002)

1.1 What do we mean by the term 'environment'?

As indicated above, in its broadest sense the environment is: 'Everything, including human beings – the whole of the planet acting as a linked and interdependent whole.' That can be a challenging definition for many people, so in an attempt to be specific, from a business management perspective, the International Standards Organisation (ISO) defines the environment as: 'Surroundings in which an organisation operates including air, water, land, natural resources, flora, fauna, humans and their inter-relationships.'

However we choose to define it, pressures on the 'environment' are increasing because of continuing social developments, for example, the enhanced production levels of food and material things as a consequence of the great increase in the world population, which passed the 7 billion mark in 2011. The world population has more than doubled since the 1950s when it was only approximately 2.6 billion, and if current trends continue, it is expected to exceed 9 billion by 2050 (Population Reference Bureau, 2013). The combination of population growth, technological development and the higher human aspirations for material wealth, raises the pressure on the

3

Earth's ecosystems, with increasing demands for resources and increasing production of materials to be disposed of as waste.

1.1.1 Ways of describing the environment

The environment is often described in terms of three separate but intimately linked component parts as shown in Figure 1.1. The interaction between these three component parts comprises the local 'environment' wherever we look. An impact on one of the three elements will normally have 'knock-on' consequences or 'secondary impacts' in one or more of the other elements. For example, the discharge of effluent into a river has an impact on the chemical qualities of the water (physical environment), which, if significant, may lead to a fish kill (biological environment) which may mean local angling groups lose their recreational resource (human environment).

In addition, the natural world is described in terms of ecosystems or habitats. The *habitat* of any particular species is the combination of physical and biological conditions in which it exists. Without such conditions the species cannot exist – hence we read about the threat of extinction to animals such as tigers arising, not only from direct pressures such as poaching, but also through 'loss of available habitat'.

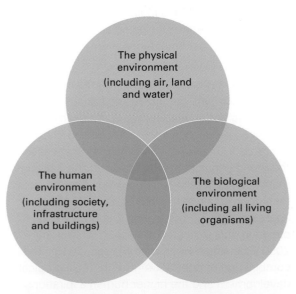

Figure 1.1 Components of the environment

Ecosystems comprise communities of interdependent organisms and the physical environment that they inhabit. Although cyclical change is frequently observed in natural ecosystems, human interference (through pollution and/or resource depletion) threatens their sustainability in many parts of the world. Examples of ecosystem types include woodlands, lakes, rivers, moorland, coral reefs, estuaries, etc.

Finally, the term *biosphere* is sometimes used to describe the 'zone of terrestrial life'. The term includes the Earth's land and water surfaces, plus the lowest part of the atmosphere and the upper part of the soil and water layers and all living things within these zones. The *Gaia Theory*, proposed by scientist James Lovelock in the 1970s, which is reminiscent of the passage from the *Tao Te Ching* quoted earlier, compares the whole of the biosphere to a single living organism. It argues that each component of the biosphere, whether considered at a local or global scale, contributes to and depends upon, the existence and stability of the rest of the system.

1.1.2 Categories of environmental impacts

At the simplest level, environmental impacts fall into two broad categories:

▶ pollution;
▶ resource depletion.

Each may be defined as follows:

▶ *Pollution* – the introduction by man of substances or energy into the environment that are liable to cause hazards to human health, harm to ecological systems, damage to structures or amenity, or interference with legitimate uses of the environment.
▶ *Resource depletion* – the consumption of natural resources that are either finite in their existence (non-renewable resources) or are managed in such a way as to permanently deplete potentially renewable biological or physical resources. Examples of non-renewable resources include fossil fuels and mineral ores. Examples of resources which may be

1 Environmental and sustainability principles

indefinitely renewed if managed/consumed appropriately include fish stocks and timber.

It should be noted that environmental impacts may be considered positive or negative and that while we often tend to focus on the negative impacts, actions may be taken which result in positive consequences in the human, biological or physical environment.

1.1.3 The process of pollution: the source–pathway–receptor model

When considering the effects of pollution, it is useful to classify the process of pollution as shown in Figure 1.2. Without the presence of all three elements of the source–pathway–receptor model, pollution cannot occur. Traditionally pollution control techniques have focused on interrupting the pathway between source and receptor, e.g. acoustic cladding around noisy plant. More recently there has been an increase in interest in addressing the sources of pollution as being a more permanent and effective form of control, e.g. the replacement of solvent-based inks by water-based inks, thereby eliminating completely the source of air pollution.

The effects of pollutants on the receptors can be further classified as acute or chronic. These are characterised as shown in Table 1.1.

Table 1.1 Acute and chronic pollution impacts

Acute pollution impacts	Chronic pollution impacts
Follow fairly immediately after exposure	Develops sometime after the exposure (often long term)
Generally clear-cut (sometimes fatal)	Less clear-cut (may ultimately cause death, otherwise leads to impairment in growth function or behaviour)
Rarely reversible	
	May be reversible

The scale of pollution can also vary from highly localised to truly global. For example, emissions of oxides of nitrogen from vehicle exhausts may cause localised air quality reduction with attendant human health problems. The same pollutants, however, also contribute to global climate change as part of the cumulative result of fossil fuel combustion. It is therefore important when considering pollution to consider the pollutant not only in relation to local ecosystems but also as part of an overall regional or even global burden upon the environment.

1.1.4 Resource depletion

It is important to note that pollution is not the only way in which human activity can impact upon the environment. Consumption of natural resources such as groundwater or surface water, forests, animal species and soils may all result in both acute and chronic environmental effects.

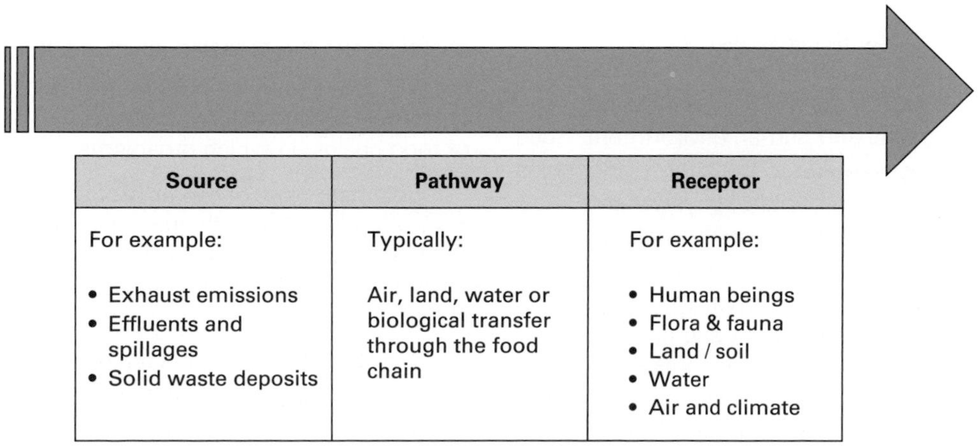

Source	Pathway	Receptor
For example:	Typically:	For example:
• Exhaust emissions • Effluents and spillages • Solid waste deposits	Air, land, water or biological transfer through the food chain	• Human beings • Flora & fauna • Land / soil • Water • Air and climate

Figure 1.2 The source–pathway–receptor model

As indicated in the definitions above, the issue here is not the consumption of resources per se – that is a natural and necessary part of existence. The issue is the scale and sustainability of rates of consumption. It is not feasible to manage non-renewable resources such as fossil fuels, which are produced over geological timescales, in such a way as to enable them to replenish at the same rate as consumption. All we can attempt to do is to use such resources in the most efficient manner possible with a consumption perspective that is appropriate to the period of their production. Such a perspective would surely call into question the wisdom of using fossil fuels in such applications as one-trip packaging.

As far as renewable resources are concerned, impact is created by the pattern of consumption and/or management of the resource in question. Fish stocks harvested at a level that allows populations to remain constant may be considered a neutral environmental impact. Over-fishing, on the other hand, will eventually lead to population decline and even species extinction with many knock-on food chain impacts. This example illustrates that, while all biological resources, as well as physical resources such as water and soil, are 'potentially renewable', only those that are managed appropriately will actually be 'renewable'.

1.2 Natural cycles and systems

1.2.1 The importance of linkages

When considering environmental impacts, it is critical to recognise the potential for 'knock-on effects' (sometimes called secondary impacts) between different parts of the local or wider physical, biological and human environment. Such impacts are possible due to the inter-relationships between different elements of physical and biological systems. These systems include the natural processes that recycle nutrients in various chemical forms in a cyclic manner from the non-living environment to living organisms and back again. The resultant impacts may be on the system itself or have wider implications for human or natural communities. Consider the examples shown below.

1.2.2 The hydrological cycle

Figure 1.3 shows the cyclical movement of water through different elements of the environment that is referred to as the hydrological cycle. Transfer rates within the cycle vary widely from very slow, e.g. groundwater flows, to very fast, e.g. rainfall. Storage times may be long, e.g. groundwater, or short, e.g. atmospheric water vapour. Changes in one part of the cycle can have significant consequences in others – consider the effect of deforestation in the higher reaches of the river catchment. With reduced vegetation cover, precipitation is able to run off into the river at a faster rate. Also less water is removed from the run-off process through plant absorption and transpiration. The result – increased erosion in the upper catchment and increased downstream flood levels.

1.2.3 The food chain/energy cycle

A similar series of linkages exists within ecosystems in the form of food chains (see Figure 1.4). An impact on one part of the system can have profound knock-on impacts elsewhere in the ecosystem. Such impacts may simply be to do with the availability of food resources, thus, reducing a population of plants will affect herbivores reliant on them as a food source, with knock-on consequences in carnivore groups, and so on. In addition, bioaccumulation is the build-up of certain pollutants in the bodies of organisms, often in specific locations such as bones, fatty tissues and organs such as the liver and kidneys. The impact of bioaccumulation varies with the chemical and the organism involved, but it can prevent reproduction or increase susceptibility to disease or even cause death. The dynamics of food chains, in which organisms at a specific level in the chain consume large numbers of the organisms in the preceding level, ensure that the products of bioaccumulation are amplified (biological amplification). Thus the chemical levels in the tissues of predators at the head of a food chain may be as many as 10 million times that in tissues of the producer organisms at the base of the chain.

One of the earliest recognised examples occurred in birds of prey populations in the southern USA

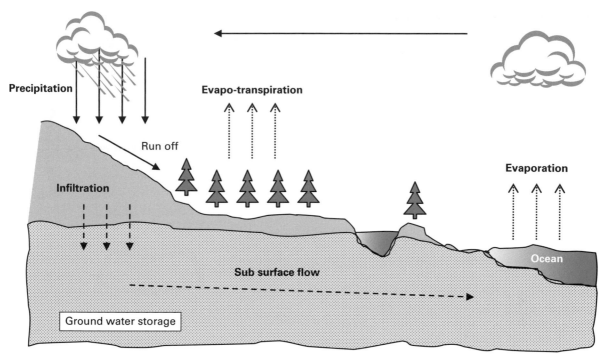

Figure 1.3 The hydrological cycle

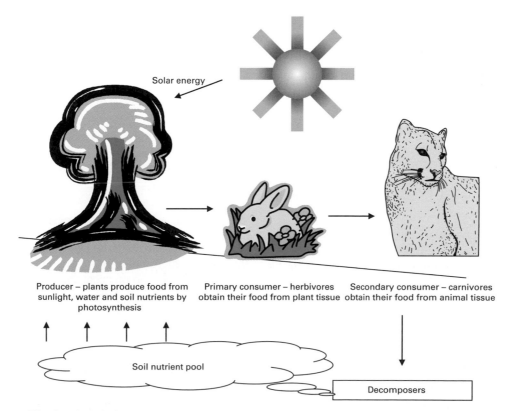

Figure 1.4 The food chain/energy cycle

in the 1960s, due to bioaccumulation of DDT pesticides, which though they remained at sub-lethal doses in the birds, severely impeded their ability to reproduce and hence led to a severe decline in populations. Although DDT has been eliminated in most parts of the world, other persistent pollutants remain a problem, as they are stored in the body tissues of organisms exposed to them and can therefore pass up the food chain, e.g. heavy metal concentrations in tuna.

1.2.4 The carbon cycle

As with water in the hydrological cycle and energy within the energy cycle described above, carbon moves between the physical and biological environment in a natural flow. In the energy cycle, in addition to the transfer of energy between different elements within the system, carbon is also transferred. Plants fix atmospheric carbon (in the form of carbon dioxide) during the process of photosynthesis. This carbon is then stored in the biomass of the plants during their life. Through the food chain, herbivores obtain carbon by eating plants and carnivores by eating herbivores. When the plants and animals within the food chain die, decomposition leads to the release of carbon to the atmosphere in the form of carbon dioxide and methane.

An equivalent carbon cycle also occurs within the water environment (particularly the world's oceans). Carbon dioxide is soluble in sea water which absorbs it from the atmosphere. A marine life cycle equivalent to the land-based version described above then cycles the carbon within the oceans.

Other key processes within the carbon cycle are respiration and combustion, both of which release carbon in the form of carbon dioxide. In both the marine and land-based ecosystems, plant respiration and animal respiration add carbon dioxide to the ocean or atmosphere. Combustion involves the conversion of carbon-based material (whether biomass, e.g. wood, or fossil fuels, e.g. coal) into energy (heat), water and carbon dioxide.

To simplify the processes described above, we may think of the carbon cycle as being represented by a number of interconnected

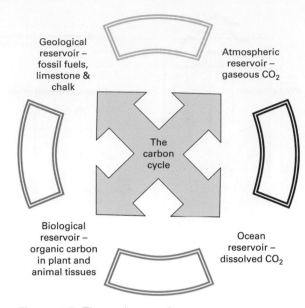

Figure 1.5 The carbon cycle

'reservoirs', each holding a quantity of carbon as shown in Figure 1.5. This cycle is normally considered to be self-regulating over geological time scales. However, in the short term, the release of carbon from the geological and biological reservoirs, through human activities in the form of deforestation and the combustion of fossil fuels, appears to be creating a significant increase in concentrations in the atmospheric reservoir. This is because combustion processes are releasing carbon dioxide faster than it can be absorbed into the biological, oceanic and geological reservoirs. The consequences of such an imbalance are discussed in the climate change sections.

1.2.5 The nitrogen cycle

The nitrogen cycle is the term used to describe the storage and transfer of nitrogen between different parts of the physical and biological environment. Figure 1.6 shows a simple representation of the key components or reservoirs.

Nitrogen is a key component of the Earth's atmosphere and in this reservoir appears as nitrogen gas. It is a key component in the cells and tissues of living organisms (the biological reservoir). It is also found extensively in soils

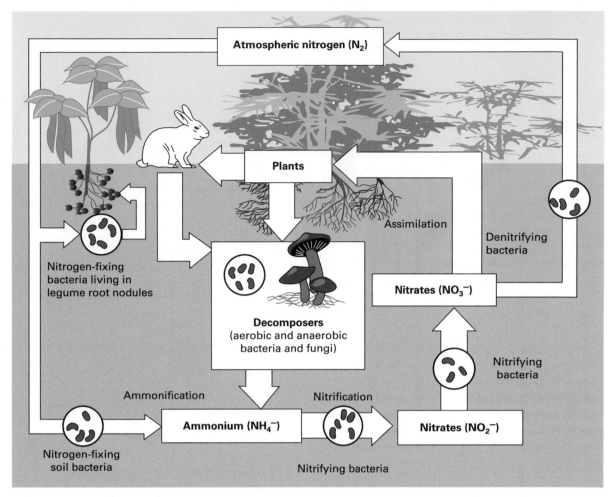

Figure 1.6 The nitrogen cycle
Source: US Environmental Protection Agency.

and the surface and groundwater environment in various forms, including ammonium, nitrites and nitrates. Transfer between the different reservoirs is essentially driven by the biological processes of assimilation and decomposition in the food chain.

Human activity can affect the nitrogen cycle in a number of ways including:

▶ *Pollution impacts in the aquatic environment*, e.g. the run-off of nitrate fertilisers from farmland into rivers and streams, or the discharge of sewage effluent from sewage treatment works both lead to increased nitrates in the water environment.
▶ *Soil nitrogen depletion* through forest clearance and poor agricultural practices.
▶ *Atmospheric and aquatic pollution* through the release of oxides of nitrogen from combustion

processes and the conversion to nitric acid (a key component in acid precipitation).

1.2.6 The phosphorus cycle

A key difference with the phosphorus cycle compared to the other biogeochemical cycles mentioned is that there is no atmospheric phase or reservoir as phosphorus rarely exists in a gaseous form. The main storage reservoir for phosphorus in the ecosystem is in sedimentary rocks in the form of phosphates. Phosphorus is an essential micro-nutrient without which living cells cannot function. Indeed, in many ecosystems (notably freshwater ecosystems), phosphorus availability is the key limitation to the productivity of plants. In both aquatic and

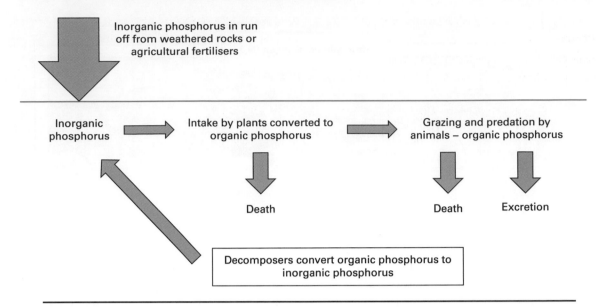

Figure 1.7 The aquatic phosphorous cycle

terrestrial ecosystems it is plants that absorb phosphates from soils and water. Phosphorus (in the form of organic phosphates) is then passed through the food chain before returning to the soils or water via breakdown by microbial activity (the decomposers in the food chain). The aquatic version of this cycle is illustrated in Figure 1.7.

The natural rate of release of phosphates from the weathering of rocks (the geological reservoir) into the water course or soils is extremely slow and explains the relative scarcity of phosphorus in ecosystems. If this rate of release is accelerated beyond this trickle, the result is a surge in plant productivity which often leads to significant imbalances in ecosystems. This is exactly what happens when an algal bloom occurs as a result of the run-off from agricultural land into water courses of high concentrations of man-made phosphate-based fertilisers (which are produced using phosphate-rich rocks obtained by mining). The plants best able to capitalise on the sudden abundance of nutrients are algae with their simple structures and rapid reproductive capacity. As agricultural fertilisers are often nitrogen–phosphorus blends, and nitrogen (in the form of nitrates) is also a key plant nutrient, the effect is even further magnified.

The consequences of such algal blooms are discussed in more detail elsewhere but encompass: secondary organic pollution arising from the decomposition of algal cells, resulting in oxygen depletion, the toxicity of blue green algae to some animal species, and the smothering and light-blocking effects of the floating algal blanket on plant and animal species below, etc.

1.3 Activities, aspects and impacts

1.3.1 Definitions and descriptions

The terminology used by ISO 14001 is useful in understanding the cause and effect links between an organisation's activities and the environmental changes that may occur as a result. The following definitions are provided by the ISO 14001 Standard:

> *Aspects* –'those elements of an organisation's activities, products or services that do or could result in an environmental impact.'

> *Impacts* –'environmental changes that occur, wholly or partly, as a consequence of an organisation's environmental aspects.'

The links between an organisation's activities and its aspects and impacts are shown in Table 1.2. Although illustrative, it should be recognised that the examples in Table 1.2 are over-simplistic in that each activity may give rise to several actual

Table 1.2 Example activity–aspect–impact linkages

Activity	Aspect	Impact
Vehicle washing	Effluent discharge	Water pollution and/or soil contamination
Paint spraying	VOC emissions	Air pollution/human health effects
Product distribution by road	Exhaust emissions	Air pollution/human health effects

and potential aspects which may have several impacts. Consider the vehicle washing example in Figure 1.8.

1.3.2 **An aspects checklist**

As indicated, a single activity may have several aspects, which in turn may have several impacts. When assessing the environmental aspects associated with a particular activity, it is useful, therefore, to have a checklist of categories to consider. A sample list is shown below:

▶ Raw material usage
▶ Effluent discharges
▶ Air emissions
▶ Land usage
▶ Hazardous material usage
▶ Energy usage
▶ Nuisance
▶ Solid waste.

This is the REALHENS checklist discussed in Chapter 6. Individual activities may have aspects associated with one or several of these categories.

1.3.3 **Environmental receptor groups**

Environmental impacts are also often grouped in terms of the receptors affected. The UK government's *Good Practice Guide to the Preparation of Environmental Statements* suggests the headings shown in Table 1.3.

As implied above, an environmental aspect may cause impacts under more than one heading, e.g. effluent discharge may cause changes to water quality (water) with secondary impacts on plant and animal communities (flora and fauna) and human health (human beings). For this reason, there is often an additional main environmental impact heading added to the left-hand column, namely indirect/secondary effects, and these show the types of impacts included under these receptor headings.

1.3.3.1 **Human beings**

Human impacts range from those that relate to a single individual to those that affect whole communities or populations. Take, for

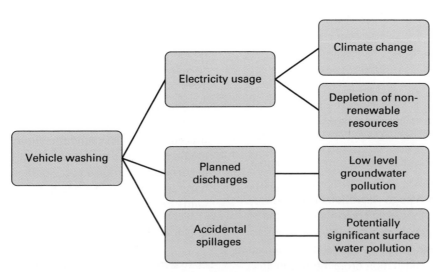

Figure 1.8 Aspects and impacts associated with vehicle washing

Table 1.3 Environmental receptor groups

Environmental receptor	Impact type sub-headings*
Human beings	Population
	Housing
	Noise and vibration (nuisance)
	Air/water/land pollution
	Infrastructure and services
	Land use
	Agriculture
	Recreation
	Forestry
	Mineral resources
	Waste disposal
Flora and fauna	Habitats
	Plant and animal communities
	Individual species
	Biodiversity
Soil	Geology
	Geomorphology
	Agricultural land quality
Water	Hydrological cycle
	Surface water
	Ground water
	Coastal/estuarine
Air and climate	Heat, chemical, odorous and gaseous emissions
	Particulate matter
Landscape	Landform/topography
	Land use
	Land cover
	Landscape character
	Landscape quality
Cultural heritage and material assets	Architectural interest
	Archaeological interest
	Historic interest
	Ancient monuments

*Note: This list is not intended to be exhaustive; nor will all sub-headings be relevant in every case.

example, the construction of a major industrial development.

Changes may relate to migration into or out of an area (e.g. to supply construction or the operational phase workforce), modifications to the demographic structure of the local population (typically in a construction project through an influx of predominately single males), etc. Such changes are likely to have secondary impacts in terms of the demand for housing and services in the region. Increases in demand can lead to requirements for additional housing, for example. This raises associated issues such as land allocation, the environmental consequences of infrastructure and housing construction, the ability of existing hospitals, schools, recreation facilities, etc. to cope with the increased population. Although primarily an issue in relation to large-scale developments, depending on the extra capacity available in an area, even smaller developments can have a significant impact.

The population impacts associated with developments often reflect the temporary and permanent impacts in terms of employment opportunities. Opportunities are normally split between construction and operational phases. In this respect, the timing and nature of associated migration are important. Initial migration associated with the construction phase of a development is likely to comprise a high proportion of young, single males who make relatively light demands on education, health or recreation facilities, and may be housed in temporary accommodation on or near the site. Such migration is of a short-term nature. A second phase of migration comprising those people directly employed in the development may take place when the development's final use commences, and is likely to include a higher proportion of family groups. A third phase of migration is less well defined and is harder to predict, comprising people who are moving to find work in the ancillary industries and expanding local companies which are supported by the new development.

On a less dramatic scale, new developments or existing activities may impact individuals or small groups of local residents on either a short- or long-term basis through nuisance-type effects associated with noise and vibration, dust, traffic hazards or even direct health effects associated with pollution.

1.3.3.2 Flora and fauna

Flora and fauna are generally considered under the combined heading of ecology. Ecology is thus the study of living organisms. It may be useful to note that habitats are comprised of both living (vegetation) and non-living elements such as climate, soil type, aspect, etc.

Commercial and industrial processes can have both direct and indirect impacts on the natural environment. The most obvious direct effect is the physical removal of soils and vegetation, and the obliteration or substantial modification of existing habitats. This could be caused by a new development or its access routes. Roads may affect the movement of wildlife if they cut across habitats, especially if traffic is heavy. If populations of scarce species are separated, their viability and hence their survival may be reduced. Habitats may also be threatened by construction routes and working areas, or by the laying of services or pipelines to supply the main development.

Less immediate extensive impacts to flora and fauna, which may affect individuals or communities, include:

▶ *Pollution impacts* – Pollutants in soil, water or air may affect individual plants and animals via complex pathways that are often difficult to trace. Their effect is often cumulative and can be influenced by other environmental factors such as temperature and sunlight. Factors such as bioaccumulation and the secondary conversion of substances may all play an important part in determining the overall impact of pollution.

▶ *Microclimate* – Major urban developments may produce micro-climatic changes sufficient to alter the health and or growth rates of some plants and animals. More frequently encountered, however, are impacts relating to the release of heat from cooling water into a receiving water body. Such releases can cause significant changes at a local level, especially in enclosed systems such as reservoirs or lakes.

▶ *Groundwater* – Impacts that cause changes in groundwater regimes may have profound consequences for habitats dependent upon the water-table.

▶ *River regimes* – Increases or reductions in the natural rate of flows are one of the

primary ways that aquatic ecosystems may be affected by development. Flash flooding may be caused by greater run-off associated with large areas of hard surfacing. Reduced flows, leading to siltation, may be directly related to increased water demand, or use of by-pass channels. In addition, changes brought about by flood alleviation schemes may affect surrounding habitats by increasing or decreasing the water levels and sediment deposition.

▶ *Public pressure and disturbance* – New developments, especially those associated with recreation and residential use, may place surrounding or other nearby habitats under increasing pressure from the new residents or visitors. This can have a number of adverse effects on the natural environment including the disturbance of animals, the physical destruction of ground flora by horse or bike riding, and the increased risk of accidents, such as fire, which may lead to major habitat destruction.

▶ *Changing relationship between habitats* – New developments may disrupt established relationships between habitats through the removal of 'buffer zones' surrounding important areas, or the severance of routes linking sleeping or roosting areas to feeding grounds. Fragmentation of a single large habitat may reduce plant or animal populations below viable levels. Such a consequence in a single species is likely to have secondary impacts for other species dependent on the first for food, shelter or other essential functions.

▶ *Loss of habitats at a distance* – In the case of some developments, habitats may be destroyed on sites indirectly associated with the development. Most commonly this occurs in relation to the winning or disposal of materials, e.g. the extension of quarries to provide building materials or landfill sites to accommodate wastes arising from the development.

As well as the immediate local or regional concerns relating to flora and fauna impacts of the type described above, there is also a growing recognition of the need to preserve the variety of life forms that inhabit the Earth. This variety of life is known as **biodiversity** and involves habitat diversity, plant and animal species diversity within the various habitats and the genetic diversity of individual species.

1.3.3.3 Soil and geology

Soil and geology play an important part in determining the environmental character of an area. The nature and alignment of rocks have a major influence on landform. Rocks and drift deposits provide the parent material from which soil is created and they influence the rate at which soil is formed. Soil chemistry and structure strongly influence the type of vegetation which occurs naturally in an area. The soil also has a considerable influence on the types of agricultural, horticultural and forestry practices an area can support. Many types of development will have direct impacts on soil, while geological impacts are especially associated with major civil engineering projects, mineral extraction and disposal of wastes to landfill.

Geological impacts may be wide-ranging and, in the case of certain types of development, can include the triggering of ground instability such as subsidence or landslides or the alteration of groundwater levels and flows.

Most soils have taken thousands of years to develop their current physical and chemical characteristics, through both natural and human (principally agricultural) processes. Soil is not an unlimited resource and may be affected in a number of ways:

▶ *Loss/destruction*: the destruction of natural soils is the inevitable consequence of many development activities. The extent of loss may, however, be limited by prior soil stripping, careful handling and storage and subsequent replacement, as in the case of mineral extraction, or reuse elsewhere, e.g. in the restoration of derelict land.

▶ *Physical damage*: in their natural state most soils have well-developed horizontal layering. The physical characteristics (texture, structure and porosity) of these layers depend on weathering processes, the activity of soil invertebrates and plant roots and the activities

of man. Such characteristics determine the soil's ability to support the growth of crops and other vegetation as well as to absorb and release water. The movement of air and water within the soil is very important to plant growth, while the movement of water through the soil is important to the flow regimes of many rivers and streams as well as in the recharge of groundwater.

Construction can have a number of adverse impacts on soil. The use of heavy machinery on site, particularly in wet weather, can cause severe soil compaction. Soil compaction can greatly restrict root growth and adversely affect the drainage characteristics of the soil which in turn can reduce plant growth, result in increased surface run-off and thereby increase the risk of erosion and the transfer of pollutants to surface waters. Deterioration in the physical quality of soil can also occur through mixing contrasting soil materials or layers (e.g. topsoil with subsoil), through contamination by other materials (e.g. rock or wastes) and by careless reinstatement.

▶ *Chemical damage*: soils may be contaminated by a wide range of industrial and construction activities. Such soil pollution may occur from materials deposited directly onto or into the soil, from the deposition of airborne pollutants or by material carried in groundwater. Categories of contaminants may include inorganic compounds – of particular concern are elevated levels of heavy metals such as lead, cadmium, mercury and copper. Although low concentrations may occur naturally, high concentrations are in the main the result of activities such as the direct disposal to land of wastes from mineral workings, industrial processes or the water industry (e.g. sewage sludge) or the deposition of particulates from the atmosphere. High concentrations are toxic to plants, fauna and humans.

Organic wastes are substances of vegetable or animal origin and do not generally cause significant damage to soils, though mobile components such as nitrates and hydrocarbons can cause pollution of surface and groundwater resources.

Landfill gases, such as methane, may be produced by the breakdown of organic materials, such as food waste, paper and timber, in anaerobic conditions. These situations are frequently found on landfill sites. The resultant gas can inhibit plant growth if it is allowed to enter the soil layers and can present an explosive hazard if allowed to build up in concentrations in an enclosed space. Radioactive wastes are produced by the nuclear industry and to a lesser extent by medical and research facilities. In sufficient quantities, such materials can be extremely toxic.

1.3.3.4 Water

Any major development has the potential to affect the water environment, both directly on site and indirectly in the wider catchment (including potentially the estuarine and marine environment). Impacts can relate to both hydrological matters and water quality. Impacts can also affect utilisation of the water environment, e.g. fisheries, navigation, recreation.

1.3.3.5 Air and climate

The evaluation of air quality impacts can be one of the most complex elements in environmental assessment due to the range of pollutants that may be emitted, their combination to form secondary pollutants, the effects of local topography and climate and the variety of different scales at which the effects may be experienced. Some pollutants may have very local, mainly health-related impacts. Others may produce effects at a national or international scale.

1.3.3.6 Landscape

Landscape is a product of the interaction between a range of physical and biological characteristics and the cultural heritage. It encompasses not only the physical features of landform and surface pattern but also the way in which these features are perceived, and the values which are attached to scenery by people. This approach recognises that landscape is a fundamental component of the wider environment and is not just associated with

a limited number of designated areas of particular scenic value.

When considering landscape impacts, it is necessary to evaluate both the direct effects on landscape resources (e.g. the removal of hedgerows) and the public perception of landscape change (e.g. the construction of an industrial building in an area of outstanding natural beauty).

1.3.3.7 Cultural heritage

Cultural heritage is the collective term used to describe aspects of the environment, which reflect the history of human activities, ideas and attitudes. It is not limited to material and economic aspects of life, but also reflects spiritual and intellectual value. The term embraces the subject areas of history, archaeology, architecture and urban design and in many cases is closely tied to the rural or urban landscape. Cultural heritage is generally irreplaceable and should be viewed in the same light as other finite or non-renewable resources.

Cultural heritage sites may include:

▶ archaeological sites, both land-based and maritime;
▶ historic buildings;
▶ conservation areas;
▶ historic landscapes, whether industrial, urban or rural;
▶ parks and gardens;
▶ historic battlefields.

Impacts on cultural heritage resulting from developments may include:

▶ *Loss/destruction*: The most dramatic negative impact is the direct loss or destruction of an element of cultural heritage. For example, this may include the demolition of an historic building or the disturbance of an archaeological feature during the construction of a development.
▶ *Visual intrusion*: An unsympathetic development may impinge on the character and appearance of an area though inappropriate siting or design, directly affecting conservation areas, historic buildings, ancient monuments, areas of archaeological importance, historic landscapes and settlements.

▶ *Physical damage*: There are a number of impacts which may potentially cause damage to the physical fabric of archaeological remains, historic buildings or historic landscapes. These include:

  ▶ air pollution – causing damage to historic buildings and ancient monuments;
  ▶ water – water table fluctuations may affect archaeological remains otherwise preserved by waterlogging;
  ▶ vibration – may cause damage to buildings, ancient monuments and archaeological remains;
  ▶ recreation pressure – which may occur as a result of improved access or by directly attracting visitors. This may cause physical damage and change the intrinsic character of the feature.
  ▶ ecological damage – flora and fauna are an important component of heritage features, particularly historic landscapes. The impacts described in Section 1.3.3.2 may therefore have secondary impacts in terms of cultural heritage.

1.3.3.8 Indirect/secondary impacts

An indirect environmental impact may be defined as:

> Any change in the environment, whether adverse or beneficial, that is caused by third parties acting on behalf of, or in support of, an organisation's activities, products or services, e.g. power generation impacts associated with the supply of energy to an organisation for use in its manufacturing process.

A secondary environmental impact may be defined as:

> Any change in the environment arising as a direct consequence of an initial change caused by an organisation's activities, e.g. fish kill resulting from water pollution caused by a discharge by an organisation.

Examples of indirect and secondary impacts have already been referred to in the individual impact categories above. However, the potential importance of this type of impact makes it worthy of separate mention. Examples of potentially significant impacts that should be considered in relation to a development project might include:

▶ indirect impacts associated with traffic (road, rail, air, water) related to the development or process;
▶ indirect impacts associated with the extraction and consumption of materials, water, energy or other resources associated with the development or process;
▶ indirect impacts associated with other developments associated with the project, e.g. new roads, sewers, housing, power lines, pipelines, telecommunications, etc.;
▶ cumulative impacts (direct, indirect and secondary) that may arise from the interaction of impacts associated with other existing or proposed activities/developments;
▶ secondary impacts resulting from the interaction of separate direct effects listed above, e.g. water abstraction and interference with drainage patterns causing a falling water table and hence secondary impacts in relation to land use or archaeological sites.

II AN OVERVIEW OF KEY ENVIRONMENTAL PROBLEMS

Section 1.3.3 on receptors began to consider the types of impacts that human activity can have on the environment. The following issues have been subject to increasing attention in recent years and certainly include some of the more urgent global environmental problems. It may be noted that each is either a pollution issue or a resource depletion issue. It is also worth noting that many issues overlap or interact for reasons touched upon in the linkages section earlier. For accessibility, the issues have been presented under four main headings:

▶ Air quality
▶ Water quality and availability
▶ Management of natural resources
▶ Human community issues.

1.4 Air quality

Four overlapping issues have been grouped under this heading:

- climate change
- ozone depletion
- acidification
- photochemical smog.

Air pollution respects no boundaries and for that reason the first three of these issues require coordinated action by the whole of the international community to deal with them effectively. Photochemical smog is a more localised problem but one of increasing concern in today's increasingly urbanised global society.

Before we begin an exploration of the key problems, however, we need to consider the key atmospheric pollutants.

1.4.1 Key atmospheric pollutants

1.4.1.1 Oxides of nitrogen (NOx)

NO_x is a collective term used to refer to the seven oxides of nitrogen. The group includes three important air pollutants, namely, nitric oxide (NO), nitrogen dioxide (NO_2) and nitrous oxide (N_2O).

Globally, quantities of nitrogen oxides produced naturally (by bacterial and volcanic action and lightning) far outweigh man-made emissions, however, the local concentrations and growing volumes of anthropogenic sources give cause for concern.

Man-made NO_x emissions are mainly due to fossil fuel combustion from both stationary sources, i.e. power generation (24 per cent) and mobile sources, i.e. transport (49 per cent). Other atmospheric contributions come from non-combustion processes, for example, agriculture (use of fertilisers and urea from livestock), nitric acid manufacture, welding processes and the use of explosives.

Impacts associated with NO_x may be summarised as follows:

- NO_2, in particular, can irritate lungs, causing bronchitis and pneumonia and lowering resistance to respiratory infections. It has also been highlighted as a potential sensitiser making susceptible people more reactive to a wide range of allergens.
- NO_x is one of the essential ingredients in photochemical smog, reacting with VOCs (see below) to form ozone which also has direct human health impacts.
- NO_2 will react with water molecules in the Earth's atmosphere to form nitric acid (HNO_3), which corrodes metal surfaces and is a significant contributor to **acid rain**. In this form, it is harmful to plants but paradoxically may also contribute to **eutrophication** (which stimulates plant growth), particularly in naturally nitrogen-poor water bodies.
- Nitrous oxide (N_2O) is a greenhouse gas and a contributor to climate change via the greenhouse effect.

1.4.1.2 Sulphur dioxide (SO2)

SO_2 is a colourless gas. It is soluble in water and can react with airborne water droplets to form sulphuric acid. It is thus a key constituent in acid rain which causes damage to both agricultural crops and natural vegetation by affecting plant membranes and inhibiting photosynthesis. In humans, it can lead to respiratory problems especially when in combination with moisture and particulate matter when inhalation leads to the formation of sulphuric acid in the lungs.

The most important sources of SO_2 are fossil fuel combustion, smelting, the manufacture of sulphuric acid, the conversion of wood pulp to paper, and the incineration of refuse. Coal burning is the single largest anthropogenic source of SO_2, accounting for about 50 per cent of annual global emissions, with oil burning accounting for a further 25–30 per cent.

1.4.1.3 Carbon monoxide (CO)

Carbon monoxide is a colourless, odourless, tasteless gas that is slightly lighter than air. CO is an intermediate product through which all carbon species must pass when combusted in oxygen. In the presence of an adequate supply of O_2, most CO produced during combustion is immediately oxidised to CO_2. This is not the case, however, in vehicle petrol engines, especially under idling and deceleration conditions. Thus, the major source of atmospheric CO is vehicle emissions. Smaller contributions come from processes involving the combustion of organic matter, for example, in power stations and waste incineration. Some sources suggest that people who spend all day in or near heavy traffic, e.g. on busy motorways or in congested city centres, may inhale as much carbon monoxide as a heavy smoker.

Carbon monoxide has direct human health impacts ranging from hypoxia (effectively, suffocation) to drowsiness and fertility impacts, depending on concentration. Those most at risk include young children and the elderly, pregnant women and people already suffering from heart or lung disease. Emissions levels have been reduced in recent years, particularly through the introduction of catalytic converters on cars.

1.4.1.4 Ozone (O3)

O_3 is the tri-atomic form of molecular oxygen. It is a strong oxidising agent, and hence highly reactive. At ground level, ozone is a pollutant with significant effects. It is an irritant to the eyes, nose and throat and can lead to respiratory and other health problems, including asthma. It also increases sensitivity to allergens (e.g. pollen) and is thought to affect the human immune system. Similar to sulphur dioxide, it interferes with photosynthesis in plants.

Most O_3 in the troposphere (lower atmosphere) is formed indirectly by the action of sunlight on nitrogen dioxide. As a result of the various reactions that take place, O_3 tends to build up downwind of urban centres where most of the NO_x is emitted from vehicles. Small quantities are also contributed from direct sources such as welding and photocopying.

1.4.1.5 Particulate matter

Particulate matter is a blanket term used to describe a range of solid and liquid air pollutants. Coarse particulates can be regarded as those with a diameter greater than 2.5 micrometres, and fine particles less than 2.5 micrometres. Coarse particles usually contain materials from the Earth's crust often in the form of fugitive dust from roads and industries. Fine particles contain the secondarily formed aerosols, combustion particles and re-condensed organic and metallic vapours.

Particulate matter is emitted from a wide range of sources, the most significant sources being road transport (25 per cent), non-combustion processes (24 per cent), industrial combustion plants and processes (17 per cent), commercial and residential combustion (16 per cent), and public power generation (15 per cent). Natural sources are less important; these include volcanoes and dust storms (though of course these can be individually highly significant). Particulate matter can also be formed by the transformation of gaseous emissions such as oxides of sulphur and nitrogen and VOCs.

Impacts depend on particle size, material and associated pollutants. Of key concerns from a human health perspective are the smallest particles (below 10 microns) known as PM_{10}, which pass into the deepest reaches of the lungs and may be irritants or carcinogens. Such particulates may also act as carriers of other pollutants such as acid droplets formed through the combination of acid gases with water vapour. In such a role, they not only have an important impact on the respiratory health of humans, but also exacerbate damage to structures caused by the combined abrasive/corrosive effect of particles carrying acidic droplets.

1.4.1.6 Volatile organic compounds (VOCs)

VOCs comprise a very wide range of individual substances, including hydrocarbons, halocarbons and oxygenates. All are organic compounds and of sufficient volatility to exist as vapour in the atmosphere. Methane is an important example of a VOC, its environmental impact being principally related to its contribution to global warming and to the production of ozone in the troposphere (photochemical smog). Other important VOCs are benzene, butadiene and toluene, all of which have been linked to human health impacts.

Sources of VOCs include hydrocarbons emitted from petrol evaporation and incomplete combustion, and from the leakage of natural gas distribution systems. Methane is generated in significant quantities from agricultural processes and also during the breakdown of organic materials in landfills. The evaporation of solvents, used in paints, adhesives and industrial degreasing processes, causes a release of hydrocarbons, oxygenates and halocarbons into the atmosphere.

VOCs are involved in a variety of impacts, including stratospheric ozone depletion, the production of photochemical smog and direct human health impacts. Two specific examples follow: benzene and butadiene:

▶ *Benzene*: Principal sources are from vehicle emissions and from evaporation losses during the handling, distribution and storage of petrol. It is a known human carcinogen, particularly associated with leukaemia.
▶ *Butadiene*: A hydrocarbon used in the production of synthetic rubber for tyres, as well as being emitted in vehicle exhausts. It is a suspected human carcinogen.

1.4.1.7 Lead

Until the widespread adoption of lead-free petrol in transport, the main source of lead emissions to air was from vehicles using fuels with alkyl lead additives. Currently the primary sources arise from lead manufacturing processes and during the incineration of plastic packaging materials. Lead inhaled from particles in the air is a serious health hazard with effects on the human body that include:

▶ damage to the nervous system;
▶ increased likelihood of high blood pressure and heart attack;
▶ problems of mental development.

1.4.1.8 Ozone-depleting substances

Ozone-depleting substances are those substances that, through their attributes of long environmental persistence and the ability to bind with oxygen molecules, are responsible for ozone depletion in the stratosphere. They include chlorofluorocarbons (CFCs), hydrochlorofluorocarbons (HCFCs), halons, 1,1,1 trichloroethane, carbon tetrachloride and bromochloromethane. These substances are most commonly associated with refrigeration, air conditioning, foam-blowing agents, solvents and fire suppression systems.

1.4.1.9 Carbon dioxide (CO_2)

Carbon dioxide is a naturally occurring constituent of the Earth's atmosphere. It is produced by the complete combustion of carbon-rich substances (e.g. fossil fuels), by the aerobic decay of organic material, by fermentation and by the action of acid on limestone. Its atmospheric concentration is increasing and this is thought to be largely due to the human sources, primarily associated with fossil fuel consumption. This, and the fact that by volume, carbon dioxide is by far the most important greenhouse gas, make it a highly significant pollutant on a global scale.

1.4.2 Climate change

First, it is important to note that climate change is a natural phenomenon and we have evidence of wide variations in global climate over geological time with ice ages and extended warm periods visible through geological formations and the fossil records. However, it is the rate of current change that is of concern and an increasing body of evidence points towards the principal cause of current trends being the result of human activity:

> We appear to be influencing the planet's key life support system in a manner that is unprecedented in the Earth's history.

Any review of this subject quickly gets caught up in a mire of political comments, claims and counter-claims. So, as far possible, we will confine ourselves to the basic facts.

1.4.2.1 What is the greenhouse effect?

The greenhouse effect (see Figure 1.9) is the term used to describe the selective absorption by the atmosphere of different types of radiation. Incoming short-wave radiation from the sun passes through largely unhindered, while long-wave radiation reflected back from the Earth's surface is to some degree absorbed by the so-called greenhouse gases. The ability of the greenhouse gases to prevent the passage of long wavelength radiation effectively traps the sun's heat close to the Earth – hence the term 'greenhouse effect'.

Of about 20 gases responsible for the greenhouse effect, carbon dioxide (CO_2) is the most abundant

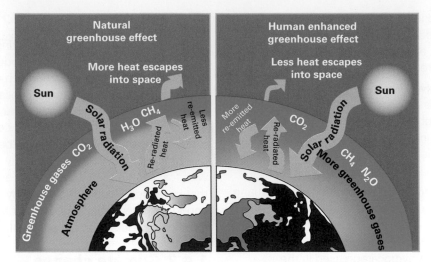

Figure 1.9 The greenhouse effect
Source: US National Park Service.

but methane (CH_4), nitrous oxide (N_2O), and the chlorofluorocarbons (CFCs) and low-level ozone also make significant contributions. Water vapour is also a key greenhouse gas but is less an issue in terms of direct emissions from human activities. Increasing temperatures arising from more long-lived greenhouse gases such as CO_2, however, may result in increased evaporation from the Earth's surface and hence increased water vapour concentration, thus compounding the warming effect.

Without this phenomenon the surface of the Earth would be significantly colder and much less hospitable to life. The current concern, however, comes from the apparent exaggeration of this effect as a result of human-generated air pollution.

1.4.2.2 How do humans influence the greenhouse effect?

The principal greenhouse gases and their human activity sources are shown in Table 1.4. The addition of these gases exaggerates the natural greenhouse effect of the Earth's atmosphere, as shown in Figure 1.9.

1.4.2.3 What the scientists think

In July 2001, the United Nations Intergovernmental Panel on Climate Change (IPCC), consisting of hundreds of the world's leading climate scientists, published their third report giving unqualified support to the view that global warming is a real and current issue and that the release of man-made greenhouse gases is largely responsible for the present trend of increasing global temperatures.

The IPCC fourth report, published in 2007, and the fifth report in 2013/14 made even stronger cases that human-influenced climate change is happening now. The fifth report combines evidence from more than 9,000 peer-reviewed papers and states that there is at least a 95 per cent chance that the current changes are a function of human-related greenhouse gas emissions.

The fifth report generated some further compelling findings:

▶ Warming of the climate system is unequivocal, and since the 1950s, many of the observed changes are unprecedented over decades to millennia. The atmosphere and ocean have warmed, the amounts of snow and ice have diminished, sea levels have risen and the concentrations of greenhouse gases have increased.
▶ Each of the past three decades has been successively warmer at the Earth's surface than any preceding decade since 1850.
▶ Levels of greenhouse gases in the atmosphere are unprecedented in the past 800,000 years.

Table 1.4 Sources of human greenhouse gas emissions

Sector	Activities	Greenhouse gases	% of global total
Energy	Fossil fuel combustion for electricity	Carbon dioxide, methane, nitrous oxide	26
Industrial activities	Fossil fuel combustion Emissions from non-energy processes	All	19
Forest	Harvesting and deforestation Burning	Carbon dioxide, methane, nitrous oxide	17
Agriculture	Rice paddies Animal husbandry Fertiliser use	Carbon dioxide, methane, nitrous oxide	14
Transport	Road, rail, air and marine fossil fuel-based transport	Carbon dioxide, methane, nitrous oxide	13
Waste management	Landfill and waste water treatment Incineration	Methane, nitrous oxide, carbon dioxide	3
Commercial and residential buildings	Heating and cooking (electricity usage is covered in the energy sector)	Carbon dioxide, methane, nitrous oxide	8

Source: International Panel for Climate Change, quoted by the US Environment Protection Agency.

▶ Concentrations of CO_2 have risen by 40 per cent since pre-industrial times, primarily from fossil fuel emissions; 2,000 gigatonnes of carbon (GtC) have been added to the atmosphere since 1751.

▶ The largest contribution to total radiative forcing (global warming) is caused by the increase in the atmospheric concentration of CO_2 since 1750.

▶ In the northern hemisphere, 1983–2012 is likely to have been the warmest 30-year period in the past 1,400 years.

▶ Human influence has been detected in the warming of the atmosphere and the ocean; in changes in the global water cycle; in reductions in snow and ice; and in global sea level rise. This evidence for human influence has grown since the fourth IPCC report. It is *extremely likely (95 per cent certain)* that human influence has been the dominant cause of the observed warming since the mid-twentieth century.

▶ The rate of sea level rise since the mid-nineteenth century has been larger than the mean rate during the previous two millennia. Sea level rise will be between 26cm and 82cm by 2100.

The IPCC fifth report provided a number of scenarios predicting average global temperatures over the twenty-first century. Projections for changes in global surface temperature by the end of the century range from 0.3°C to 4.8°C compared with 1985–2003 levels. The report notes that warming over land will be greater than over the ocean.

It has previously been suggested that a 2°C increase represents the threshold beyond which 'the risks to human societies and ecosystems grow significantly'. Such an increase is considered probable if atmospheric concentrations of carbon dioxide exceed 400 parts per million (ppm). The 400 ppm concentration was recorded at several monitoring stations around the globe during 2013. The United Nations Environment Programme expects the global average to reach 400 ppm by 2015 or 2016.

1.4.2.4 So where does the climate change debate come from?

Unfortunately the clear picture of scientific consensus described above has been muddied and confused by statements made from the platforms of a variety of interested parties, including political and business interests, media speculation and even, ironically, environmental lobby groups. Bob May, President of the Royal Society and a former chief scientific advisor to the UK government, put it this way in 2005:

On the one hand we have the International Panel on Climate Change, the rest of the world's major scientific organisations and the Government's Chief Scientific Advisor all pointing to the need to cut emissions. On the other hand, we have a small band of sceptics, including lobbyists funded by the US oil industry, a sci-fi writer and certain elements of the tabloid press who deny the scientists are right. It is reminiscent of the tobacco lobby's attempts to convince us that smoking does not cause cancer. There is no danger this lobby will influence the scientists. But they don't need to. It is the influence on the media that is so poisonous.

Since this statement was made, however, the debate seems to have begun to shift away from 'whether or not there is a problem' and focuses more on 'what we should do about it'.

1.4.2.5 What might be the consequences of climate change?

The consequences of climate change are extremely difficult to predict with accuracy, although the IPCC fifth report reports increasing confidence in the climate change models being used by various research facilities. As indicated above, it is thought that global temperature increases could lead to impacts such as sea level rise with consequent flooding of low lying coasts (and many of the world's major cities), extreme weather events (droughts, storms) and shifts in vegetation zones and hydrological systems with consequent impacts in terms of biodiversity, desertification, agricultural productivity and associated human impacts. We are already observing evidence of some of these consequences in parts of the world. Many examples of such current changes have been documented, including the following:

▶ widespread melting of permafrost in Alaska, Canada and Siberia with catastrophic effects on Arctic wildlife;
▶ coral bleaching (as a result of increased sea temperatures) on the Great Barrier Reef in Australia;
▶ glacial retreat in the Peruvian Andes and other areas with knock-on human and environmental consequences, e.g. 7 million people in Lima, the capital of Peru, rely on glacial meltwater as the primary source of water supply;
▶ dust storms and desertification in Mongolia and Sub-Saharan Africa;
▶ spread of tropical diseases such as malaria into temperate zones;
▶ increasing numbers, severity and geographical distribution of hurricanes.

1.4.2.6 A growing sense of urgency

In late 2006, the publication of the *Stern Review on the Economics of Climate Change*, commissioned by the UK Chancellor of the Exchequer, provided a significant alternative presentation of the issue from an economic viewpoint. In a nutshell, the review concluded that 'the benefits of strong, early action on climate change outweigh the costs'. Coming from a leading economist, the report attracted a lot of attention internationally and may yet prove to be a significant marker in the global response to climate change.

The Stern Report concludes: 'Uncertainty is an argument for a more, not less demanding goal [in emissions reductions], because of the size of the adverse climate change impacts in the worst case scenarios.'

1.4.3 Ozone depletion

In the stratosphere, 15–30 km above the ground, intense ultraviolet (UV) radiation from the sun forms ozone directly from the oxygen in the air. This process absorbs quantities of UV radiation and hence the so-called 'ozone layer' protects plants and animals from dangerous levels of UV radiation at ground level. Most importantly, it absorbs the portion of ultraviolet light called UVB. UVB has been linked to many harmful effects, including various types of skin cancer.

Some chemicals, such as the synthetic chlorofluorocarbons (CFCs), break down ozone more quickly than it forms naturally. Scientists have shown that this has thinned, or depleted, the ozone layer to such an extent that an 'ozone hole' has formed above the Antarctic in spring/summer, and that the same processes are taking place too in the Arctic, though to a lesser degree.

1.4.3.1 CFCs and ozone

For over 50 years, chlorofluorocarbons (CFCs) were thought of as miracle substances. They are stable, non-flammable, low in toxicity, and inexpensive to produce. Over time, CFCs found uses as refrigerants, solvents, foam-blowing agents, and in other smaller applications. Other key ozone-depleting substances include:

▶ hydrochlorofluorocarbons (HCFCs);
▶ methyl chloroform – a solvent;
▶ carbon tetrachloride – an industrial chemical;
▶ halons – extremely effective fire extinguishing agents;
▶ methyl bromide – an effective produce and soil fumigant.

All of these compounds have atmospheric lifetimes long enough to allow them to be transported by winds into the stratosphere. CFCs are so stable that only exposure to strong UV radiation breaks them down. When that happens, the CFC molecule releases atomic chlorine. Ozone is destroyed by reaction with chlorine, bromine, nitrogen, hydrogen and oxygen gases. Due to the catalytic function of these gases in their interaction with ozone, an individual chlorine atom can, on average, destroy nearly a thousand ozone molecules before it is converted into a form harmless to ozone. The net effect is to destroy ozone faster than it is naturally created.

Numerous experiments have shown that CFCs and other widely used chemicals produce roughly 85 per cent of the chlorine in the stratosphere, while natural sources contribute only 15 per cent. In the ozone 'hole' over Antarctica during each spring the ozone levels fall by over 60 per cent during the worst years.

In addition, research has shown that ozone depletion occurs over the latitudes that include North America, Europe, Asia and much of Africa, Australia and South America. Over the USA ozone levels have fallen 5–10 per cent, depending on the season. Thus, ozone depletion is a global issue and not just a problem at the Poles.

1.4.3.2 The consequences of ozone depletion

These are thought to be more far-reaching than may be initially obvious and include:

▶ reduced photosynthesis by marine phytoplankton with knock-on effects on the food chains throughout the world's oceans;
▶ human health impacts – increased sunburn, skin cancers, cataracts and immune system suppression;
▶ plant and animal health impacts;
▶ reduced agricultural crop yields and associated socio-economic impacts.

1.4.3.3 Signs of improvement?

However, on a positive note, studies in the Antarctic provide evidence that since 2001 there has been some improvement in ozone levels, perhaps indicating that the phase-out of key ozone-depleting substances under the Montreal Protocol is beginning to take effect. In 2012, the ozone hole over Antarctica was the smallest in a decade. However, scientists agree that it will be at least 30 years and the continued commitment to the phase-out of ozone-depleting substances before we can say with certainty that the problem is completely solved. That said, the signs are good that this is an example of where international agreement and action can yield positive results in halting and reversing environmental degradation.

1.4.4 Acid rain

Air pollution from the burning of fossil fuels is the major cause of acid rain (see Figure 1.10). The main chemicals in air pollution that create acid rain are sulphur dioxide (SO_2) and oxides of nitrogen (NO_x). Acid rain usually forms high in the clouds where sulphur dioxide and oxides of nitrogen react with water, oxygen and oxidants. This mixture forms a mild solution of sulphuric acid and nitric acid. Sunlight increases the rate of most of these reactions. Rainwater, snow, fog and other forms of precipitation containing those mild solutions of sulphuric and nitric acids fall to earth as acid rain.

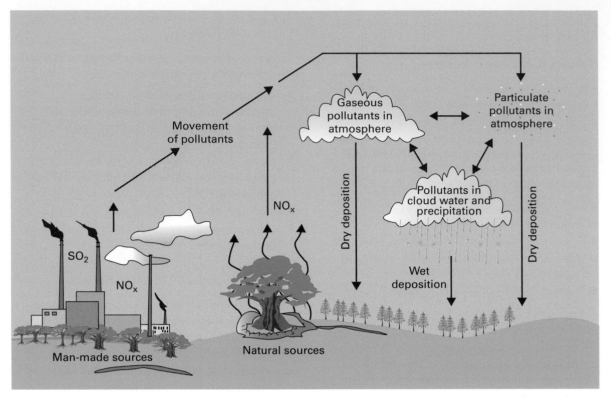

Figure 1.10 Acid rain
Source: Adapted from US EPA.

Because rain travels over long distances in clouds, acid rain is a regional as well as a local problem. In Britain, the prevailing winds come in from the Atlantic, so they are unpolluted. This means that 87 percent of the sulphur dioxide in the air in Britain is produced here (only about 1 percent is produced naturally). Other countries are much less fortunate, especially Scandinavia and the central European countries. In Sweden alone, 20,000 lakes have been affected by acidification, causing imbalances in nutrient levels which affect fish and other aquatic life.

The effects of acid rain are not confined to Europe. Every country with a power station or a significant number of road vehicles helps to generate the gases which cause acid rain. In recent years the massive increase in coal-fired power generation in China has resulted in increasing acid gas production in the Far East. Acid rain is a global problem, and needs a global solution.

1.4.4.1 The consequences of acid precipitation

Rainwater is naturally slightly acidic because of the presence of carbonic acid arising from atmospheric carbon dioxide dissolving in water vapour in clouds prior to precipitation. However, the processes described above can lead to significantly increased acidity (lower pH). The consequences of this vary somewhat depending on capacity of the receiving environment to neutralise or buffer the effect. In areas with limestone geology and relatively neutral or even alkaline soils and water courses, the acid precipitation has much less of an impact than in areas with acidic soils and/or vegetation, e.g. upland heath, coniferous forests, granite soils. In the latter, impacts arising from acid rain include the following:

▶ contact damage to plants, leading to foliage loss and hence reduced photosynthesis;
▶ reduced soil pH can lead to inhibition of soil nutrient breakdown and hence to reduced soil fertility;

▶ reduced soil water pH can lead to the mobilisation of naturally occurring metal ions (especially aluminium). The metals react with the acidic solution and are then taken up by terrestrial plants or washed into water courses. These metals are 'secondary pollutants' and are often toxic to plants or aquatic organisms.

The knock-on consequences from these types of impacts can have far-reaching effects through food chains and at its worst can lead to forest die-back and the decimation of aquatic ecosystems.

In addition to this well-documented form of acidification, there is also growing evidence to suggest that, as global atmospheric levels of CO_2 increase, there is a corresponding increase in ocean acidity as greater quantities of CO_2 are absorbed. There are concerns that this increasing acidity will have negative consequences for marine ecosystems.

1.4.5 Photochemical smog

Certain conditions are required for the formation of photochemical smog (see Figure 1.11). These conditions include:

▶ a source of oxides of nitrogen and volatile organic compounds. High concentrations of these two substances are associated with industrialisation and transportation and in particular the combustion of fossil fuels.
▶ the time of day is a very important factor in the amount of photochemical smog present though primarily because of variation in meteorological and traffic conditions;
▶ several meteorological factors can influence the formation of photochemical smog, e.g. high pressure, lack of wind, temperature inversions;
▶ topography is another important factor influencing how severe a smog event can become, e.g. valleys help slow the dispersal of pollutants and may also increase the likelihood of temperature inversions which help hold the pollutants near the generation source.

An abundance of nitrogen oxides and volatile organic compounds in the atmosphere and the presence of these particular environmental conditions primarily determine the development of photochemical smog. In essence, the chemical process of photochemical smog requires the following conditions:

▶ sunlight and relatively still air;
▶ oxides of nitrogen (NO_x) being produced;
▶ volatile organic compounds (VOCs) being produced;
▶ temperatures greater than 18°C.

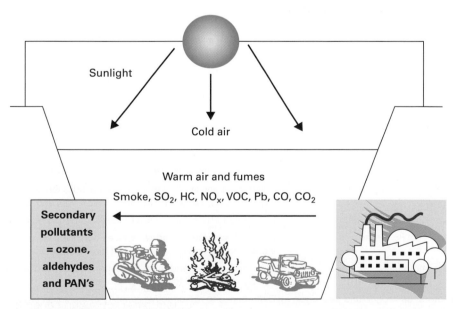

Figure 1.11 Photochemical smog formation

The environmental conditions allow for the interaction of the pollutants without rapid dispersal and this leads to a series of human health impacts. Oxides of nitrogen are directly linked to increased respiratory problems (including asthma) in sensitive individuals. In addition, several reactions will occur, producing further toxic chemical constituents of photochemical smog, including:

▶ ozone (harmful to plants; irritant to eyes, nose and throat);
▶ aldehydes (smelly, poisonous and irritant to eyes, nose and throat);
▶ peroxyacetyl nitrate (PAN) (toxic; irritant to eyes, nose and throat).

It is important to make a distinction between the ozone in photochemical smog at ground level, where high concentrations create serious pollution, and ozone in the upper atmosphere, where its presence in the 'ozone layer' is critical to protect the Earth from excessive ultraviolet radiation.

1.5 Water quality and availability

Water bodies, whether fresh or saline, surface or groundwater, are often particularly sensitive receptors when it comes to pollution. They have the ability to transfer or spread pollution, as well as accumulate pollutants from a variety of sources. They are, of course, also often valuable biological and human resources. Groundwater represents a particular concern from a water supply perspective because of the difficulty of cleaning it up once contaminated.

1.5.1 Pollutant categories

Water pollution is any chemical, biological, or physical change in water quality that has a harmful effect on living organisms or makes water unsuitable for desired uses. Pollution may arise as point sources, such as discharges through pipes, or may be more diffuse, such as from run-off from streets and buildings, or agricultural nutrients lost from fields. There are several categories of water pollutants, as summarised in Table 1.5.

The scale of impact associated with the discharge of water pollutants varies, depending on the nature and concentration of pollutants present, the quantity and frequency of discharge and the characteristics of the receiving water body. However, the following examples cover many of the principal categories of impact.

1.5.1.1 Toxicity effects

Pollutants individually or in combination may cause direct health impacts on aquatic organisms. Detailed impacts vary from corrosive/irritant through to metabolic toxicity. In some instances,

Table 1.5 Categories of water pollutants

Pollution	Effect
Nitrates and phosphates: run-off from intensive agriculture using nitrate-based fertiliser	Can promote eutrophication leading to excessive algae growth which reduces light to plants below the surface. Dissolved oxygen levels can be reduced through bacteria living on the dead plant.
Organic wastes: including untreated sewage, food industry wastes, etc.	Reduces dissolved oxygen levels which causes changes in aquatic flora and fauna. Sewage contains bacteria and other pathogens harmful to humans.
pH: industrial discharges	Excessive acidity or alkalinity can be toxic to fish.
Oil: accidental seepage from storage tanks, or deliberate dumping	Poisons aquatic life. Prevents oxygen absorption so the dissolved oxygen levels will decline and may inhibit flora and fauna.
Suspended solids: coal washing, china clay	Increased turbidity reduces light levels. Smothering of river beds.
Thermal: industrial cooling water	Elevated water temperature reduces dissolved oxygen levels while increasing the rate of chemical and biological activity.
Toxic compounds: industrial	Toxicity to aquatic and other life depending on levels. Toxic materials can be stored up in fish and concentrated.
Endocrine disruptors: oestrogen and oestrogen mimicking substances such as phthalates	Disturbance to reproductive capacity of individual organisms or whole species. Mutagenic effects. Human health impacts.

persistent pollutants may bio-accumulate and lead to knock-on effects through the aquatic food chains, and may even cross over into the terrestrial food chains and perhaps affect humans.

Pathogens are the biological version of toxic chemicals causing health impacts in susceptible organisms and in some cases linked back to human receptors via aquatic disease vectors.

1.5.1.2 Tolerance effects

Some pollutants may generate sub-lethal impacts on aquatic organisms that may nonetheless have significant impacts through the reduction in the ability of organisms to photosynthesise (e.g. suspended sediments), to reproduce (oestrogen-mimicking compounds), or to stay healthy via robust immune responses (temperature or pH changes).

1.5.1.3 Oxygen availability

Lack of oxygen is a particular type of tolerance effect but one to which many aquatic organisms are particularly sensitive. Whether through physical blocking of air–water interchange (e.g. surface coating by hydrocarbons), or through biological or chemical reactions associated with pollutants in the effluents, oxygen concentrations in a receiving water body may be reduced to such an extent that aquatic organisms (both plants and animals) effectively begin to suffocate.

1.5.1.4 Physical impacts

Some pollutants such as hydrocarbons or sediments literally coat aquatic organisms, smothering them in the case of plants or impeding their ability to move and behave normally, as in the case of oil contact on the plumage of birds.

Sediments may also result in downstream flow changes, either clogging up abstraction points or navigation channels, covering spawning grounds or changing erosion patterns.

1.5.1.5 Nuisance impacts

In some instances pollutants may cause visual and/or odour nuisance to human receptors downstream of the discharge. Even where these are not overtly present, impacts on fish populations or water quality may limit the utility of the water course for amenity or recreation purposes.

1.5.1.6 Eutrophication

There are several naturally occurring groups of algae, which can be identified by colour. They provide essential food for animal plankton and hence form part of the aquatic food chain, eventually providing food for fish and even non-aquatic animals such as herons or otters. However, certain species of blue-green algae can produce toxins, which are potentially poisonous to mammals and fish. And in the presence of certain pollutants, algal growth can become magnified to a level where it ceases to be a natural element in the food chain and instead becomes a 'secondary pollutant' in its own right.

In freshwater, phosphorus is regarded as the most important limiting nutrient, though nitrogen and silica also play major roles in controlling the growth of algae. The growth rate of algae is limited by the amount of nutrient available, even when the physical conditions such as light, temperature and stability are optimal. Waters, in which the level of nutrients has increased such that the growth of algae is no longer limited, can be described as eutrophic (meaning 'well feeding'). Such waters can support a large population of algae. Eutrophication of a lake or reservoir can arise naturally or as a result of activities such as discharges from sewage treatment works, industry and agriculture.

Problems associated with excessive algal growth in lakes and reservoirs include the de-oxygenation of still waters and the reduction in light levels on the bed, both of which can result in the long-term disruption of the ecosystem. Excessive growths of blue-green algae can lead to surface scum and the production of toxins, which are potentially poisonous to fish and mammals (including humans).

1.5.1.7 Endocrine disruptors

There is a large range of chemicals with endocrine-disrupting properties. These include naturally occurring hormones such as oestrogen, and synthetic chemicals like pesticides, fungicides and insecticides, drugs such as oral contraceptives, and industrial chemicals like tributyltin (TBT), polychlorinated biphenyls (PCBs), polyaromatic hydrocarbons (PAHs) and dioxins.

Of these, the damaging effects of TBT, once widely used in antifouling paint on boats, are perhaps the most widely documented. TBT is known to cause imposex (female growth of male sex organs) and intersex (hermaphrodite condition) in marine invertebrates such as dog whelks, oysters and mussels. TBT will leach from boat hulls treated with TBT-based anti-foulant paint, and the devastating effect on commercial shell fisheries in the 1980s led to a European ban on TBT use for boats less than 25 metres in length in 1987. A total European ban came into effect on 1 January 2008.

Research on endocrine disruptors has also concentrated on the effects of synthetic oestrogens discharged into the sewer system as a result of the widespread use of oral contraceptives.

In humans, reduced sperm counts in men have been attributed to oestrogen contamination in drinking water, and its feminising effects on fish, both in the laboratory and in freshwater environments, have been well documented.

In the UK, alarm bells over oestrogen pollution were first rung by researchers who reported hermaphrodite fish in the settlement lagoons of sewage treatment works. Further research found that discharged sewage effluent contained high levels of oestrogen and several UK estuaries including the Mersey, Tee and Tyne were severely contaminated. The effect on coastal and offshore marine life is still largely unknown.

1.5.2 Groundwater pollution

Groundwater is a term used to describe water located beneath the ground surface in soil pore spaces and rock fractures. An underground area is called an **aquifer** when it can yield a usable quantity of water. The depth at which soil pore spaces or fractures and voids in rock become fully saturated with water is called the *water table.* Groundwater is recharged, and eventually flows to the surface naturally; natural discharge often occurs at springs, seeps and streams. Groundwater is also often withdrawn for agricultural, municipal and industrial use by constructing and operating extraction wells. The study of the distribution and movement of groundwater is hydrogeology.

Groundwater is of particular concern as a pollution receptor because it is:

▶ difficult and expensive to clean up;
▶ linked to surface water, providing a diffuse and difficult-to-control source of pollution;
▶ valuable as a human water supply resource;
▶ out of sight and therefore out of mind! Groundwater pollution is often not obvious from the surface and can therefore go undetected for long periods, with large areas contaminated before the issue comes to light.

The Love Canal incident is one of the most widely known examples of groundwater pollution. In 1978, residents of the Love Canal neighbourhood in upstate New York noticed high rates of cancer, and an alarming number of birth defects. This was eventually traced to organic solvents and dioxins from an industrial landfill that the neighbourhood had been built over and around, which had then infiltrated into the water supply and evaporated in basements to further contaminate the air. Some 800 families were reimbursed for their homes and moved, after extensive legal battles and media coverage.

More recently, debate has begun about the potential for serious groundwater contamination arising from the controversial practice of 'fracking', which involves the injection of water and chemicals into geological formations to enhance the recovery of gas (or, to a lesser extent, oil). The lack of good studies to date makes any conclusions contentious, but serious independent studies, such as that conducted by the Pacific Institute in 2012, indicate that there are good grounds for concern and that the current lack of sound independent evaluation should be addressed.

1.5.3 Freshwater availability

More than a billion people – almost one-fifth of the world's population—lack access to safe drinking water, and 40 per cent lack access to basic sanitation, according to the second UN World Water Development Report, published in 2006.

A growing global population that is becoming increasingly industrialised and dependent on intensive agricultural practices is creating an increasing demand for dependable supplies of good quality freshwater. Such supplies are threatened by both over-abstraction and the pollution of water sources. Especially in areas of the world where supplies are naturally scarce, an increase in demand coinciding with a decrease in quality as a result of pollution can produce adverse effects in relation to:

▶ aquatic and wetland eco-systems;
▶ the amenity and biodiversity value of such resources;
▶ human health and welfare.

From a resource depletion perspective, water shortages are at the very least disruptive in developed nations but in developing nations, such problems may be devastating both in human and ecological terms.

From a pollution perspective, water is a receptor in its own right but is also a key pathway linking pollution sources not only to aquatic organisms but also to terrestrial and marine ecosystems and human communities that depend on water for life and livelihood.

1.6 Management of natural resources

A number of overlapping and varied issues will be considered in this section, namely:

▶ population growth
▶ energy
▶ land contamination
▶ availability of productive land
▶ biodiversity and ecological stability
▶ chemicals exposure
▶ waste.

1.6.1 Population growth

In the study of ecology, we are very familiar with the idea of species populations exploding to take advantage of available resources and collapsing when those resources become scarce. Such

terminology is rarely used in relation to our own species. Indeed, discussion of human population numbers often seems to have become something of a politically incorrect topic. This is a theme to be considered further during the sections of the manual looking at 'sustainability', but for now let's consider the numbers:

▶ The world population reached 7 billion in 2011 (Population Reference Bureau, 2013).
▶ Human population numbers are now expected to reach 9 billion by 2040.
▶ A population almost the size of Germany's (about 75 million) is being added to the planet each year, with the equivalent of one new city added every single day.

Big numbers – but still only a partial picture! One of the reasons that population pressure has become a 'difficult topic' is that it has been used by some as a way of passing the blame for poverty and environmental degradation onto the developing world, i.e. to the people who experience the effects of these issues most keenly. A number of authors have attempted to present the issue in a more balanced light by looking at the effects of population in different contexts rather than simply the raw numbers of people. For example, it has been estimated that a child born in the USA creates 13 times as much environmental damage over the course of its lifetime than a baby born in Brazil and 35 times as much as an Indian baby.

Put simply, it is not just the number of people that determines the scale of the impact but also their rate of consumption of the Earth's resources. Using this idea of 'equivalent children' it becomes clear that the problem of population growth is not confined to the developing world where absolute growth rates average 1.5 per cent per annum but also to the developed world where absolute growth rates average only 0.1 per cent but where the relative impact of any increase is so much higher (statistics from the Population Reference Bureau, 2013, 2006 World Population data sheet). Factor in the rising annual consumption levels (and hence environmental impact) in *all* population groups and we begin to really appreciate the scale of the problem.

1.6.2 Energy

We are a 'carbon society' with global economies dependent on energy supplied by fossil fuel combustion. This is a problem for two reasons previously discussed:

▶ Fossil fuels are a finite *non-renewable resource* – sooner or later they will run out.

▶ Fossil fuel combustion is the source of a variety of pollution issues discussed elsewhere.

In the UK, the current energy breakdown is shown in Figure 1.12.

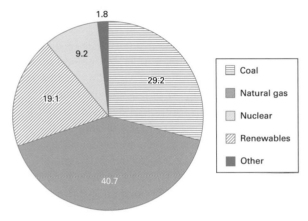

Figure 1.12 UK energy mix by fuel type, 2011–12
Source: Based on data held on the Department for Energy and Climate Change website.

There has been an increase in the UK in the use of renewables over the past few years as Figure 1.13 illustrates, but there is clearly still a long way to go before the problems associated with fossil fuel dependence are addressed.

It must be acknowledged that these figures do not reflect the potential capacity for renewables generation in the UK, which is one of the highest in Europe. Other countries have been much more progressive in increasing the contribution made by renewables to energy production as Figure 1.14 demonstrates.

The 2011 renewable energy review by the UK Department for Energy and Climate Change (DECC) confirmed that 65 per cent electricity generation from renewables is technically feasible by 2030. Overall, the potential percentage of renewables across electricity, heating and

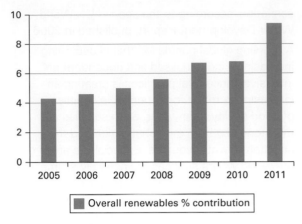

Figure 1.13 Overall renewables contribution to UK energy supply (%)
Source: DEFRA.

transport is said to be 45 per cent by 2030. This data supports the view that potentially the UK has one of the highest renewables capacities in Europe.

1.6.3 Land contamination

This is an issue of particular relevance to countries with a long industrial heritage. The UK, for example, which is considered the birthplace of the Industrial Revolution, has many sites throughout the country with enhanced levels (concentrations) of a wide range of hazardous substances. These usually resulted from minimal environmental controls and a lack of understanding of environmental issues. Contamination typically has been caused by the dumping of solid or liquid wastes, or from the spillage or leakage of materials and wastes.

Key issues associated with such sites include:

▶ the hazardous nature and variety of the substances (contaminants, e.g. heavy metals, cyanide, solvents, oils, PCBs);

▶ the risk of substances reaching receptors (humans, environment-specific organisms) to cause harm;

▶ the ability of substances to migrate along pathways to receptors;

▶ contamination of the aquatic environment, particularly groundwater (since once this is contaminated, it is difficult to clean up).

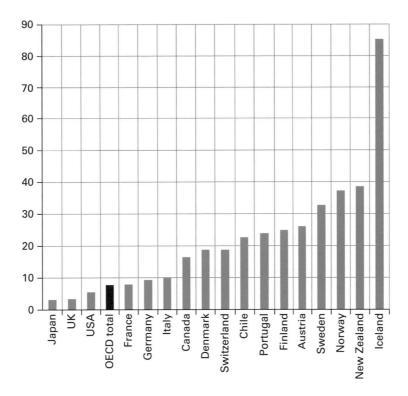

Figure 1.14 Renewables contributions for OECD countries, 2010
Source: DEFRA.

While some of the examples given in Table 1.6 clearly highlight historical contamination, many of the others relate to current industries and sites. Pollution control standards have improved significantly in the past few decades but it must be acknowledged that land contamination is not only associated with historic heavy industry sites but also with current operations in a variety of sectors.

Contaminated land may pose a threat to a wide variety of receptors or may simply constitute a 'loss of amenity' with valuable land areas previously used for industrial purposes being unfit for residential or municipal development without significant clean-up operations.

1.6.4 Availability of productive land

Pollution is not the only threat to the availability of productive land. Desertification resulting from climate change, salinisation resulting from ill-managed irrigation schemes, deforestation (particularly in the tropics), soil erosion and increasing urbanisation all represent major threats on a global scale to the availability of productive land.

Nearly all of the world's productive land is already exploited. Most of the 'good bits' have been in agriculture for a long time. Some of the remaining land surface, such as tropical rainforests, is highly productive under native vegetation (hence not much use for feeding people), but not very productive when converted to agriculture. In fact, cropland per capita is declining world-wide, as agriculture land is degraded, or urbanised. Increasing the yields from available farmland appears to be the key to increased food production.

The Green Revolution of the 1960s and 1970s produced a dramatic increase in productivity through the introduction of high-yielding crop varieties, but with some significant environmental consequences, e.g. the increase in fertiliser and pesticide usage and irrigation systems all present

Table 1.6 Examples of sites and contaminants

Industry	Example sites	Likely contaminants
Chemical	Acid/alkali works Dyeworks Fertiliser and pesticide manufacturing Pharmaceuticals Paint works Wood treatment plants	Acids, alkalis, metals, solvent (e.g. toluene, benzene) phenols, specialised organic compounds
Petrochemicals	Oil refineries Tank farms Fuel storage depots and filling stations Tar distilleries	Hydrocarbons, phenols, acids, alkalis, asbestos
Metals	Iron and steel works Foundries and smelters Electroplating, anodising and galvanising works Ship building and breaking Scrap yards	Metals, particularly iron, copper, nickel, chromium, zinc, cadmium and lead, asbestos
Energy	Gas works Power stations	Combustible substances (e.g. coal and coke dust), phenols, cyanides, sulphur compounds, asbestos
Transport	Garages, vehicle builders and maintenance workshops Railway depots	Combustible substances, hydrocarbons, asbestos
Mineral extraction and land restoration (including landfill sites)	Mine and spoil heaps Pits and quarries Landfill sites	Metals (e.g. copper, zinc, lead), gases (especially methane), leachates
Water supply and sewage treatment	Water works Sewage treatment plants	Metals (in sludges) Micro organisms
Miscellaneous	Docks, wharfs and quays Tanneries Rubber works Military land	Metals, organic compounds, methane, toxic flammable or explosive substances, micro organisms, asbestos

a threat to the local and regional environment. In the light of the population pressures discussed above, the management of productive land and the development of high efficiency food production without magnification of environmental degradation are key challenges to human society in the future.

1.6.5 Biodiversity and ecological stability

Biodiversity may be defined as the variety of life in any given area – it may be expressed as the variety of habitats, species or genetic variation within species. Many scientists consider species extinction to be the most important environmental challenge facing us today. We are all aware of the threats to high profile species such as the giant panda and the Siberian tiger, but these represent the tip of the iceberg. It has been estimated that the extinction rate associated with tropical forest clearance is of the order of 27,000 species every year or 3 per hour! These losses represent a rate of extinction 1,000–10,000 times faster than the 'normal' rate documented in the fossil record, which goes back hundreds of millions of years.

These rates translate into some disturbing statistics – over half the world's primates and one in eight plant species are threatened with extinction (UNEP's Global Biodiversity Assessment and World Conservation Union global survey). Activists and scientists try to convey humanity's deep dependence on other species by appealing to people's self-interest – pointing out, for example, that 9 out of the top 10 pharmaceuticals in the United States are derived from natural sources. For humans to eliminate other species thus forecloses agricultural, medical and other breakthroughs that cannot yet be imagined.

Yet cures for cancer and the aesthetic pleasure of watching creatures in the wild are perhaps the least of the benefits that humans derive from other species. A basic tenet of ecology is that everything is connected to everything else, and nowhere is this more apparent than in the interplay of the planet's species; without it, phenomena such as the pollination of plants, purification of air and water and the decomposition of waste would be severely impeded and in many cases non-existent.

E.O. Wilson describes this relationship in his (1987) essay, 'The little things that run the world':

> The truth is that we need invertebrates but they don't need us. If human beings were to disappear tomorrow, the world would go on with little change . . . but if invertebrates were to disappear, it is unlikely that the human species could last more than a few months . . . The soil would rot. As dead vegetation piled up and dried out, narrowing and closing the channels of the nutrient cycles, other complex forms of vegetation would die off, and with them the last remnants of the vertebrates.

A dramatic thought perhaps, but no less startling than the permanent loss of three species per hour as a consequence of human activity. Clearly biodiversity matters and is an urgent environmental problem. The UK statistics are similarly disturbing – since the Second World War 97 per cent of our wildflower meadows have been destroyed, half our ancient woodlands, 75 per cent of our heathland, and 98 per cent of our unique lowland raised bogs (dug up for garden peat!).

1.6.6 Chemicals exposure

This is a catch-all category that relates to both wide-scale and highly localised exposure to a variety of chemical substances. The concern, in a nutshell, is that we now produce and use a huge quantity and variety of man-made chemicals compared with society only a few decades ago. The European Environment Agency (EPA) suggests that in 1930 the global production of organic chemicals was approximately 1 million tons a year. Today it is about 400 million tons a year. Perhaps surprisingly this increase in quantity is accompanied by an apparent lack of knowledge of the impact of many of these chemicals either in isolation or combination on human beings and the environment. Of the 100,000 or so chemicals thought to be in production today – very little or no information is available on 85 per cent of them.

The immediate concern in many people's minds is what impact exposure to such chemicals may have on humans. Two groups of chemicals of particular concern are *persistent organic pollutants* (POPs) which will accumulate in body tissues and *endocrine disruptors* which affect reproduction, as humans are top of the food chain and often most directly in contact with domestic and industrial chemicals. Chemical accumulation in our body tissue (known as bio-accumulation) is thus a very real possibility. Numerous cases have been recorded including the following landmark example:

> The group of chemicals known as 'brominated flame retardants' is used for preventing fire in computers, televisions, etc. A sub-group of these chemicals has been found in women's breast milk. In 1998 a Swedish study showed that the concentration of PBDE (polybrominated diphenyl ethers) has increased 50 times in Swedish women's milk over the last 25 years. These industrially made chemicals have numerous effects in humans and in the environment, and it has been shown that miscarriages, learning disabilities and changes in the immune system are just some of the side-effects of flame retardants when tested on apes.
>
> (Darnerud, *et al.*, 1998)

Box 1.1 Waste in the UK: a case study

Waste generation in the UK is typically divided into three main sectors: construction and demolition, commercial/industrial and household waste. Waste production in the UK has been gradually declining in recent years reaching 288.6 million tonnes in 2008, down from some 335 million tonnes of waste in 2004. Of the 2008 total, 45 per cent was recovered while 48 per cent was deposited onto or into the land. Figures 1.15 and 1.16 illustrate the breakdown in both source and disposal routes.

Although the last few years have seen significant improvements, a significant proportion of UK waste still goes to landfill for disposal. There are a number of concerns (which are set out in the UK Government's Waste Strategy), associated with society's current reliance on landfill as a disposal option. These are:

▶ the risk of contaminating water with leachate;
▶ the risk of contaminating land and making it unsuitable for some uses (e.g. housing);

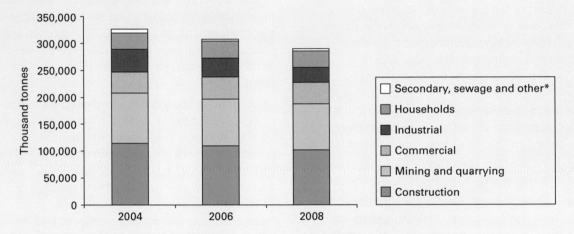

'Other' includes healthcare wastes, batteries and accumulators, and wastes containing PCB.

Figure 1.15 Breakdown of UK waste arisings
Source: DEFRA - data at end 2012.

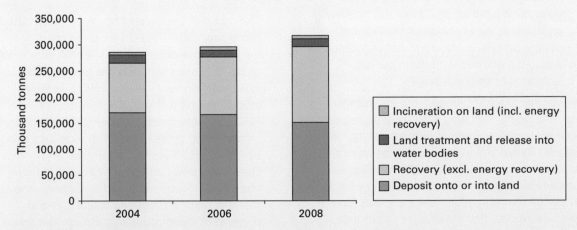

Figure 1.16 Breakdown of disposal routes
Source: DEFRA - data at end 2012.

▶ landfill gas can be dangerous and its methane content is an important greenhouse gas;

▶ suitable disposal space is running short in some areas.

These concerns do not mean that a modern well-engineered and well-managed site is an unacceptable environmental risk and landfill will continue to be an important disposal option for the foreseeable future. The UK government, however, is concerned that the versatility and convenience of landfill make it less attractive for waste producers to be innovative in the way that they deal with their waste.

Hazardous brominated flame retardants are used in many of the products we use every day, for example, computers, fax machines, textiles, toys, building materials, etc. The problem with flame retardants is that they do not stay in the products. Flame retardants are able to ooze from computers, televisions, etc. and via the air they enter our bodies and our blood and nervous system. Moreover, consumers ingest the flame retardant PBDE through fatty foods. There are still no absolute certainties on the toxic effects of flame retardants in human bodies, but scientists agree that it is extremely alarming that those chemicals are found in humans and especially in women's breast milk from where they will be passed on to infants.

The adverse human reactions posed by individual chemicals or complex mixtures range from sensitisation and associated allergic responses, through infertility to cancers. This is not to say that all chemicals are bad – undoubtedly they bring a plethora of benefits. However, the problem is that for many chemicals and chemical mixtures, we simply do not know what the potential risks are. Pre-release testing, bio-monitoring of accumulation of substances like PBDEs in human populations and correlations with health impacts would seem to be the appropriate response and in some parts of the world this is beginning to happen, e.g. the European REACH programme and the US bio-monitoring programme established by the Centers for Disease Control and Prevention in Atlanta.

1.6.7 Waste

As with water resources, the issue of waste is a 'double-headed monster'! First, we have the resource depletion issues associated with the waste generated. Second, we have the direct and indirect pollution issues associated with the disposal of the waste.

In short, to reduce the associated environmental consequences, we should produce less waste in the first instance and improve the way we manage any waste that is produced so as to minimise the associated pollution and maximise resource utilisation through reuse and recycling.

As a benchmark of UK performance in relation to other countries, the graph in Figure 1.17 relates to 2009 data for municipal waste in the European Union.

1.7 Human community issues

There are a wide range of issues relating to human communities and the impacts that can occur in relation to them. They range widely in scale from impacts facing one person or a household through to the regional to national and even global scale impacts (e.g. cultural heritage loss). Three key areas are outlined below.

1.7.1 Nuisance

Nuisance can be caused by a number of different factors and results in a range of undesirable outcomes. Common sources of nuisance arising from industrial sites include:

▶ noise disturbance

▶ odour

▶ light disturbance – excess at night, shading during the day

▶ visual intrusion

▶ dust

▶ traffic movements.

Examples of the consequences arising from such sources might be:

▶ disturbed sleep patterns or inability to use outside space;

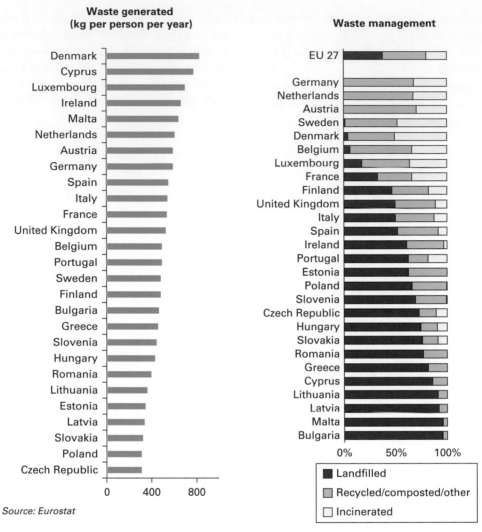

Figure 1.17 Eurostat data for waste totals and disposal route by country
Source: © European Union, 1995–2014.

▶ visual amenity loss, i.e. 'ruining the view';
▶ damage to or devaluation of property (permanent or temporary);
▶ congestion on transport routes.

A sometimes confusing quirk of terminology is that both source and consequence are often termed 'nuisance'.

1.7.2 Infrastructure/amenity impacts

This category includes a variety of 'nuisance-type' issues that tend to have a wider-scale impact than those listed above. For example, increased traffic associated with a large construction site may have a local 'nuisance' impact on households at key locations along the access route. However, it may also affect traffic flows over a wider area if hold-ups and congestion affect through traffic. On a large construction project this may last a considerable period of time. With a major new industrial site the change may be 'permanent'.

Similarly with access to amenities, whether schools, doctor's surgery, temporary or permanent accommodation, etc. Increased long- or short-term demand on such amenities in an area will affect all those relying or using the amenities in question.

1.7.3 **Cultural heritage impacts**

Loss or damage to cultural heritage may take many forms, both physical and social. The most obvious example might be the destruction of archaeological remains during the clearance of a site for a construction project. But losses may also take less dramatic forms, e.g.:

▶ pollution or vibration damage to historic buildings;

▶ devaluation associated with inappropriate nearby development e.g. building a McDonald's alongside Stonehenge!

In addition, changes to language and traditions associated with demographic change in an area, e.g. an influx of new population from different cultural groups which become dominant either by dint of cultural characteristics or simply numbers, leading to abandonment of the pre-existing language and/or traditions. Examples of this exist in distant locations like the Amazon but also closer to home, e.g. in some Welsh and Scottish villages.

III THE PLACE OF HUMAN BEINGS IN THE 'ENVIRONMENT'

The question of 'who are we and why we are here?' is undoubtedly beyond the scope of this book! However, the question 'what is our place on this Earth?' is squarely at the centre of our inquiry into 'environmental issues'. However, paradoxically, it is a question that is rarely voiced and yet might even be considered to constitute the heart of the matter. The idea of human beings being somehow separate from the living and non-living systems that make up our planet is perhaps the root of much of our current predicament. Here are some obvious, if often ignored, examples of our utter link to and dependence on natural systems:

▶ We can survive for only a matter of minutes without breathing. Breathing is an exchange of gases from within our bodies with the atmosphere. Every breath we take involves an interrelationship with the Earth's atmosphere – a taking in of oxygen and a giving out of carbon dioxide. This oxygen that we absorb drives the whole of the life process within us – our thoughts, feelings, acts are all utterly dependent on this exchange. In a sense, we are the air that we breathe!

▶ We can survive for only a matter of days without drinking – freshwater systems are essential for human life. In fact, our bodies are made up of more than 80 per cent water – in a sense we are 'clean water'!

▶ We can survive for only a matter of weeks without eating – everything we eat, whether animal or vegetable, involves a dependence on the ecological processes of the Earth. Even food produced from intensive farming relies on 'wild' insect pollination, microbial breakdown of wastes and a gene pool of plant and animal species that we continue to adapt to our needs. In the words of the old adage, we are what we eat and we eat the living planet!

It is a quirk of the human intellect that we can often behave individually and collectively as if none of these obvious statements were true. The truth of the matter is that the answer to the question of 'what is our place on Earth?' is obvious to anyone who looks:

> We are beings fundamentally dependent on and linked to the physical and biological processes that comprise the universe in general and planet Earth in particular. Individually and collectively we have some capability of influencing and changing those physical and biological processes. However, as with all other species on Earth, we have a range of needs essential to life which can only be met by our connection to the rest of the planet.

From this perspective our collective interest in the health and relative stability of physical and ecological processes is clear.

1.8 **Sustainability and sustainable development**

An inevitable consequence of the relationship described above is that everything we do has an environmental impact. When we breathe, we change the composition of the air that passes through our lungs, reducing the oxygen content and increasing the carbon dioxide. Change in the physical characteristics of the environment

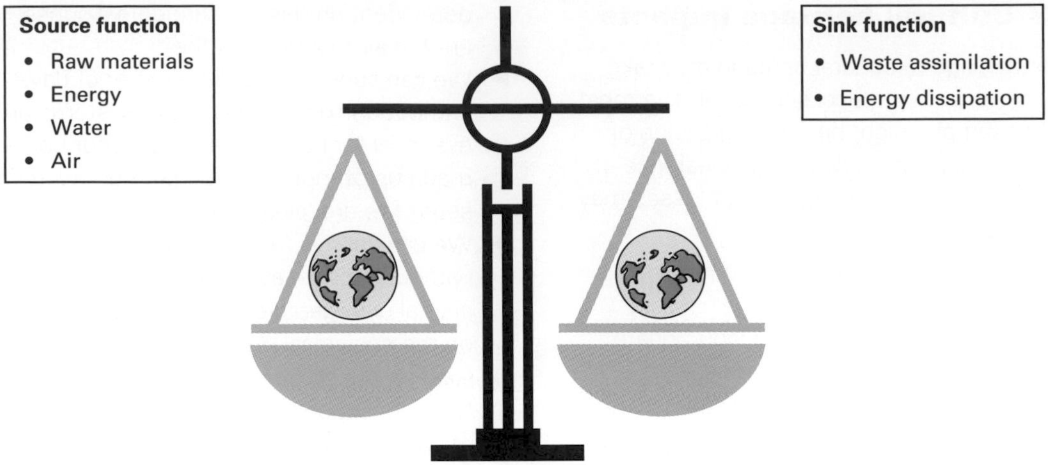

Source function
- Raw materials
- Energy
- Water
- Air

Sink function
- Waste assimilation
- Energy dissipation

Figure 1.18 The source and sink functions provided by the planet to humankind

is a natural and inevitable part of life. Indeed, the physical and ecological systems of the Earth are based on change in the form of transfer of energy/food between different levels in the living and non-living systems. There is a limit, however, to the level of change that can be sustained. A fish population can remain static even though 100,000 fish are removed annually, because the rate of removal does not exceed the rate of replacement in the form of birth and growth to adulthood. If, however, only 90,000 fish reach adulthood annually, then the population will begin to decline. Fishing has reached an *unsustainable* level.

Human existence depends on the sufficient availability of natural resources to satisfy our basic needs (see Figure 1.18). With a global population that continues to grow, the importance of the concept of sustainable resource management is crucial. However, the provision of resources is only part of what we depend on the Earth to provide. Human activity also depends on those natural resources to fulfil a 'sink function' to assimilate the wastes we produce, whether solid, liquid, atmospheric or energy. As long as these source and sink functions are used within their capabilities, then the environment (either locally or globally) will remain in equilibrium. From an environmental perspective, this is the basis of sustainability.

Essentially all the environmental impacts that an individual or organisation may have will relate to one or both of these functions. Sustainability, however,

has become recognised as a concept that cannot be dealt with from such an environmental perspective alone. The argument is that, unless social and economic conditions are acceptable, individuals and society as a whole cannot be expected to consider change linked to environmental sustainability regardless of whether detrimental consequences will occur in the long term. Short-term priorities dominate people's thinking. This tripartite view of sustainability (involving environmental, social and economic issues) will be explored further below. It may be useful to distinguish between two terms before we get involved in the politics and approaches to sustainability:

▶ *Sustainability* is a description of environmental balance (including human society's place in that balance) – sometimes the term 'environmental sustainability' is used to denote this meaning.

▶ *Sustainable development* is the process of transforming human society into a form that allows 'sustainability' to exist. It involves consideration of social and economic issues (as well as the environmental ones) that will need to be addressed along the way. However, it is not a catch-all phrase used to describe improvements in relation to a miscellany of social, ethical and environmental issues. Nor is it a vicarious and slightly woolly principle – it is the process of achieving sustainability, i.e. getting back within environmental limits as the only way of meeting human needs into the future.

In practice, the two terms are often used interchangeably but it is worth reminding ourselves of this distinction on occasion. The tripartite vision of sustainable development was never intended to imply that the end goal is anything other than ecological balance and should not be seen as a justification for environmental degradation.

1.9 Ecosystem products and services

Ecosystems services is a way of thinking about the source and sink functions described above in a rather more detailed way that even allows conversion into monetary values to facilitate incorporation into 'cost benefit analysis' style decision-making.

As a reminder, an ecosystem may be defined as: 'A community of interdependent organisms and the physical and chemical environment that they inhabit.' Ecosystem services may thus be defined as: 'Those attributes of ecosystems that enable or facilitate human life and well-being.'

Although the concept of ecosystem services has existed for a long time in academic circles, it received an increase in interest in corporate and policy-making circles following the publication of the UN Millennium Ecosystem Assessment. This was the culmination of a five-year study aimed at providing a state-of-the-art scientific appraisal of the condition and trends in the world's ecosystems and the services they provide. The study also provided an evaluation of the scientific basis for action to conserve and use those ecosystems sustainably.

Subsequently The Economics of Ecosystems and Biodiversity (TEEB) study was launched as

> a major international initiative to draw attention to the global economic benefits of biodiversity, to highlight the growing costs of biodiversity loss and ecosystem degradation, and to draw together expertise from the fields of science, economics and policy to enable practical actions moving forward.

The intention of the Millennium Assessment and the follow-on TEEB study was to demonstrate in ecological and economic terms the dependency of human society on natural ecosystems and biodiversity and to highlight the threat that on-going degradation of such resources represents to humankind.

The 2005 Millennium Assessment report outlined four basic categories of services provided by ecosystems to humanity. These have become known as Ecosystem Services and they provide another way for us to consider the type of impacts caused by human activity:

- *Provisioning services:*
  - food (including seafood and game), crops, wild foods, and spices
  - water
  - minerals (including diatomite)
  - pharmaceuticals, bio-chemicals, and industrial products
  - energy (hydropower, biomass fuels).

- *Regulating services:*
  - carbon sequestration and climate regulation
  - waste decomposition and detoxification
  - purification of water and air
  - crop pollination
  - pest and disease control.

- *Supporting services:*
  - nutrient dispersal and cycling
  - seed dispersal
  - primary production.

- *Cultural services:*
  - cultural, intellectual and spiritual inspiration
  - recreational experiences (including ecotourism)
  - scientific discovery.

The on-going goal of TEEB and those who see ecosystem services as a particularly useful concept, is to find ways of quantifying the economic benefit of ecosystem services so that environmental protection can be assigned a monetary value. The aim is to find a way for decision-makers to relate to environmental damage or protection that presents a clear and comparable picture to the fiscal data that forms the language of decision-making in most organisations.

1.10 The international commitment to sustainable development

The Bruntland Report of the World Committee on Environment and Development gave us the following definition of sustainable development as long ago as 1987: 'Development which meets the needs of the present without compromising the ability of future generations to meet their own needs.' The UK government expanded this definition in its 1990 White Paper, *This Common Inheritance*:

> Sustainable development means living on the earth's income rather than eroding its capital. It means keeping the consumption of renewable natural resources within the limits of their replenishment. It means handing down to successive generations not only man-made wealth, but also natural wealth, such as clean and adequate water supplies, good arable land, a wealth of wildlife, and ample forests.

1.10.1 The Earth Summit, 1992

In June 1992, close to 200 heads of state met in Rio de Janeiro for the Earth Summit (the United Nations Conference on Environment and Development). At this meeting four key agreements were signed and a United Nations (UN) Commission was established. One of these agreements was *Agenda 21*, which is the agenda, or plan of action, for the twenty-first century to bring about sustainable development. The UN Commission for Sustainable Development (UNCSD), established at Rio, is to oversee the implementation of Agenda 21. Agenda 21 is not a legally binding document but a strategy covering a wide range of areas, including business and industry. It embraced the idea that environmental or natural resource management issues could not be dealt with in isolation, and that factors such as social and economic needs form an important part of the equation. The challenge to the world governments is to achieve a high 'quality of life' for all, in a way that remains within the limits of our natural systems to sustain indefinitely. Hence the introduction of the term sustainable development, which allows for economic growth aimed at achieving high living standards for all, but in a way that does not exceed the capacity of the environment to sustain indefinitely.

The bottom line of this argument is that if people do not have, or are struggling to obtain, access to the basic needs of life, they cannot be expected to be concerned about long-term environmental viability. Few would argue with this premise, however, the idea that economic growth should form a central part of a long-term sustainability strategy is much more contentious.

Proponents of this view claim that as countries become more affluent, they move away from polluting manufacturing industries and towards a service sector, with heavy reliance on modern technology (which is less pollution-intensive). This, in combination with the increased awareness that high standards of education bring, leads to increased spending on environmental protection and enhancement. Therefore, the argument runs – economic growth equals increased environmental protection.

In recent years, however, a counter-argument has been receiving increasing support. This runs as follows: the idea that shifting manufacturing away from any particular country leads to a reduction in overall impact is a fallacy as increasing affluence has in the past always led to increased consumption. This in turn means increased manufacturing, albeit often in other countries. If such countries have environmental protection standards that are lower than in the original 'wealthier' country, there may even be a net increase in environmental impact. Second, the idea that technology always leads to reduced pollution may be flawed, as long-term implications of key technological changes remain unclear (consider the nuclear industry and genetic engineering – day-to-day impacts undoubtedly reduced but with an increased risk of major catastrophic impact). Third, the assumption that increased wealth automatically leads to an interest in environmental protection in individuals (and society) is open to challenge – especially where some environmental protection measures may have financial consequences or lead to some kind of infringement of an individual's rights to enjoy the freedom that such increased wealth brings.

1.10.2 The Earth Summit II, 1997

During the Earth Summit II in June 1997, the UN General Assembly concluded that the 'overall trends with respect to sustainable development are worse today than they were in 1992 (at the original Earth Summit)'. In response to this observation, they stated:

> Although sustained economic growth is essential to the economic and social growth development of all countries, in particular developing countries . . . unsustainable patterns of production and consumption, particularly in the industrialised countries, are the major cause of continued deterioration of the global environment.

A couple of other voices seem relevant here:

> You cannot solve a problem with the mindset that created it.
> (Albert Einstein)

And a more detailed statement by the late, great Jacques Cousteau:

> Sustainable development is wishful thinking. If development means growth, then you cannot have sustainable development on a limited planet. It is a principle that is impossible. I think we should change the wording to a 'sustainable future'. That could be achieved, *if* we use our efforts to grow in wisdom, culture, education and those things that are unlimited. But if our development is expressed in electricity and motor cars, then it is impossible.

A diplomatic view would perhaps conclude that while economic growth can be used to reduce adverse social and environmental impacts, it does not automatically follow that such outcomes will occur.

This raises some difficult questions about the rate and direction of change initiated by world governments in relation to the sustainability challenge since the agreement to pursue the principle at the 1992 Earth Summit.

1.10.3 Rio+10, 2002

Few real answers to such doubts were provided during the +10 Sustainable Development Conference which took place in Johannesburg in late 2002. Critics panned the conference because of its inability to produce concrete global targets. The only quantitative target agreed was: 'to halve the number of people lacking clean drinking water and basic sanitation by 2015'. This commitment at once epitomises the disparities in wealth and power between different parts of the world and highlights the gap between need and action in the pursuit of sustainability. The fact that the third Global Sustainability Conference should make as its single concrete commitment the provision of something that to most in the developed world is seen as a basic human right, may have raised international awareness of the social aspects of sustainability in a similar way that the 1992 Earth Summit did for climate change. This is perhaps further emphasised by subsequent estimates that suggest that the cost of providing clean water and basic sanitation to everyone currently lacking it would amount to less than the total annual marketing spend by US companies in one year alone!

Since the global financial crisis commenced in 2008, this whole debate has been brought to the fore once more. As governments around the world struggle to find ways to 'get us back on track', the question about 'how we gauge prosperity and economic/social success' should perhaps be re-examined. This is a big topic and not an easy one to grapple with from an individual or company perspective. However, ignoring the wider context in which we live and work is probably not sensible in the longer-term strategy of sustainability. For those interested in furthering their understanding of this debate, a good place to start is the Sustainable Development Commission's (2009) report, *Prosperity without Growth: The Transition to a Sustainable Economy*, available for download at www.sd-commission. org.uk/publications.php.

1.10.4 Rio+20, 2012

The most recent Earth Summit took place in June 2012, again in Rio de Janeiro. The published objectives of Rio+20 were:

▶ to secure renewed political commitment for sustainable development;

▶ to assess the progress to date and the remaining gaps in the implementation of the outcomes of the major summits on sustainable development;

▶ to address new and emerging challenges.

The outcome of the conference, a document entitled *The Future We Want*, was received by reactions that ranged from 'less than enthusiastic' to 'strongly hostile'. Although the principles and philosophies of Agenda 21 and the original Earth Summit were reaffirmed, commentators seem to broadly agree that little in the way of break-through or significant next steps had been agreed. One key outcome, however, was that from 2013 the UN Sustainable Development Commission was replaced by a 'High Level Political Forum' whose goals include the 'reinvigoration of international commitment to sustainable development'.

1.11 Sustainable development: an economic, social and environmental goal

For all the debate on individual policies described above, it is generally accepted that the three elements of economic, social and environmental health must be part of any sustainability strategy. The model of the three-legged table shown in Figure 1.19 represents the balance that must be achieved for 'sustainability' to exist.

Figure 1.19 The three-legged table model of sustainability

The model suggests that only when the table top is level (i.e. equal emphasis is allocated to the three legs) will sustainability be possible. What the model fails to convey, however, is what the view from the top of the table might be! In a society that is currently dependent on fossil fuel consumption and committed to economic growth, it is difficult to formulate a view of what such sustainability might look like in terms of the way business operates and people live.

As Kofi Annan, ex-UN Secretary-General put it: 'Our biggest challenge in this new century is to take an idea that seems abstract – sustainable development – and turn it into a reality for all the world's people.'

1.11.1 Relative sustainability

The way in which many individuals, companies and governments have dealt with this lack of clarity of the 'end goal' of sustainability is to approach the problem from the opposite direction. It is much easier to identify what is *not* sustainable about a company or society and in doing so we can usually identify actions and/or indicators that would make the company or society *more* sustainable. Adrian Henriques in his book, *Sustainability – A Manager's Guide* (2001), talks about this concept of 'relative sustainability' (the path), as being the basis of an approach to sustainability that gets around the difficulties of grappling with 'absolute sustainability' (the goal). The approach fits well with the evolutionary development of organisations responding to multiple pressures and has also been adopted by governments (see resources in Table 1.7) to track nationwide progress.

1.11.2 Absolute sustainability

There have also been a number of attempts to define absolute sustainability with one of the more cogent examples coming from the Sigma project (a UK government-supported initiative aimed at producing a systematic approach to sustainability for industry). Described in Henriques' book as the 'system conditions', for any activity to be sustainable, the following conditions should be satisfied:

- the environmental conditions, according to the Natural Step, see Table 1.7:
  - substances from the Earth's crust must not systematically increase in nature;
  - substances from society must not systematically increase in nature;
  - the physical basis for the productivity and diversity of nature must not be systematically diminished.
- the social conditions according to the Sigma Project, see Table 1.7:
  - organisations practise stakeholder dialogue and accountability, recognising the needs and values of stakeholders;
  - acceptable social, economic and environmental impacts are stakeholder-defined and equitable.
- the economic conditions, according to the Sigma Project, see Table 1.7:
  - scarce resources are used efficiently;
  - levels of economic activity are stable;
  - scarce resources are effectively used at all scales – from local to global.

IV FURTHER RESOURCES

Table 1.7 presents some resources for further study.

Table 1.7 Further information

Topic area	Further information sources	Web links (if relevant)
Aspects and impacts	ISO 14001:2004 Environmental management systems – requirements with guidance for use	www.iso.org
Sustainability	United Nations sustainable development knowledge platform	www.sustainabledevelopment.un.org/
	The Sigma Project	www.projectsigma.co.uk/
	The Natural Step	www.naturalstep.org/
	Forum for the Future	www.forumforthefuture.org/
Sustainability indicators	DEFRA publication – *UK Sustainable Development Indicators* 2013	www.gov.uk/defra
	Welsh Assembly publication – *Sustainable Development Indicators* 2013	wales.gov.uk

CHAPTER 2

Environmental policy

Chapter summary

This chapter covers three distinct areas, namely:

- *Principles* – this section explores the agreed 'fundamentals' on which international and most national policy relating to environment and sustainability are based. These include the polluter pays principle, the preventative approach, the precautionary principle, best available techniques, producer responsibility, lifecycle thinking and best practicable environmental option.

- *Instruments* – this section explore the tools or approaches used by the international community and, more commonly, national governments to translate policy into action. Four somewhat overlapping categories of instruments are considered – legal, fiscal, market-based and education/awareness measures. The section concludes with a consideration of the pros and cons of non-legal instruments.

- *Policy examples* – while international, European and UK policies are referenced in some detail in Chapter 3, some examples of policy initiatives are summarised here in relation to key global environmental challenges.

I ENVIRONMENTAL POLICY: AN OVERVIEW

The *Oxford English Dictionary* defines a policy as: 'a course or principle of action adopted or proposed by an organization or individual'. Increasingly environmental issues are being recognised as a shared challenge, caused by the whole of human society and trends in populations, economies and lifestyle aspirations. In response to this recognition, many attempts have been made to produce a standard benchmark of environmental principles to which all governments, institutions and individuals could refer. While no definitive international benchmark has been achieved to date, we have a reasonable attempt in the form of

Agenda 21, which is the internationally agreed sustainable development strategy drawn up at the 1992 Earth Summit in Rio de Janeiro. Most of the principles described in this chapter come from Agenda 21 with additional principles drawn from European law and organisational best practice.

II GENERIC PRINCIPLES OF ENVIRONMENTAL POLICY AND LAW

As indicated above, a number of key principles and approaches have emerged that form the basis of our current approach to environmental protection. The following list is not definitive and it is to be hoped that, as we progress, further principles will emerge but for now these are the key concepts built into many policies and laws at both international and national levels.

2.1 The polluter pays principle

This principle assumes that those responsible for the creation of pollution, in whatever form, should meet the cost of controlling and mitigating the resulting environmental impacts. Such costs may include the regulator's time, abatement techniques and monitoring, as well as remediation expenses.

2.2 The preventative approach

This principle assumes that it is preferable to prevent the creation of pollution or nuisances at source, rather than subsequently seeking to counteract their effects. It is epitomised by seeking to avoid the generation of an unwanted impact rather than mitigate the effects later.

In many countries, this approach is linked to the land use planning system though it can also be applied to enforcement of noise control and emissions of grit and dust from new installations. The Seveso Directive (which operates under the Control of Major Accident Hazards Regulations 1999 in the UK) also requires operators of certain industrial sites to prepare a major accident prevention policy and emergency plans for potential major accidents.

2.3 The precautionary principle

This principle confirms that: 'Even without scientific information or certainty that any particular level of emission causes harm, the level of emission should always be reduced to the minimum that technology can achieve.'

Or, to put it in another way: 'The lack of irrefutable proof in relation to the cause of an environmental impact should not be used as a reason to take no action against the "probable" cause of impact.' As with the polluter pays principle, the precautionary principle was first introduced to the international community as part of Agenda 21 in 1992.

2.4 Best available techniques (BAT)

As a general principle, Best available techniques (BAT) imply that anyone involved in the generation of an environmental impact, whether of a pollution or resource consumption type, should employ best practice to either minimise the likelihood of occurrence or ensure appropriate control and mitigation. In many ways, this is a development of the precautionary principle but with application to a particular industrial sector or activity.
The principle is at the heart of the European Integrated Pollution Prevention and Control regime which aims to ensure that all operators of highly polluting or potentially polluting industrial processes work to agreed best practice standards for their industrial sector.

2.5 Producer responsibility

In general terms, this principle looks similar to the polluter pays principle, however, it is normally used specifically in relation to waste. For example, in UK law it appears in two rather distinct areas, namely the 'duty of care' and the 'producer responsibility regime'. The 'duty of care' relates specifically to the responsibilities of those generating or managing waste to ensure that it is dealt with appropriately to minimise the scope for pollution. The 'producer responsibility regime' is part of an EU-driven legal framework that is aimed at manufacturers and sellers of goods and products that constitute key waste streams when disposed of at the end of their productive life. These 'producers' have 'responsibility' to ensure that at least a proportion of such wastes are recovered at end of life for reuse, recycling or waste to energy purposes. The principle of producer responsibility could be expanded to cover any product or service provision.

2.6 Lifecycle thinking and Best practicable environmental option (BPEO)

Over the years there have been many examples where an action taken to reduce or eliminate one environmental impact has simply shifted the impact in another direction. Sometimes this is done intentionally and is based on well-researched and informed decisions. However, often a decision is made without adequately considering the wider implications of the 'improvement' vs the 'original'. Lifecycle thinking implies that the consequences of an improvement or control method should be considered in a wide sense to avoid making changes that may, for example, reduce the air pollution generated at a particular location but perhaps lead to an overall increase in the amount of material usage or hazardous waste generated in the wider context.

Best practicable environmental option (BPEO) implies that the final design of a product, process or activity has considered the wider lifecycle impacts and represents the overall best option, given current standards of control and understanding. Clearly BPEO in this context has some overlap with BAT.

III POLICY INSTRUMENTS: DRIVING CHANGE AND IMPROVEMENT

Policies of themselves do not create change. They require mechanisms of implementation that bring them to the attention of appropriate parties and cultivate desirable behaviour while discouraging undesirable behaviour. Once such positive and negative perceptions are adopted as a widely held social norm, change can be said to have occurred.

Policy implementation measures aimed at driving change and improvement in both society at large and in organisations in particular may be broadly grouped under the following headings:

▶ Legal instruments
▶ Fiscal measures
▶ Market-based measures
▶ Education and awareness schemes.

2.7 Legal instruments

Perhaps the most immediately important mechanism from both a company and government perspective is the development,

application and enforcement of laws that require some or all sectors of society to behave in a particular way. This is the subject of extensive further coverage in Chapter 3. Suffice to say here that the success of legal instruments to generate change depends on the will and resources available to enforce them. In countries where regulators are well resourced and afforded significant operational powers, legislation is often a key driver to initially mobilise organisations into a process of change. In states with poor regulatory structures or in the international community where compliance is often based on a 'name and shame' basis, legal instruments may be less dependable.

2.8 Fiscal measures

As with legislation, fiscal measures or economic instruments aim to curb discharge of pollutants and/or consumption of environmentally damaging resources. They exist in many different forms as the examples below illustrate.

2.8.1 Taxation

Taxation may be both negative and positive. UK examples of negative taxation include:

▶ *Landfill tax*: payable on every tonne of commercial/industrial waste disposed of to landfill.
▶ *Climate change levy*: payable by all commercial/industrial users of energy in the form of electricity, gas or fuel oil for heating, lighting and power.
▶ *Aggregates tax*: payable on each tonne of primary aggregate extracted for commercial/industrial use.
▶ *Fuel duty*: payable on every litre of fuel used for motor vehicles.

A UK example of positive taxation in the form of a tax relief scheme is the contaminated land remediation relief applied to the clean-up costs associated with the redevelopment of brownfield sites, i.e. a site previously occupied by industrial activities.

2.8.2 Grants or incentive schemes

These are essentially the opposite of taxation and can be created to encourage the adoption of technology or practices that in some way generate an improvement in pollution control or resource efficiency. Funds are given or loaned to individuals or organisations for use in a specific way. UK examples include government grants for home insulation and other energy-efficiency measures.

2.8.3 Charges for licensing or administrative time relating to particular activities

Often given as a direct example of 'the polluter pays principle', charges for the regulator's time involved in ensuring that the most polluting or potentially polluting activities are managed appropriately can be used to encourage organisations to find alternative processes or raw materials to avoid the requirement to submit to expensive compliance assurance measures.

2.9 Market-based measures

These are an interesting area where we can perhaps expect to see many further developments in coming years. They come in a variety of forms but all essentially provide an incentive for organisations to consider environmental performance as part of their management priorities and strategy. For example, the European Union emissions trading programme is based on the allocation of tradable emissions quota to key industries to provide a positive incentive for companies to reduce their overall greenhouse gas emissions. Each individual company will consider its own position and make a commercial choice, based on a wide range of market factors, as to whether to invest in energy-efficiency improvements that enable it to sell excess quota or to continue to operate as normal and buy in any additional emissions quota required. The market essentially decides where improvements make most commercial sense rather than requiring everyone to improve in a similar manner.

In a completely different manner, regulating to push organisations to report on key environmental performance metrics can provide an incentive for them to improve because of the public visibility of their impact and the ability to compare products or overall performance with competitors. Customers can use such information to create market forces by selecting high performers and avoiding low performance organisations. This is the basis of reporting requirements in relation to things like new vehicle emissions.

2.10 Education and awareness schemes (voluntary initiatives)

In both the domestic and the commercial/industrial sector, education and awareness-raising programmes tend to be an important part of any long-term social change strategy. Whether they take the form of media-based advertising programmes or more direct sector-based initiatives, all take the stance that being better informed means that individuals and organisations will be more likely to make choices that lead to reduced impact.

In the UK, there are some long-standing government-funded best practice programmes aimed at sharing the benefits and ideas of a wide range of organisations in relation to improvements in waste reduction, energy and water efficiency and resource consumption in more general terms. Perhaps the best known examples in terms of assistance are now offered by WRAP (waste and resource efficiency) and The Carbon Trust (energy efficiency and greenhouse gas emissions reduction initiative). For further information, see www.wrap.org.uk and www.carbontrust.com.

2.11 The pros and cons of non-legal instruments

Fiscal and market mechanisms are increasing in relation to environmental issues and there seems to be increasing interest in further development. However, it is worth considering the relative advantages and disadvantages of this approach. Table 2.1 provides a summary.

Table 2.1 Advantages and disadvantages of non-legal policy instruments

Advantages of non-legal instruments	Disadvantages of non-legal instruments
• Allow individual flexibility and choice of action based on personal circumstances • Taxation can raise revenues to be invested in causes supporting the environmental issue in question • Has a greater tendency toward 'self regulation' • More likely to result in active engagement in solution finding / performance improvement (the more we do the more we save)	• May lead to organisations simply 'paying to pollute' or ignoring best practice guidance • If trading prices, arrangements or tax levels are set too low, there may be little incentive to change behaviour • If trading prices, arrangements or tax levels are set too high, there may be trade disadvantages created compared with operations in jurisdictions with no similar mechanisms • Can be cumbersome to administer with costs associated with accounting and assurance that reduce the benefits of any revenues raised.

IV EXAMPLES OF POLICIES AND IMPLEMENTATION INSTRUMENTS

International, European and UK policies are referenced in some detail in Chapter 3. However, Table 2.2 provides some examples of policy initiatives in relation to key global environmental challenges.

V FURTHER RESOURCES

Table 2.3 presents some resources for further study.

Table 2.2 Examples of environmental policies, principles and implementation measures

Environmental issue	Example of related policy	Key principles incorporated	Examples of implementation instruments
Climate change	The UK Low Carbon Plan, 2011	Polluter pays principle, Precautionary principle	The Climate Change Levy The Green Deal The carbon reduction commitment (CRC) scheme Mandatory greenhouse gas reporting for plcs
Ozone depletion	The Montreal Protocol on substances that deplete the ozone layer	Producer responsibility	EC Regulation 744/2010 on ozone-depleting substances
Trans-boundary air pollution	The convention on long-range trans-boundary air pollution	Polluter pays, Producer responsibility, BPEO	EU Integrated Pollution Prevention and Control Regime, 1999
Water pollution	The International Convention for the Prevention of Pollution from Ships (Marpol Convention)	Polluter pays principle and Producer responsibility	The UK Merchant Shipping (Prevention of Pollution by Sewage and Garbage from Ships) Regulations, 2008
Biodiversity loss	The Convention on Biological Diversity	The precautionary approach	National enforcement of the Convention on International Trade in Endangered Species UK local and regional biodiversity action plans
Pollution associated with hazardous waste disposal	The strategy for hazardous waste management in England, 2010	Polluter pays principle, preventative approach, lifecycle analysis and BPEO	The Hazardous Waste (England and Wales) Regulations, 2005

Table 2.3 Further resources

Topic area	Further information sources	Web links (if relevant)
Generic principles	Agenda 21	www.sustainabledevelopment.un.org/
Specific policy examples (national)	UK government policy web portal	www.gov.uk/government/policies
Specific policy examples (international)	UNEP Global Outlook on Sustainable Production and Consumption policies	www.unep.org/resourceefficiency/

CHAPTER 3

Environmental law

Chapter summary

This chapter explores the important area of environmental law. It is written from the perspective of the UK, and in most cases even more specifically, English law, but attempts to place this in the context of international and European law and policy. The layout of the various sections in this chapter is such that readers from outside the UK can dip into the sections on legal structures, international agreements and European law, without immersing themselves in the details of UK-specific legislation.

The chapter begins with Section 3.1 considering the basic structures of the UK legal system, including an overview of the courts, the regulators and the development of environmental law through the twentieth and into the twenty-first century.

There then follow eight sections exploring distinct areas of environmental legislation: permitting, air pollution, water pollution, waste management, contaminated land, nuisance control, development control (including conservation law) and hazardous substances. Although some UK legislation covers more than one of these areas, in terms of understanding legal requirements, this has proven to be a more accessible way of approaching an introduction to environmental law.

We then consider civil liability issues and the key case law relevant to environmental issues. The chapter concludes with a section on further information sources.

3 Environmental law

I UK ENVIRONMENTAL LAW: AN INTRODUCTION

3.1 Overview of legal systems in the UK

Law can be defined simply as a body of rules that aim to regulate the behaviour of society. In the UK and in many other parts of the world there are two sources of law:

▶ common law (civil or case law)
▶ legislation (statute law).

3.1.1 Common law (civil or case law)

British common law consists of a body of principles built up on judicial precedents (i.e. court rulings) dating back to the twelfth century, and they in turn superseded a network of courts that existed in Anglo-Saxon days. The expression 'common law' can also be used to denote our 'case law'. The decisions of courts are regarded as precedents, sometimes called judicial precedents, which subsequent courts are bound to follow when they are called upon to determine issues of a similar kind.

3.1.2 Legislation (statute law)

Since the eighteenth century, much use has been made of legislation or statute law. Legislation comprises Acts of Parliament and delegated legislation made by subordinate bodies or persons given authority by an Act of Parliament. Examples of delegated legislation are ministerial orders, regulations (statutory instruments), Local Authority bye-laws and court rules of procedure. Often, but not always, delegated legislation requires the approval of Parliament.

The most direct way to make a new law is by Act of Parliament, sometimes called a statute. New Acts are the result of a long process of discussion and negotiation with interested organisations, and the Act must pass through five stages in the House of Commons, five stages in the House of Lords and, finally, royal assent.

3.1.3 Civil claims and criminal cases

An important distinction between civil claims and criminal cases is the outcome expected in each category. The difference between these two systems is not the event or circumstances that give rise to the legal action, but the purpose for which the legal process is initiated namely punishment (crime) or compensation (civil). The same incident or occurrence may constitute a 'crime' and a 'civil wrong' (or 'tort') and hence be subject to parallel proceedings in both the criminal and civil courts.

3.2 Sources of legislation

Criminal law is usually enacted through Parliament in the form of statutes. Such legislation may be brought into UK law from a variety of sources, as shown below.

3.2.1 International agreements and conventions

These typically commit the UK and Europe to certain policy objectives, for example:

▶ the UN Montreal Protocol on substances that deplete the ozone layer (1989);
▶ the Basel Convention on the control of transboundary movements of hazardous wastes and their disposal (1989);
▶ the United Nations Convention on biological diversity (1992).

3.2.2 EU legislation

A significant proportion of UK environmental legislation has its origins in European law and for the foreseeable future this is likely to be the direction from which new initiatives arrive, either in policy or legislative format. The different forms of European law are shown in Table 3.1.

The most common form of European environmental legislation is the Directive, and the requirements of the Directive must be transposed into the national law of member states before they apply to organisations or individuals operating within the individual countries.

Box 3.1 Example of parallel proceedings

If the driver of a van in which you are travelling drives dangerously, resulting in an accident in which you are injured, both civil and criminal actions may follow. The police (actually the Crown Prosecution Service) may prosecute the driver in the criminal court. This would usually be the magistrates court. The result of a conviction might be a fine, a driving ban or imprisonment. This is intended to deter individuals from behaving in a way that society has democratically decided is unacceptable.

You as an individual may well seek compensation for the losses (damages) that you have suffered as a result of the negligent actions of the driver. You would sue the individual by taking out a civil action, possibly in the County Court or the High Court (depending on the amount of money you are claiming). Your action would be entirely independent of the criminal case, though if the driver is convicted of dangerous driving it will certainly help your case. Even if the driver is not charged, or is acquitted, you might still be able to prove that he was liable for compensation or 'damages'.

Civil cases are becoming increasingly important in relation to environmental issues, particularly in the areas of nuisance and contaminated land. However, on the majority of issues, statute is the key form of law shaping the way organisations and individuals behave in relation to environmental matters.

Table 3.1 Forms of European Union law

Type of EU law	Relevance to member states
Regulation	Approved by the Council of Ministers and apply directly to all member state governments. They come into effect on the date of publication but are seldom used due to their lack of flexibility.
Directive	Addressed to member states and binding with regards the result to be achieved but flexible as to means of implementation. Implementation timetable is normally specified and is achieved by the passing of national legislation. Failure to meet deadlines may lead to penalties from the European Court of Justice. There are two types of Directive: Framework Directives – deal with policy and general objectives Daughter Directives – specify limits, targets and standards.
EU Council Decisions	Applicable directly to whom they are applied (member states, groups or individuals), often administrative and used to ratify international agreements. They are legally binding and may be directly effective. They are only occasionally used in environmental law.
Recommendations and opinions	Not legally binding and only have a persuasive effect.

3.2.3 The UK legislative framework

UK legislation consists of laws made by Parliament directly, or under the authority of Parliament via the Secretary of State.

3.2.3.1 Primary legislation

These are the Acts of Parliament which tend to set out general requirements and provide a legal framework for detailed regulations to be introduced by the Secretary of State.

3.2.3.2 Delegated or secondary legislation

These are the detailed regulations made under the Primary Act and issued by the relevant Secretary of State. For example, the Environmental Permitting (England and Wales) Regulations 2010 were brought into force under the Pollution Prevention and Control Act 1999.

3.2.3.3 Guidance on legislation

Guidance is produced by government departments, the Environment Agency and other similar bodies. Such guidance does not carry the force of law and cannot be used in a court of law. However, it provides practical guidance and interpretation on Regulations. An example of this would be the Technical Guidance Notes issued by the Environment Agency.

3.2.4 Main UK environmental legislation

It is worth noting that since the mid-1990s most UK legislation has followed or been linked to legal developments in Europe. Furthermore, since devolution in the late 1990s, UK legislative coverage has, in some instances at least, followed a regional pattern (particularly in relation to secondary legislation) with variation in implementation arrangements between England, Wales, Scotland and Northern Ireland. However, particularly noteworthy UK environmental legislation includes:

> The Environmental Protection Act 1990
>
> The Water Resources Act 1991
>
> The Water Industry Act 1991
>
> The Clean Air Act 1993
>
> The Environment Act 1995
>
> The Pollution Prevention and Control Act, 1999 and the Environmental Permitting (England and Wales) Regulations, 2010.

The details of all this legislation will be covered in the subsequent chapters.

3.3 The evolution of environmental law from a UK perspective

In legal terms there has, perhaps unsurprisingly, been a huge increase in environmental coverage since the 1970s, though there are several examples of controls that were introduced long before this. Although this is perhaps an over-generalisation, we can observe shifts of focus in what might be thought of as the evolution of environmental control.

3.3.1 Pre-1960s: a focus on public health and property rights

The earliest legal references to what would be now considered to be pollution control appeared in civil law cases. The *Rylands v Fletcher* ruling from 1867 refers to responsibility of individuals to safeguard the release of substances from their property that might cause harm or nuisance to someone else.

In addition, there were early pieces of statute law relating to public health protection that still retain some relevance today. The Alkali Act of 1863 introduced a requirement for all alkali works which produced soda ash (a substance which was used extensively in the glass, textiles and paper industries of the day) to use 'best practicable means' to prevent the discharge of all noxious or offensive gases. The same standard of control is still referred to today in defences related to cases under nuisance law.

More recently the Clean Air Act of 1956 was enacted in response to the public health problems created by the urban smogs (the 'peasoupers') generated by the very widespread combustion of coal in domestic and industrial settings.

3.3.2 1960–1990: Issue specific legal controls and the emergence of planning control

The 1960s brought environmental issues into the spotlight in a number of ways but landmark publications such as *The Silent Spring* by Rachel Carson (1962) began to highlight the very real threats to not just human health, but also to the wider environment caused by human-generated pollution. As a result, specific legislation began to emerge, often prompted by individual concerns or problems. The Control of Pollution Act 1974 was a key piece of UK legislation that expanded on the concept of 'best practicable means' to noise control. The same control standard was applied under the Health and Safety (Emissions into the Atmosphere) Regulations 1983 which introduced the idea of a range of polluting processes that had to have a 'licence to operate' in order to facilitate regulatory control.

By the mid-1980s the emergence of the 'prevention is better than cure' approach was beginning to make its mark though conservation legislation such as the Wildlife and Countryside Act 1981 which established protected species and areas. Planning law was also being used to help control new sources of pollution and nuisance. The adoption in 1985 of the European Directive on Environmental Impact Assessment set the foundations for the 'risk assessment' and 'mitigation through planning' approach that continues in that area of planning law today.

In addition to these more general planning and permission style laws, specific legislation was emerging in relation to key issues of the time which included acid rain, the use of lead in petrol and ozone depletion. As an example, the Montreal Protocol was agreed by the international community in 1989 and formed the basis of legislation right up to the present day, requiring the phase out of ozone-depleting substances.

3.3.3 The 1990s: the beginnings of 'joined-up thinking'

In both the international arena and in the UK, the 1990s brought some major shifts in legal controls, beginning the process of considering wider environmental issues and requiring solutions or controls that did not simply mean moving a pollutant from one discharge route to another. The polluter pays principle was finding its way into legal controls in a variety of ways but perhaps is best epitomised by the 'Duty of Care' requirements in relation to waste introduced under the Environmental Protection Act 1990. Similarly, the Producer Responsibility regime introduced under the Environment Act 1995 brought requirements for organisations involved in the manufacture and sale of goods that ultimately became priority waste streams. The establishment of the Environment Agency in 1996 also brought into immediate being an 'environment-specific' regulator. The Agency's powers were equivalent to the Health and Safety Executive which had taken years to develop.

International policy was forming and negotiations commenced around a global response to the emergent threat of climate change which would require action in multiple areas and still represents perhaps our biggest challenge in terms of control and social change.

3.3.4 Into the twenty-first century: dealing with historic damage and looking to the future

In more recent years, issue-specific legislation has continued to emerge as our understanding of the threats posed by particular pollutants or to particular receptors has grown. However, there has also been some interesting signs of further trends in legislation towards consideration of the clean-up of pollution through legislation such as the Contaminated Land Regulations, 2001, and more recently the Environmental Damage Regulations, 2010, which afford considerable powers to the regulators to find and hold accountable 'appropriate persons' to clean up recent and historic pollution.

Legislation and financial measures are being increasingly used not just to curb pollution but also to protect natural resources though improving efficiencies in energy and materials usage. Concepts such as sustainability and corporate social responsibility are being nudged along through international and national policy with occasional legislation being used to make firm minimum requirements with regards availability of information and accountability for supply chain issues. An example of the latter might be the 2013 commencement of mandatory greenhouse gas reporting for publicly listed companies.

The UK Climate Change Act 2008 perhaps represents the shift in emphasis best with its 40-year commitments to radical change by requiring a formula of planning and review linked to interim and long-term targets. Such legislation is in some ways more akin to establishing a framework for widespread social change than to simply regulate a problem pollutant or activity.

Also relevant in this period has been the emergence of separate legislation under the devolved administrations of Wales, Scotland and Northern Ireland. However, while this appears to provide greater variation in legal requirements, at the same time we have moved firmly into an era when changes in environmental law are almost always negotiated and brought in under regional (European) or international structures. This again reflects the growing sense of recognition that many environmental issues are not confined within national borders and that most are ill suited to isolated legal responses.

3.4 Regulatory bodies and the courts

3.4.1 Government regulators

Environmental legislation is enforced by a number of regulatory bodies in the UK. They are summarised in Table 3.2 and then described in more detail below. DEFRA and DECC have a UK remit, while others operate on a devolved administration basis. Table 3.2 applies to England while Wales, Scotland and Northern Ireland are considered separately below.

3.4.1.1 Department for Environment, Food and Rural Affairs (DEFRA)

DEFRA is the central government department with responsibility in England for a wide range of matters including:

Environmental protection	Agriculture
Conservation	Rural issues
Water	Food safety
Hunting Animal welfare	Pesticides

The Welsh Assembly, the Scottish Parliament and the Northern Ireland Executive deal with policy-making on such matters in their relevant areas but DEFRA is responsible for negotiating on the UK's behalf at European and international policy levels.

The central negotiation role and the fact that the main policies of DEFRA and the devolved institutions are based on European initiatives and policies mean that there are typically significant similarities in policy and legislation adopted throughout the UK. Specific responsibilities of DEFRA include:

▶ drafting and implementing legislation;
▶ negotiating national agreements;
▶ formulation of policy;

Table 3.2 Environmental regulators in England

Legislative body	Areas of responsibility
Department for Environment, Food and Rural Affairs (DEFRA)	Drafting and implementing legislation
	Negotiating national agreements
	Formulation of policy
	Provision of guidance to other regulatory bodies
	Provision of scientific advice
Department of Energy and Climate Change	The UK Climate Change programme
Environment Agency	Environmental Permitting
	Water
	Waste
	Producer responsibility
	Radioactive substances
	Contaminated land (special sites)
Local Authorities	Environmental Permitting
	Clean air
	Air quality
	Nuisance
	Contaminated land
Sewerage undertakers	Effluent to sewer
Natural England	Conservation and land management issues

▶ provision of guidance to other regulatory bodies;

▶ provision of scientific advice.

3.4.1.2 Department of Energy and Climate Change

This department was formed in 2008 from the elements of DEFRA and the Department of Business Enterprise and Regulatory Reform (BERR), previously responsible for energy and climate change policy.

3.4.1.3 The Environment Agency

The Environment Agency (EA) has six regions in England with a head office in Bristol. Within each region they are further divided into areas, which can be seen as the first stop shops for regulated businesses. The Head Office is responsible for overall policy and relationships with national bodies including government. The organisation has wide-reaching statutory objectives and responsibilities as it was established under the Environment Act 1995. In general terms, it is responsible for enforcing legislation and prosecuting individuals and companies who fail to comply. EA officers have a range of generic powers to enable them to carry out their duties:

▶ right of entry at any reasonable time to any premises which they believe it necessary to enter (in an emergency this is any time and by force if necessary);

▶ rights to carry out investigations, including taking measurements and photographs;

▶ rights to take and remove samples and use them in evidence;

▶ rights to seize and render harmless any article or substance which appears to be the cause of imminent danger of pollution or harm to health.

In 2013, the two Welsh regions of the Environment Agency merged with the Countryside Council for Wales and the Forestry Commission in Wales to form a new regulatory body known as *Natural Resources Wales*. All of the previous responsibilities of the Environment Agency now fall under this body in Wales.

3.4.1.4 Local government

Local government consists of councils at different levels within an area. The main two are the County Council and the District Council, though in some areas these are combined as a Unitary Authority. County Councils are responsible for larger-scale planning consents and for the co-ordination of emergency plans for specified industrial activities.

District Councils determine local planning applications, produce local development plans and must appoint an Agenda 21 Officer who is responsible for the promotion of sustainable development issues at a local level.

District Councils also have Environmental Health Departments with Environmental Health Officers (EHOs) responsible for regulating the legislation within their remit and taking appropriate enforcement action. EHOs are responsible for:

▶ emissions from works not falling under the remit of the Environment Agency;

▶ local air quality under statutory nuisance provisions;

▶ enforcement of noise legislation;

▶ limiting smoke, grit and dust emissions;

▶ designating smoke control areas;

▶ monitoring and reviewing ambient air quality;

▶ monitoring drinking water quality of private and public supplies;

▶ identifying and initiating the assessment and remediation of contaminated land.

3.4.1.5 Sewerage undertakers

Although these are now private companies in England and Wales, they still carry out a regulatory role in terms of licensing discharges to public/municipal sewer systems.

3.4.1.6 Natural England

In Natural England's own words, their responsibilities include:

▶ managing England's green farming schemes, paying nearly £400million/year to maintain two-thirds of agricultural land under agri-environment agreements;

▶ increasing opportunities for everyone to enjoy the wonders of the natural world;

▶ reducing the decline of biodiversity and licensing of protected species across England;

▶ designating National Parks and Areas of Outstanding Natural Beauty;

▶ managing most National Nature Reserves and notifying Sites of Special Scientific Interest.

3.4.2 Regulators in Scotland

In Scotland, the Scottish Environment Protection Agency (SEPA) fulfils a similar function to the EA. Scottish Natural Heritage fulfils a similar role in Scotland to Natural England in England.

3.4.3 Regulators in Northern Ireland

In July 2008, the Northern Ireland Environment Agency (NIEA) was launched, encompassing many of the roles filled by the Environment Agency and Natural England in England. The Northern Ireland Environment Agency is subject to the overall direction and control of the Secretary of State for Northern Ireland; legislation-making staff have remained at the Department of the Environment (NI).

3.4.4 The courts

Figure 3.1 illustrates the structure of the England and Wales court system. Scotland and Northern Ireland have separate lower court systems but converge at the Supreme Court which has UK jurisdiction.

3.4.4.1 Courts of criminal jurisdiction

Minor criminal cases may be tried in a magistrates court who can issue fines up to a value of £20,000 and under certain circumstances a prison sentence of up to six months. For more serious offences, a formal charge (indictment) is issued in the magistrates court and the case referred for trial by jury in the Crown Court. Crown Courts can issue unlimited fines. Appeals against rulings follow the chain shown in Figure 3.1.

3.4.4.2 Courts of civil jurisdiction

In civil cases, the decision on which court deals with a case is based simply on the level of compensation claim that is being pursued.

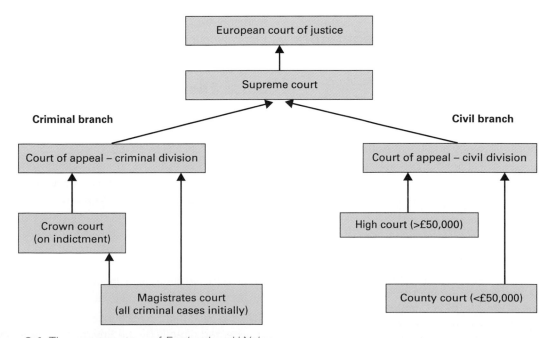

Figure 3.1 The court system of England and Wales

Claims less than £50,000 are normally heard in the County Court while those in excess of this amount are heard in the High Court. A similar chain of appeal exists as with the criminal branch.

3.4.4.3 The UK Supreme Court

In October 2009, the UK Supreme Court replaced the Appellate Committee of the House of Lords as the highest court in the United Kingdom. This court hears appeals from both the criminal and civil lower courts. Under European law, member states' courts should always make their rulings according to principles laid down in relevant decisions by the European Court of Justice (ECJ).

If the Supreme Court is considering a case where interpretation of an ECJ decision is unclear, the Justices can refer the question to the ECJ for clarification. They will then base their own decision on this answer.

In cases relating to the European Convention on Human Rights, and where human rights principles seem to have been breached, it may be possible to appeal to the European Court after all avenues of appeal in the United Kingdom have been exhausted.

3.5 The powers of the regulators

In addition to the generic powers described in Section 3.4, the environmental regulators have specific powers granted under specific statute. These are described in more detail in the relevant section of this chapter but those for the Environment Agency and Local Authorities are summarised in Tables 3.3 and 3.4.

3.5.1 Rights of appeal

3.5.1.1 The Environment Agency

In certain circumstances operators can appeal against a decision of the Environment Agency to the Secretary of State. The *Environmental Protection (Applications, Appeals and Registers) Regulations 1991* specify the conditions under which appeals procedures are to operate.

Appeals may be lodged with the Secretary of State by:

(a) a person who has been refused an authorisation;
(b) a person aggrieved by conditions attached to an authorisation;
(c) a person who has been refused a variation of an authorisation;
(d) a person whose authorisation has been revoked;
(e) a person on whom a variation, enforcement or suspension notice has been served.

3.5.1.2 Local Authority

An appeal against a Local Authority statutory nuisance abatement notice must be made within 21 days and there are several grounds for appeal:

(a) the notice was served on the wrong person;
(b) the time allowed for compliance is insufficient;
(c) the event does not constitute a nuisance;
(d) the requirements of the notice are imprecise;
(e) the requirements are unreasonable;
(f) there is some other material defect in the notice.

In addition, appeals can also be made against planning decisions with referral to the Secretary of State and public inquiry allowable under the Town and Country Planning Act.

3.5.2 The Environmental Civil Sanctions (England) Regulations 2010

This legislation has been introduced in an attempt to give the Environment Agency (and other regulators) an alternative to prosecution to secure binding and effective action to prevent or remediate pollution. Civil sanctions do not replace any other enforcement tools. The EA can and will still prosecute serious offenders, but *for a specified list of offences,* they are able to use alternative sanctions with legitimate businesses who are trying to do the right thing.

Unlike prosecution, civil sanctions are imposed or accepted directly by the Environment Agency. There are six types of civil sanctions:

Table 3.3 Legislative powers of the Environment Agency

Relevant legislation	EA powers
Waste Regulations, 2011	Register and withdraw registration from carriers of waste
	Confiscate and in some circumstances dispose of vehicles used in the illegal dumping of waste
EPA 1990 Part II	In response to suspected pollution – enter premises, take samples and request information or documentation (failure to comply may be submitted as evidence against the offender)
	Require the removal of fly-tipped material by the owner of the premises (defence available of unknown, unauthorised tipping) or in the event of high pollution risk, remove the waste and reclaim the cost
Water Resources Act 1991	Grant abstraction licences in relation to controlled waters
	Enter premises to investigate pollution events, take samples and demand the provision of information
	Clean up pollution damage and recover the costs of the clean-up from the polluter
Anti-Pollution Works Regulations, 1999	Issue works notices which require the person served to carry out specified works to stop or prevent pollution of controlled waters
Control of pollution (oil storage) (England) Regulations, 2001	Issue improvement notices on organisations where oil stores do not meet the standards of the regulations (in accordance with the implementation timetable)
Environmental Permitting (England & Wales) Regulations, 2010	Grant Environmental Permits with conditions for Part A1 regulated facilities and advise Local Authorities in relation to applications from A2 processes where discharges to water occur
	Grant Environmental Permits with conditions for specified 'water discharge' activities
	Issue enforcement and/or suspension notices where organisations have contravened or are likely to contravene the terms of their permit
The Environmental Civil Sanctions (Miscellaneous Amendments) (England) Regulations 2010	Provide for the issue of various statutory notices that focus on investment in environmental clean-up rather than forcing offenders to pay fines. Notices may be raised in relation to a wide variety of offences covered by other legislation.

▶ *Compliance Notice*: a regulator's written notice requiring actions to comply with the law, or to return to compliance, within a specified period;

▶ *Restoration Notice*: a regulator's written notice requiring steps to be taken, within a stated period, to restore harm caused by non-compliance, so far as possible;

▶ *Fixed Monetary Penalty*: a low-level fine, fixed by legislation, that the regulator may impose for a specified minor offence (£300 for organisations and £100 for individuals);

▶ *Variable Monetary Penalty*: a proportionate monetary penalty of up to £250,000, which the regulator may impose for a more serious offence;

▶ *Enforcement Undertaking*: an offer, formally accepted by the regulator, to take steps that would make amends for non-compliance and its effects; if the offer is accepted by the Agency and is completed, then there is no further criminal or civil sanction.

▶ *Stop Notice*: a written notice which requires an immediate stop to an activity that is causing serious harm or presents a significant risk of causing serious harm – this is the only civil sanction that may be combined with a prosecution.

Most of these notices are similar in format to those issued under powers granted by the Environmental Permitting Regulations 2010 but the monetary penalties are a new regulatory feature as is the Enforcement Undertaking. Where an Enforcement Undertaking is available for the particular legal breach, the offending party must set out how they propose to put the matter right. If the Environment Agency accepts their proposals, the Enforcement Undertaking becomes a legally binding voluntary agreement. The Enforcement Undertaking must identify

65

Table 3.4 Legislative powers of the Local Authorities

Relevant legislation	Local Authority powers
Control of Pollution Act, 1974	Set up noise abatement zones. Thereafter, the authority must measure noise levels at different premises, record the levels in a register, and serve notices on the owners and occupiers of the premises concerned. After that, the persons on whom notices have been served must not exceed the noise levels in the register. The Local Authority can also serve notice on an occupier or owner of premises to reduce noise levels and, if the person served fails to do so, it may carry out works to reduce the noise level.
	Under s.60 of COPA, Local Authorities have powers to control noise (and vibration) on or from building sites. This control is by the service of a notice making requirements on the person responsible for the construction operations to observe specified controls to minimise noise. The notice can specify types of plant and machinery, permitted hours of operation, boundary noise levels, etc.
Planning (Hazardous Substances) Act1990	In consultation with the Environment Agency and the HSE, issue consents for the presence of hazardous substances on, over or under land, at or above certain quantities.
EPA 1990 Part III	Where satisfied that a statutory nuisance exists, serve an abatement notice on the 'person responsible' for the nuisance, specifying the actions, works or steps required to abate the nuisance or prevent its occurrence (or recurrence) and the timescale for implementation.
Town and Country Planning Act 1990 and associated regulations	Grant or refuse planning permission with reference to planning policy. Also to call for EIAs to be completed for specified developments as part of the application process.
Clean Air Act 1993	Establish and enforce the provisions of smoke control orders. Section 20(1) of the Clean Air Act 1993 states that if smoke is emitted on any day from a chimney of any building within a smoke control area, the occupier of the building shall be guilty of an offence.
Noise and Statutory Nuisance Act 1993	If the Local Authority is satisfied that noise from vehicles, machinery or equipment in the street is causing, or is likely to cause, a nuisance, it must serve an abatement notice on the person responsible. The Act extends the powers of entry of EPA 1990 to allow anyone authorised by the local authority to enter or open the vehicle or remove it to a safe place, for the purposes of abating the nuisance.
Environmental Permitting (England & Wales) Regulations, 2010	Grant Environmental permits with conditions for Part A2 (in consultation with the EA) and Part B regulated facilities.
	Issue enforcement and/or suspension notices where organisations have contravened or are likely to contravene the terms of their permit.

the steps the business will take to put right any harm caused and it can also include providing compensation for the local community. It must also identify what the business will do to return to compliance, in both the immediate term and long term. The use of Enforcement Undertakings is interesting in that it allows for 'action that will secure equivalent benefit or improvement to the environment, where restoration of the harm arising from the alleged offence(s) is not possible'. For example, several companies that have agreed enforcement undertakings for breaches in relation to the Producer Responsibility Obligations (Packaging Waste) Regulations, 2007 – have agreed to make monetary donations to environmental charities such as the Wildlife Trust and the Woodland Trust.

Civil sanctions can only be used for designated offences committed in England after 6 April 2010 or after 15 July 2010 in Wales. Civil sanctions are available for offences under the following regulations:

▶ Control of Pollution (Oil Storage) (England) Regulations 2001
▶ Environment Act 1995
▶ Environmental Protection (Disposal of Polychlorinated Biphenyls and other Dangerous Substances) (England and Wales) Regulations 2000
▶ Hazardous Waste (England and Wales) Regulations 2005
▶ Hazardous Waste (Wales) Regulations 2005
▶ Land Drainage Act 1991
▶ Nitrate Pollution Prevention Regulations 2008 (England only)

▶ Producer Responsibility Obligations (Packaging Waste) Regulations 2007
▶ Salmon Act 1986
▶ Salmon and Freshwater Fisheries Act 1975
▶ Sludge (Use in Agriculture) Regulations 1989
▶ Trans frontier Shipment of Waste Regulations 2007
▶ Water Industry Act 1991
▶ Water Resources (Environmental Impact Assessment) (England and Wales) Regulations 2003
▶ Water Resources Act 1991.

Natural England can also use civil sanctions in relation to offences under legislation such as the Wildlife and Countryside Act 1981 and the Protection of Badgers Act 1992.

3.6 Penalties for legal non-compliance

Non-compliance with environmental regulation can give rise to significant liabilities for an organisation in a variety of different ways.

3.6.1 Prosecution

In the event of a breach of statutory requirements an organisation may be subject to prosecution and, certainly in the case of serious pollution incidents, may find themselves in receipt of significant fines. Many environmental statutory offences carry fines of up to £50,000 and the threat of prison sentences up to 12 months. More serious offences that proceed to the Crown Court may receive unlimited fines and custodial sentences of up to two years.

The custodial offences referred to above apply where an individual, typically a company Director or Manager, is found personally liable for the offence in question.

3.6.2 Statutory notices

In addition to prosecutions, the regulatory authorities have a range of administrative tools at their disposal which might be employed instead of or as well as court proceedings. Civil sanctions, works notices and environmental damage notices may be raised by the Environment Agency requiring remediation or reinstatement works to

be completed or even compensation payments to be made to appropriate parties. Local Authority abatement notices, and Agency enforcement/ prohibition notices may require specific action to be taken to deal with the offending activity and may even require operations to cease until the problem has been resolved. All such improvement notices may impact upon the organisation's ability to conduct its operations in both the short and longer term through restrictive practices.

3.6.3 Clean-up costs

The financial liabilities associated with pollution clean-up costs may also be very significant. Contaminated land remediation costs frequently run into millions of pounds, while a major incident affecting a vulnerable river, stream or aquifer can also be hugely expensive in terms of clean-up and reinstatement. In addition to specifying works that must be completed by the polluter, the regulators are also able to seek repayment of any costs incurred by them in dealing with the incident.

Depending on the nature of the incident there may even be long-term implications in terms of land values if the pollution incident requires significant clean-up operations.

3.6.4 Civil liability

Parties affected by a pollution incident may seek compensation for losses incurred through the civil courts. Such claims can also be significant if there is a long-term loss of amenity or perhaps the potential for human health impacts.

3.6.5 Reputation damage

Finally, the reputation damage arising from a significant non-compliance event and the accompanying negative publicity may result in difficulties with a wide range of stakeholders. These might include local regulators (planning and permitting), customers (supplier selection criteria), neighbours (infringement of trust and hence increased likelihood of complaints and/or objections to planning or permit applications), shareholders (direct pressure from current shareholders or via ethical investment fund managers), lending institutions (lending criteria exclusions) and/or insurance companies (increased premiums or refused coverage).

3.7 Environmental liability in the European Union and the UK

Under Part IX of the UK Environmental Protection Action 1990, the legal principle of personal liability for environmental damage was established:

> Where an offence under any provision of this Act committed by a body corporate is proved to have been committed with the consent or connivance of, or to have been attributable to any neglect on the part of, any director, manager, secretary or other similar officer of the body corporate or a person who was purporting to act in any such capacity, he as well as the body corporate shall be guilty of that offence and shall be liable to be proceeded against and punished accordingly.
>
> (EPA 1990 Part IX, section 157[1])

There are also many examples of incorporation of the 'polluter pays principle' into UK legislation, e.g. the Environmental Permitting Regulations, Contaminated Land Regulations, etc.

In 2004, the EU produced *Directive 2004/35/ CE on Environmental Liability with Regard to the Prevention and Remedying of Environmental Damage*. The Directive provides a legal framework for three of the key legal principles introduced in Chapter 2: 'the polluter pays', 'prevention' and 'the precautionary principle'. The aim is to cause business to behave more responsibly in relation to environmental damage.

The Directive has been enforced in England and Wales via the *Environmental Damage (Prevention and Remediation) Regulations, 2009*. Similar legislation also applies in Scotland in the form of the *Environmental Liability (Scotland) Regulations 2009*. Under the Regulations, environmental damage is described as:

▶ serious damage to surface or groundwater;
▶ contamination of land where there is a significant risk to human health;

▶ serious damage to EU protected natural habitats and species or damage to Sites of Special Scientific Interest (SSSIs).

The Environmental Damage (Prevention and Remediation) Regulations, 2009 are applicable to both the public and private sector and require any organisation whose activities are causing or are threatening to cause 'environmental damage' to do the following:

▶ take steps to prevent the damage (or further damage) occurring;
▶ inform the appropriate regulatory body of the threat or incident.

The Regulations are enforced in England by the regulators listed in Table 3.5.

The appropriate regulator may raise a statutory 'remediation notice' if a satisfactory response is not taken by the responsible party. Any costs of assessment, pollution prevention or remediation incurred by the regulators are recoverable from the operator. Essentially the Regulations make an organisation responsible for pollution prevention and remediation, regardless of any other legal offences that may have been committed.

In addition, persons 'with a sufficient interest' can request action from the appropriate regulator if they perceive a threat of environmental damage to something of concern to them. This could include NGOs, residents associations, anglers, ramblers, birdwatchers, etc.

Damage that occurred prior to the Regulations coming into force or if associated with operations that ceased prior to the Regulations is exempt.

Table 3.5 Regulatory enforcement of the Environmental Damage (Prevention and Remediation) Regulations, 2009

Environment Agency	Permitted activities and water damage/ biodiversity damage in waters other than the sea
Natural England	Damage to biodiversity
Local Authorities	Land damage
Secretary of State	Damage at sea

3.8 Access to environmental information

There are international, European and national levels of commitment to providing public access to environmental information.

The so-called *Aarhus Convention* was adopted in 1998 in Aarhus, Denmark, by the members of the United Nations Economic Commission for Europe (UNECE) and came into force in October 2001. Its aims are to strengthen the role of members of the public and environmental organisations in environmental protection issues through the provision of greater access to information held by public bodies.

The EU met the terms of the Aarhus Convention relating to environmental information in the form of the *2003 Directive on Public Access to Environmental Information*. The Directive had to be implemented by member states by February 2005. It has a very wide definition of environmental information which in effect means that any *publicly held information* on the state of the environment or on processes that could or do affect the environment must be made available to the public unless reasons of genuine commercial sensitivity or national security exist. As well as national, local and regional government and public administrations (e.g. the Environment Agency), the definition of public authorities includes those carrying out functions or providing services directly or indirectly related to the environment, e.g. water companies, waste disposal companies. Information must be made available on request as soon as possible and at least within one month (2 months for complex requests). A formal refusal and appeals procedure is also defined.

The Directive is met in general terms in the UK through the *Freedom of Information Acts* (2000 – England, Wales and Northern Ireland, 2002 – Scotland) and specifically via the *Environmental Information Regulations 2004* (separate but similar regulations exist for Scotland). These Regulations define public bodies and environmental information and set out procedures and a code of practice for dealing with requests, timescales, charging, appeals and exemptions.

As demanded by the Directive, if a request for information is refused, the enquirer must be given a reason for the refusal in writing. Requests for information can be refused if the disclosure would adversely affect:

- ▶ international relations, defence, national security or public safety;
- ▶ justice or disciplinary reasons;
- ▶ commercial or industrially confidentiality;
- ▶ the interests of the person who provided the information;
- ▶ the protection of the environment to which the information relates.

In all these cases, however, information can only be withheld if the public interest in doing so outweighs the public interest in disclosure. An Information Commissioner has been established to oversee the implementation of the Regulations by the various bodies affected.

3.8.1 Chemical release inventory

In addition to the requirements relating to public requests for information, there is a significant amount of information required to be put into the public domain by individual pieces of legislation. Examples include the announcement of planning applications and the public availability of both applications and supporting environmental impact assessments under planning law. Under the Environmental Permitting regime a pollutant-specific register (known as the chemical release inventory or CRI) is maintained by the Environment Agency with the aim of providing information on releases of pollutants which will help enforcing authorities to identify environmental black spots. It also provides information to the public and environmental interest groups in an understandable format.

The information is presented by substance, by industry and by Local Authority area, there is commentary on the actual data in terms of emission and the potential environmental effects. It is held by the Environment Agency on an electronic database which is partially accessible through the EA website.

At a European level, as part of the IPPC Directive, member states, from June 2003, have had to submit national pollutant release data to the European Commission. The data is used to collate a Europe-wide Pollutant Release and Transfer Register. This is also available online as an interactive database: http://prtr.ec.europa.eu/MapSearch.aspx, providing individual site locations, licence details and emissions summaries (as with the CRI in the UK). In addition, it is possible to search by pollutant and to view aggregate totals of emissions to air, land and water for individual countries or the EU as a whole.

II ENVIRONMENTAL PERMITTING

3.9 The issue

In the UK, we have a control regime that covers what may be considered the most polluting or potentially polluting industrial activities through a system of permits. Such permits require operators to consider and control their emissions to air, land and water and in many cases, to meet defined standards in relation to efficiency of resource use.

In a sense the environmental permitting regime provides a licence to operate system that demands best practice standards from those activities with the greatest potential to cause environmental harm.

3.10 International agreements

There are no international agreements directly relevant to this part of UK environmental law, though the regime embodies many of the core pollution prevention principles set out in Agenda 21 at the Earth Summit in Rio de Janeiro in 1992.

3.11 European legislation

The 1998 *EC Directive on Integrated Pollution Prevention and Control* now amended as *Directive 2008/1/EC* has as its main purpose

> to achieve integrated prevention and control of pollution from a wide range of industrial and agricultural activities. This is to be done by preventing, or, where that is not possible, reducing emissions to the air, water and land,

so as to achieve a high level of protection of the environment as a whole.

This is an example of what has become known as a 'multi-media approach to regulatory control'. This approach to regulation came about in order to prevent the simple shifting of pollutants from one discharge route to another.

Annex 1 to the Directive lists those activities covered by the regime. Technical guidance documents describing 'Best Available Techniques' for each of the sectors included in the regime are prepared and published by the European IPPC Bureau on behalf of the European Commission (further details in the description of the UK regime that follows).

3.12 UK legislation overview

3.12.1 The Pollution Prevention and Control Act 1999 and the Environmental Permitting (England and Wales) Regulations 2010

The *Environmental Permitting (England and Wales) Regulations 2010*, enacted under the Pollution Prevention and Control Act 1999, provide a consolidated system of environmental permitting in England and Wales. They replace the *Environmental Permitting (England and Wales) Regulations 2007*, the system of consenting of water discharges in the Water Resources Act 1991, the groundwater permitting system in the Groundwater (England and Wales) Regulations 2009 and the system of radioactive substances regulation in the Radioactive Substances Act 1993.

Operators who were already in possession of an existing environmental permit, pollution prevention and control permit, water discharge consent, waste management licence, etc. automatically became Environmental Permit holders from April 2010. There was no requirement to reapply for or receive a new permit and all permit conditions remained the same.

3.12.2 Statutory offence and penalties

Under the terms of the Regulations:

(a) a person must not, except under and to the extent authorised by an environmental permit –

(b) operate a regulated facility; or

(c) cause or knowingly permit a water discharge activity or groundwater activity (unless exempted by the Regulations).

Failure to abide by this may lead to fines of up to £50,000 and/or prison sentences of up to 2 years if the case is heard in the magistrates court. If the case is referred by the magistrates to the Crown Court, then fines are unlimited and the maximum prison sentence increases to 5 years.

Essentially operators of 'regulated facilities' or 'water discharge activities' must hold a valid permit from the appropriate regulator (the EA, or the relevant Local Authority). It is for the regulators to determine the number and nature of conditions attached to a permit, however, each process must meet a best practice standard for all aspects of the licensed activities. Such standards are known as 'Best Available Techniques' or BAT and are defined for industrial sectors at European level with guidance published in the form of *BREF Notes* (best practice reference) or UK guidance notes for Part B processes. BAT refers to operational as well as technological factors, including such things as staff qualifications, working methods, training and supervision, design, construction and maintenance procedures, design of plant and equipment, monitoring and analysis procedures.

The regulators are also required to review the conditions of each permit regularly – roughly once every six years – and to consider any changes to BAT as part of this process.

3.13 UK regime: key points

3.13.1 Who requires an environmental permit?

Schedules to the Regulations outline the activities covered by the Environmental Permitting regime as well as exemptions. In schedule 1, under sector headings, three categories of process are identified:

▶ *Part A1 processes* – require a permit from the EA and are regulated in relation to emissions to air, land and water as well as with regard to other operational conditions such as noise, energy efficiency and waste minimisation. Processes in this group are considered those with the highest pollution potential and include such activities as large power stations, chemical plants and industrial activities using large quantities of very hazardous materials.

▶ *Part A2 processes* – require a permit from the Local Authority. They may be regulated against the same range of operational conditions as A1 processes and, if discharges to water are involved, then in England and Wales, the EA must be consulted.

▶ *Part B processes* – require a permit from the Local Authority but are only regulated in relation to their discharges to air. BAT for this group of activities is set by the UK rather than on a European basis.

Other schedules deal specifically with waste operations, mining operations, water discharge and groundwater activities and radioactive substance activities – defining those requiring a permit and those eligible for exemptions. All of these activities are regulated by the Environment Agency.

3.13.2 Types of environmental permit

There are three types of environmental permit that the Environment Agency or Local Authority can issue:

▶ Standard permits: Standard permits (SPs) are suitable for some activities and contain a standard set of compliance conditions. This type of permit is simpler and cheaper to apply for and, because they do not require individual public consultation, they are likely to be issued more quickly.

▶ Bespoke permits: Where a standard permit is not suitable, for example, where operations could have a high impact on the environment or if they are novel or complex, a bespoke permit will be required. This means that the Environment Agency or local authority will set specific conditions for the facility.

▶ Consolidated permits: For organisations with more than one permitted activity, it is possible to apply to the relevant regulator for a consolidated permit.

3.13.3 Coverage/scope of conditions associated with environmental permits

Permit coverage will vary widely but may cover:

▶ emission limit values to land, air and water (air only for part B processes);
▶ resource efficiency conditions (relating to raw materials, wastes and energy);
▶ protection measures for soil and groundwater; site reinstatement requirements;
▶ nuisance control limitations;
▶ monitoring and operational management requirements (including incident prevention and response).

The key references used to determine consent conditions for any individual permit are as follows. For standard permits, it is the general binding rules that have been produced by the Environment Agency for all activities of a similar nature. For bespoke permits, the reference is the relevant BREF note or Process Guidance note for Part B processes. These references are used by both the operator and the regulator during the permit application process.

3.13.4 Revocation, enforcement and suspension powers

The appropriate regulator may issue a partial or complete *revocation notice* which specifies a 'permanent' cessation of activity as well any steps that must be taken to prevent pollution or return the site to a satisfactory condition.

An *enforcement notice* may be issued where the regulator believes the operator is not complying with the permit. The notice states the contravention of the permit, the remedial steps to be taken and a time limit for compliance.

A *suspension notice* may be issued where the regulator believes that the operation of the regulated facility constitutes a serious pollution risk. The permit holder may not operate the installation until such time as the remedial actions specified within the suspension notice have been successfully implemented and the notice withdrawn by the regulator.

3.13.5 Definitions of BAT

As previously indicated, BAT under the Environmental Permitting regime is set at a European level, with guidance for each Part A1 and A2 process being published in the form of 'BREF notes'. BREF notes are available from the website of the European Integrated Pollution Prevention and Control Bureau. Process Guidance notes for Part B processes, which are developed and apply only within the UK, are available from the DEFRA website at www.defra.gov. uk/Environment/airquality/lapc/pgnotes/default. htm. In both cases it should be noted that Best Available Techniques refers to operational as well as technological factors and can include (but are not limited to) any or all of the following:

▶ staff qualifications;
▶ working methods;
▶ training;
▶ supervision;
▶ design;
▶ construction and maintenance;
▶ design of plant and equipment;
▶ monitoring and analysis.

3.13.6 Regulatory guidance

In addition to the process/sector BREF notes, there is considerable additional guidance on the Environmental Permitting regime. A range of general guidance documents, plus the so-called 'horizontal' guidance notes (covering issues such as monitoring, energy efficiency, waste minimisation) are available for download from the Environment Agency and DEFRA websites at www.gov.uk/environment-agency and www.defra.gov.uk.

3.14 The permit application process

3.14.1 Full/bespoke permit application requirements

The Environmental Permitting application process essentially requires the operator to prove that:

▶ they have adequately considered all environmental receptors that may be affected by their operation;

▶ they will operate their process and all related operations in accordance with the Best Available Techniques (BAT), as defined in the relevant sector BREF note, and any relevant horizontal guidance.

In order to meet the requirements of the application process, the operator must compile information on such issues as:

▶ the presence of sensitive receptors, e.g. conservation areas, watercourses, aquifers, etc.;

▶ operational controls – both technical and administrative, e.g. management systems, training and competency, etc.;

▶ existing site conditions (with particular reference to the presence of contaminants associated with the process for which the application is being made).

The completed application pack is submitted to the Environment Agency (or Local Authority for Part A2 and B processes). There then follows a period of review by the regulator and a number of statutory consultees which vary depending on the application but may include, among others:

▶ the conservation bodies, e.g. Natural England;

▶ the relevant health authority;

▶ the sewerage undertaker;

▶ the Health and Safety Executive.

There is also a requirement for a public consultation to be made by the applicant, in the form of a press notice in both appropriate local newspapers and the *London Gazette*. The lead regulator (the EA or Local Authority) acts as the coordinator of the views expressed by all parties, and must consider them in deciding whether or not to grant the Environmental

Permit. Often the regulator will ask the applicant for further information in support of the application and this is done via an official notice. The applicant then responds with additional details. This process may be repeated as many times as necessary.

Assuming the application is successful, the regulator will issue first a draft and then a final permit which becomes the site's 'licence to operate' and will include constraints relating both to emissions and in relation to any of the operational standards considered in the description of BAT in Section 3.13.5. In addition, the permit:

▶ is linked to application commitments, including operational controls, monitoring and reporting activities;

▶ may contain improvement actions over defined timescales;

▶ is subject to rolling review by regulator.

3.14.2 'Standard permit' applications

Applications for standard permits essentially follow the above process but with the following notable exceptions:

Applicants must refer to the generic assessments of risk for standard facilities to determine whether their activity is within the scope of the standard rules and, if so, they must demonstrate the adoption of suitable control measures to meet those rules.

Public consultation on applications for individual standard facilities is not required (other than for Part A installations). This reflects the fact that consultation in the development of the rules has already taken place.

3.15 EP OPRA: the Environment Agency's risk-based approach to regulation

Risk-based regulation enables the Environment Agency to systematically establish priorities and target its regulatory effort on higher-risk issues. A key component of this risk-based approach to regulation is the Environmental Protection Operator and Pollution Risk Appraisal (EP OPRA).

This is a tool for assessing the environmental risks of sites regulated by the EA and the competence of site operators.

The EP OPRA scheme is primarily a risk-screening methodology. The risk assessment framework incorporates an element of professional judgement, but the methodology itself is simple to apply and objective in nature. EP OPRA helps the Environment Agency target its regulatory effort at those activities that present the greatest risk to the environment. In line with the polluter pays philosophy, outputs from this methodology are built into the charging scheme for the Environmental Permitting regime and are reviewed annually.

3.15.1 Three tiers of regulated activities

Under the terms of the Environmental Permitting Regulations, the EA considers there to be three tiers of regulated activities:

Tier 1 – activities required to be registered under the EP regime but not subject to full-scale permits.

Tier 2 – activities for which the EA decides whether a permit will be issued or not but, where permitted, fall under a fixed charging scheme as they are simpler processes. In this category, however, there may be some consideration taken of compliance history in setting annual subsistence fees.

Tier 3 – more complicated, typically A1 activities for which the OPRA scheme is applied in full.

All Tier 3 activities are subject to OPRA approval to determine regulatory priority and annual subsistence charges to the permit holder. Although the assessment is based on a common approach, specific guidance is provided for 'installations', 'waste management facilities' and 'mining waste operations', In each case, the EP OPRA assessment is based on the determination of five attributes:

Complexity
Location
Emissions
Operator performance
Compliance rating.

Complexity: From an Agency perspective, the more complex an installation is, the more regulatory effort will be needed to understand the processes involved, their interactions and pollution potential. The Agency has assessed each sub-paragraph of Schedule 1 of the Environmental Permitting Regulations. It has placed each of the activities listed into one of five bands, A–E. An example is shown in Figure 3.2.

This classification process uses a simple scoring system and considers the following factors in determining the band rating for each process:

▶ has significant releases to one or more media;
▶ uses one or several interconnected but distinct processes;
▶ has a significant potential for accidental emissions;
▶ carries a significant inventory of potentially hazardous materials;
▶ is a specified waste activity;
▶ is of a significant size, relative to the sector in which it appears;
▶ is likely to require significant regulatory effort to assess and maintain compliance and thereby maintain public confidence.

Location: The presence or absence of key receptors that could be affected by the regulated activity is a further indication of the potential hazard of the installation and of the assessment required. As with the complexity rating, A–E bands are generated for the location attribute, with the scores produced by reference to Tables 3.6 and 3.7.

Emissions: An emission's index approach is used for emissions to air, water and land. The indices are calculated from the following:

▶ the annual load of each pollutant that would be emitted if the installation operated at the emission limit values contained within the permit to operate;
▶ the 'Emission Thresholds' for that pollutant (which vary between pollutants and reflect the potential of the substance to cause environmental harm).

Annex 1 table 1 definition of activities and ep opra complexity band scores. Reference in this Table to Part A(2), Part B, Part 2 of this Schedule is to Schedule 1 contained in the Pollution Prevention and Control (England and Wales) Regulations 2000 SI 2000 No 1973.			Band
Chapter 1 – Energy industries			
Section 1.1 – Combustion activities			
Part A (1)	a)	Burning any fuel in an appliance with a rated thermal input of –	
		(i) 50 megawatts or more	C
		(ii) 300 megawatts or more.	D
	b)	Burning any of the following fuels in an appliance with a rated thermal input of 3 megawatts or more but less than 50 megawatts unless the activity is carried out as part of a Part A(2) or B Activity –	
		(i) waste oil	C
		(ii) recovered oil	C
		(iii) any fuel manufactured from, or comprising, any other waste.	C

Figure 3.2 Sample complexity classification from the EA OPRA methodology
Source: Contains Environment Agency information ©Environment Agency and database right.

Once an overall emission index has been determined for each substance the scores are added together for each of air, water, sewer, land and waste transfers off-site and used to determine an emission band for each. The EP OPRA emission bands are shown Figure 3.3.

Operator performance: The Agency believes that effective management systems are critical to managing the risk associated with an activity and to delivering permit compliance. As the operator is responsible for managing an installation's impacts and for ensuring compliance with permit conditions, the Agency believes that the absence of a documented environmental management system will require an increased level of regulatory oversight.

The overall operator performance attribute is calculated on the basis of the management systems in place and the operator's enforcement history for the installation. Essentially, the banded score for this attribute is arrived at via the completion of a detailed questionnaire that relates to the management systems in place and the compliance history of the organisation. The extract in Figure 3.4

Table 3.6 OPRA bands for location attributes

Score	Band
1–4	A
5–8	B
9–12	C
13–17	D
18–20	E

Source: Contains Environment Agency information © Environment Agency and database right.

illustrates the positive and negative scoring mechanism that leads to an overall banding for the organisation, based on a weighted assessment methodology.

The operator performance rating is generated from the final score from the questionnaires in accordance with Table 3.8.

Compliance rating: Essentially every breach of a permit is scored over a period of a year in accordance with a Compliance Classification Scheme (CCS) which divides breaches into four different categories (with associated points allocations) based on the work generated for the Agency. The total non-compliance points are totalled over each assessment year and are converted into an OPRA band as shown in Table 3.9.

Table 3.7 OPRA methodology questions for location attribute scoring

Table 3A – Opra location attribute	Score if yes
Human occupation/presence	
a) If within 50m of the installation boundary, score 5.	
b) If greater than 50m but less than 250m, score 3.	
c) If greater than 250m but less than 1km, score 1.	
Leave blank if 1km or more	
Statutory sites designed under Habitats Directive or Countryside Rights of Way Act 2000:	
a) If "relevant" under Habitats Directive, score 3.	
Or	
b) If CRoW Act 2000 assessment required, score 2.	
Leave blank if not applicable.	
Groundwater/aquifers	
a) If on any aquifer and in a groundwater source protection zone (GPZ), score 2.	
Or	
b) Leave blank if not applicable.	
Sensitivity or receiving waters	
a) If river category 1 or 2 or estuarine, score 3.	
b) If river category 3 or 4, score 2.	
c). If river category 5, score 1.	
Direct run-off	
a) If there are direct run-offs to water (for example, via a surface water drain) without interceptors or other active control measures, score 2.	
Or	
b) If as above, but there are interceptors or active control measures, score 1.	
Air quality management zone (AQMZ)	
a) If within on AQMZ and emit a declared pollutant for that AQMZ, score 3.	
b) If within 2km of an AQMZ and emit a declared pollutant for that AQMZ, score 2.	
c) As a), but do not emit pollutants declared for the AQMZ, score 1.	
Leave blank if not applicable.	
Flood plain	
If within a flood plain, score 2.	
Total score (maximum of 20)	

Source: Contains Environment Agency information © Environment Agency and database right.

3.15.2 Compilation of the installation banded profile

The final output from each of the five attributes is one or more bands in the range A–F, where:

    A = lower regulatory oversight required and,
    F = greater regulatory oversight required.

These bands are then carried forward to generate the EP OPRA Banded Profile for the installation covered by the permit. Such a profile may look as shown in Figure 3.5.

In order to convert an OPRA profile to a score for the purposes of charging, the total OPRA charging score is calculated using the Environment Agency spreadsheets (available from the EA website). The

Media	Emission Index (EI)	Emission OPRA band
Air	1–9	A
	10–99	B
	100–999	C
	1000–9999	D
	>10,000	E
Water	1–9	A
	10–99	B
	100–999	C
	1000–9999	D
	>10,000	E
Sewer	1–9	A
	10–99	B
	100–999	C
	1000–9999	D
	>10,000	E
Land	1–9	A
	10–99	B
	100–999	C
	1000–9999	D
	>10,000	E
Off site disposals	1–9	A
	10–99	B
	100–999	C
	1000–9999	D
	>10,000	E

Figure 3.3 OPRA emission band scoring
Source: Contains Environment Agency information ©Environment Agency and database right.

spreadsheets contain the current fees schedule and automatically apply multipliers for each band in order to calculate the appropriate fee for an initial application, annual subsistence, variation or surrender. The compliance rating essentially acts as a multiplier adjustment after the operator specific fee has been calculated (see Table 3.10). The compliance rating impact only applies to annual subsistence fees but its impact is substantial, reflecting the desire that the poorest operators should improve their performance as soon as possible.

To summarise, the key stages/elements of the EP OPRA process are as follows:

▶ The EP OPRA methodology is supported by a calculation spreadsheet, which is completed by Environmental Permit applicants and existing permit holders and supplied to the Agency at time of permit application or annual renewal.

▶ Applicants are able to identify influences on their EP OPRA Banded Profiles (and corresponding charge implications).
▶ The Agency assesses the spreadsheet and amends it if required.
▶ The initial EP OPRA Banded Profile is finalised at the time of permit issue.
▶ The Agency will review the EP OPRA Banded Profile when any enforcement/prosecution action is taken.
▶ The operator may request a review of the EP OPRA Banded Profile but needs to state the grounds for the review.
▶ EP OPRA Banded Profiles are used for calculation of permit applications, subsistence, variation and surrender fees.
▶ EP OPRA Banded Profiles are used to inform Agency compliance assessment work planning (not just inspections).

		Y/N	POINTS AVAILA-BLE	Points scored	Post or group responsible for each requirement	Document reference(*) or date when systems will be in place (*see para 4.4.2)
	Operations and maintenance 20% Effective operational and preventative maintenance shall be employed on all aspects of the process where any failure could impact on the environment.					
1	Are there documented operating procedures for operations that may have an adverse impact on the environment?		2.0			
2	Is there a defined procedure for identifying, reviewing and prioritising items of plant for which a preventative maintenance regime is appropriate?		2.0			
3	Are there documented procedures for monitoring emissions or impacts?		2.0			
4	Is there a preventative maintenance programme for those items of plant whose failure could lead to impact on the environment?		1.0			
5	Does the preventative maintenance programme include regular checks and formal inspections of 'static' items such as tanks, pipework, retaining walls, bunds and ducts?		1.0			

Figure 3.4 Extract from the OPRA operator performance attribute questionnaire
Source: Contains Environment Agency information ©Environment Agency and database right.

Table 3.8 OPRA banding for the operator performance attribute

Score	Band
10.0–8.0	A
7.9–6.0	B
5.9–4.0	C
3.9–2.0	D
<2.0	E

Source: Contains Environment Agency information © Environment Agency and database right.

Table 3.9 OPRA banding for compliance rating scores

CCS points	0	0.1–10	10.1–30	30.1–60	60.1–149.9	150+
OPRA band	A	B	C	D	E	F

Attribute	Band	Number of activities within each band
Complexity	A	1
	B	0
	C	2
	D	0
	E	0
		Band
Emissions	Air	n/a
	Water	B
	Land	C
	Waste input	n/a
	Sewer	C
	Off-site waste	n/a
Location		C
Operator performance		B
Compliance rating		B

Figure 3.5 A completed OPRA profile
Source: Contains Environment Agency information ©Environment Agency and database right.

Table 3.10 Multipliers generated by the OPRA compliance rating band for use in calculation of annual subsistence fees

Compliance rating band	A	B	C	D	E	F
Adjustment to OPRA score/subsistence fees (%)	95	100	110	125	150	300

Source: Contains Environment Agency information © Environment Agency and database right.

III AIR POLLUTION

3.16 The issue

Air pollution may result in impacts on human health, ecosystems and the physical environment, at both a local and global scale. It is the range of impacts and the lack of barriers to dispersal of pollutants that make this such an important pollution medium. An overview of key pollutants and air pollution issues is provided in Chapter 1.

3.17 International agreements

3.17.1 The Montreal Protocol (ozone-depleting substances)

The 1987 Montreal Protocol (and the multiple revisions to the original agreement) relate to the phase-out of use of ozone-depleting substances, initially chlorofluorocarbons (CFCs) and, currently, halocarbons. This protocol has been implemented in member states of the European Union through a number of regulations, the latest of which is *EC Regulation 744/2010*, supported in the UK by the *Environmental Protection (Controls on Ozone-Depleting Substances) Regulations 2011*. These regulations affect users, producers, suppliers, maintenance and service engineers, and those involved with the disposal of ozone-depleting substances (ODS). Some of the key substances affected by the Regulation are shown in Table 3.11.

All listed ozone-depleting substances go through three stages to final phase-out:

Table 3.11 Examples of substances covered by the Montreal Protocol

Refrigerants	CFCs (11, 12, 13, 113, 114, 500, 502, 503)
	HCFCs (22, 123, 124)
	HCFC blends (R401a, R402a, R403a, R406a, R408a, R411b)
Solvents	CFC (113)
	1, 1, 1 trichloroethane
	Bromochloromethane (CBM)
Foam-blowing agents	HCFCs (22, 141b, 142b)
Fire-fighting fluids	(1211, 1301)

Stage 1 – All users of listed substances must prevent escape and recover as much as of the ozone-depleting substance as possible but can continue to use equipment and new ozone-depleting substances for top-up or recharge of existing equipment.

Stage 2 – Manufacturers of listed substances are no longer allowed to produce new products. Users of equipment using listed ozone-depleting substances can continue to use the equipment but any recharge or top-up must use 'recovered product only'.

Stage 3 – The use of the listed ozone-depleting substance is prohibited and all users of equipment utilising such substances are required to decommission it and recover the banned substance.

Under the terms of the EC Regulation, from Stage 1 onwards, all organisations using listed ozone-depleting substances must ensure that they are recovered, wherever practicable, for recycling, reclamation or destruction. This requirement applies both during service and maintenance of equipment, as well as prior to equipment dismantling or disposal. It is also necessary to take precautionary measures to prevent leakages during manufacture, installation, operation and servicing.

In addition to these conservation and pollution prevention measures relating to ozone-depleting substances already in use, organisations would be well advised to carefully consider the phase-out programme when commissioning new equipment. Installing equipment that uses a substance due for phase-out in the short term may seriously reduce the working life of that product or at the least increase maintenance costs once the substance enters the Stage 2 'recovery only' period of the phase-out programme, i.e. when no new product can be introduced but where recovered products can be used for top-up and servicing on existing equipment.

3.17.2 Trans-boundary pollution and European Air Quality Standards

The Convention on Long-Range Trans-boundary Air Pollution (LRTAP) was adopted in Geneva in 1979 and came into force in 1983. The Convention has its foundations in concerns from Sweden and Norway over acid rain and related problems resulting in part from certain pollutants (e.g. SO_2 and NO_x) originating in other countries. The Convention requires countries to 'endeavour to limit and, as far as possible, gradually reduce and prevent air pollution, including long-range trans-boundary air pollution'. This should be achieved through the 'use of best available technology that is economically feasible'.

3.17.3 UN Framework Convention on Climate Change (UNFCCC)

Established as part of the international agreements reached at the Earth Summit in Rio de Janeiro in 1992, the Convention gave birth to a permanent secretariat at the United Nations and a programme of on-going meetings with national representatives from 192 countries negotiating an international response to the threat of climate change. With four negotiating sessions per year, including one main annual meeting known as the Conference of the Parties (COP), there have been a number of key agreements and milestones achieved including the Kyoto Protocol, 1997.

3.17.3.1 The Kyoto Protocol, 1997

The Kyoto Protocol was produced after lengthy negotiations in 1997, and aimed at combating the problem of climate change. The detailed requirements of the protocol cannot be covered

here but essentially signatories (all are developed countries) committed themselves to reducing the total greenhouse gases emitted by their country by varying amounts by 2008–2012. A notable absentee from the list of final signatories was the USA, who refused to participate. By the end of the implementation period the signatories had reduced their combined total emissions by 14.7 per cent vs 1990 levels (International Energy Agency data). However, in the same period globally, total emissions have grown vs 1990 levels by just over 38 per cent.

3.17.3.2 The Copenhagen Accord, 2009

In December 2009, the 15th UN-led climate change conference (COP15) was held in Copenhagen. All the participants had earlier agreed that the aim of the conference was to agree a new 'post-Kyoto' global initiative that would involve both developed and developing countries in long-term action plans to address climate change. After two weeks of debate the final outcome was the Copenhagen Accord 2009, which may be summarised as follows. The signatory countries agreed to the following:

- ▶ recognised the need to limit global temperature rise to 2 degrees centigrade;
- ▶ promised to deliver $30 billion of funding to developing nations for adaption measures between 2009–2012 and outlined a goal to increase this to $100 billion per year by 2020;
- ▶ agreed an emissions verification structure for industrialised nations;
- ▶ provided for the voluntary submission of national emissions reduction targets as an annex to the Accord;
- ▶ agreed that the Accord should be subject to review by 2015.

The Accord was a very long way from the quantified plan of action that many had hoped for. Time will tell whether the 2009 conference was a turning point or a catastrophic failure to act in the face of a global problem. Since the 2009 Accord, the annual COP talks have continued and we are now looking towards a key commitment date of 2015 by which to agree a series of binding international GHG reduction commitments. After the Copenhagen conference, most national governments pledged voluntary targets to reduce their GHG emissions but at Durban in 2010, it was recognised that these pledges fell well short of the cuts required to achieve the 2050 GHG levels required to limit global warming to the agreed 2 degrees centigrade. The 2015 target is the focus for correcting this situation, increasing the level of commitment by member states and putting the international community on track to meet this critical figure, which is seen to be the point at which the worst of the climate change consequences may be avoided.

3.18 European legislation

3.18.1 Framework Directive on Ambient Air Quality Assessment and Management

In 1996, a European *Framework Directive on Ambient Air Quality Assessment and Management* was adopted. The Directive requires that 'Daughter' Directives be prepared for several key air pollutants which set air quality objectives, limit values, alert thresholds and monitoring guidance. The following pollutants have been covered by Daughter Directives:

▶ sulphur dioxide, nitrogen dioxide, particulates and lead (Directive adopted April 1999);
▶ ozone (Directive adopted February 2002);
▶ benzene and carbon monoxide (Directive adopted December 2000);
▶ polyaromatic hydrocarbons (PAHs), cadmium, arsenic, nickel, mercury (Directive adopted December 2004);
▶ volatile organic compounds (the solvents Directive, adopted December 1999);
▶ sulphur dioxide and nitrogen oxides from large combustion plant (Directives adopted 1988 and 2001);
▶ total emissions of SO_2, NO_x, VOCs and ammonia are also subject to national ceiling limits to be achieved by 2010 (Directive adopted 2001).

Member states are required to assess ambient air quality and, where limit values are exceeded, take action to reduce ambient concentrations and inform the general public accordingly. In the UK, the Framework Directive and the associated Daughter Directives are being implemented via the Air Quality Strategy (under Part IV of the Environment Act, 1995, see p. 83) and supporting regulations (including the Environmental Permitting regime) that incorporate the various standards and emission limits relating to individual pollutants and specific sources. The priority pollutants and threshold concentrations used under the UK Air Quality strategy thus reflect the European standards set by the Daughter Directives.

3.18.2 European Climate Change Policy

In 2007, EU leaders endorsed an integrated approach to climate and energy policy and committed to transforming Europe into a highly energy-efficient, low carbon economy. To kick-start this process, a series of demanding climate and energy targets to be met by 2020 have been set, known as the '20–20–20' targets. These are:

▶ a reduction in EU greenhouse gas emissions of at least 20 per cent below 1990 levels;
▶ 20 per cent of EU energy consumption to come from renewable resources;
▶ a 20 per cent reduction in primary energy use compared with projected levels, to be achieved by improving energy efficiency.

The EU leaders also offered to increase the EU's emissions reduction to 30 per cent, on condition that other major emitting countries commit to do their fair share under a global climate agreement. United Nations negotiations on such an agreement are on-going. With its 'Roadmap for moving to a competitive low-carbon economy in 2050' the European Commission is looking beyond these 2020 objectives and has set out a plan to meet the long-term target of reducing domestic emissions by 80–95 per cent by mid-century as agreed by European Heads of State and governments. It shows how the sectors responsible for Europe's emissions – power generation, industry, transport, buildings and construction, as well as agriculture – can make the transition to a low-carbon economy over the coming decades.

Key elements of programmes linked to the 2020 targets and beyond include:

▶ The EU Emissions Trading Scheme (ETS) is intended to be a key tool in driving achievement of the targets in the most efficient manner. The scheme is discussed further in Section 3.19.4.6. It currently focuses only on energy-intensive industrial sectors such as steel-making, paper mills, etc. (aviation and shipping were later additions) – the justification being that such sectors are responsible for a large percentage of total emissions.

▶ Light-duty vehicles – cars and vans – are also a major source of greenhouse gas emissions, producing around 15 per cent of the EU's emissions of CO_2. Following up on a European Commission strategy adopted in 2007, the EU has put in place a comprehensive legal framework to reduce CO_2 emissions from new light duty vehicles as part of efforts to ensure it meets its greenhouse gas emission reduction targets under the Kyoto Protocol and beyond. The legislation sets binding emission targets for new car and van fleets.

▶ For cars, manufacturers are obliged to ensure that their new car fleet does not emit more than an average of 130 grams of CO_2 per kilometre (g CO_2/km) by 2015 and 95g by 2020. This compares with an average of almost 160g in 2007 and 135.7g in 2011. In terms of fuel consumption, the 2015 target is approximately equivalent to 5.6 l/100 km (50.44 mpg) of petrol or 4.9 l/100 km (57.65 mpg) of diesel. The 2020 target equates approximately to 4.1 l/100 km (68.9 mpg) of petrol or 3.6 l/100 km (78.47 mpg) of diesel.

▶ For vans, the mandatory target is 175 g CO_2/km by 2017 and 147g by 2020. This compares with an average of 203g in 2007 and 181.4g in 2010. In terms of fuel consumption, the 2017 target is approximately equivalent to 7.5 litres per 100 km (l/100 km) (37.6 mpg) of petrol or 6.6 l/100 km (42.8 mpg) of diesel. The 2020 target equates approximately to 6.3 l/100 km

(44.84 mpg) of petrol or 5.5 l/100 km (51.36 mpg) of diesel.

▶ In the area of fluorinated gases, which contain some potent greenhouse gases, the *EU F-gas Regulation* and the *MAC (Mobile Air Conditioning) Directive*, both published in 2006, regulate the use of hydrofluorocarbons (HFCs), perflourocarbons (PFCs) and SF6 in fixed and mobile equipment. Essentially both require regular testing of equipment by competent personnel to minimise the risk of leaks. Competency standards are also set for those involved in the maintenance and top-up of such equipment. In the UK this regime is implemented under the *Fluorinated Greenhouse Gases Regulations 2008*.

3.19 UK legislation

3.19.1 Environmental Permitting (England and Wales) Regulations 2010

While the Environmental Permitting Regime is clearly very important in terms of regulation of air pollution, it has essentially been covered in Section III. In summary, Part A1 and A2 processes may incorporate regulation and monitoring of air emissions as part of a wider spectrum of permit conditions. Part B permits may be expressly written to ensure the regulation of air emissions from a particular activity, e.g. a crematorium. Essentially it is an offence to operate a 'regulated activity' without or in breach of a permit.

3.19.2 The Clean Air Act (1993)

This modern-day version of the original 1956 Act provides Local Authorities with responsibility and authority for the control of dark and black smoke from chimneys and industrial/trade premises. Legal standards of emissions of smoke refer to two types of smoke:

▶ 'dark' smoke is defined as smoke as dark as or darker than Ringlemann shade 2;

▶ 'black' smoke is defined as smoke as dark as or darker than Ringlemann shade 4.

The Ringlemann shade numbers refer to the British Standard scale that runs from clear (0) to black (7).

Key elements of the Act include the following:

▶ Part I of the Clean Air Act prohibits the emission of dark smoke from the chimney of any building (domestic or industrial), with certain exemptions for boiler start-up, etc. It also prohibits the emission of dark smoke from industrial/trade premises (as opposed to chimneys).
▶ Part II requires Local Authority notification of installation of industrial furnaces and boilers.
▶ Part III enables the Local Authority to issue smoke control orders.

The key offence under the Act is to emit dark smoke (i.e. Ringlemann shade 2 or greater) from any chimney or industrial/trade site.

3.19.3 **The Environment Act (1995) Part IV**

The Act requires UK regulators to drive air quality improvements through a framework of national standards related to key air pollutants. The objectives of the national strategy are to reduce the hazard posed to human health by air pollution in the UK and to protect vegetation and ecosystems. A number of priority pollutants and concentration limits have been set in supporting regulations and policy documents. The ambient air quality standards for the following pollutants must be monitored by Local Authorities:

▶ benzene
▶ 1,3-butadiene
▶ carbon monoxide (CO)
▶ ozone (O_3)
▶ sulphur dioxide (SO_2)
▶ particulates (PM_{10})
▶ oxides of nitrogen (NO_x), and, in particular, nitrogen dioxide
▶ lead
▶ polyaromatic hydrocarbons (PAHs).

The standards set for each pollutant link in, where appropriate, to limits or targets set by European legislation, e.g. sulphur dioxide, NO_x, particulates and lead. Where ambient concentrations are found to exceed the standards set for individual pollutants, Local Authorities are required to take appropriate action to bring pollutant levels in line. As these pollutants are particularly associated with road transport, most responses relate to traffic control measures to reduce speeds or numbers of vehicles in critical locations. While there is no operator/emitter offence under this legislation, Local Authority strategies may also affect individual organisations through planning or permitting constraints, as well as via traffic control measures.

This system of ambient air quality objectives/standards is important because it acts in relation to cumulative and non-point source emissions (particularly transport-related), rather than simply in relation to individual sources.

3.19.4 **The UK Climate Change Strategy**

Existing key elements of the UK long-term emissions reduction strategy include the following elements.

3.19.4.1 *The Climate Change Act 2008*

The Climate Change Act 2008 aims to commit this and successive governments to long-term action (and thereby remove the issue from party politics to a degree). In summary, the Act:

▶ made legally binding carbon dioxide reductions targets for 2020 (26–32 per cent) and 2050 (80 per cent);
▶ introduced a system of 'carbon budgeting' capping emissions over five-year periods – with three budgets set ahead to help businesses plan and invest with increased confidence – the first round of budgets commenced in June 2009;
▶ created a new independent body (the Committee on Climate Change) to advise on the setting of carbon budgets and to report on progress;
▶ contained enabling powers to make future policies to control emissions quicker and easier to introduce;
▶ introduced a new system of government reporting to Parliament on climate change adaptation policies.

3.19.4.2 The UK Low Carbon Plan (2011)

The UK climate change targets set out in the Climate Change Act look likely to have significant impact on the energy and transport sectors as well as increasing the pressure on industry to reduce direct emissions of greenhouse gases. The reason for these priority sectors is clear when the key sources of greenhouse gases are considered (Table 3.12).

Mechanisms for reducing greenhouse gas emissions were set out in the UK Climate Change Programmes, published in 2000 and 2006. In December 2011, a new and ambitious programme was published in the UK Low Carbon Plan. The plan sets out a package of integrated measures and policies to achieve the national carbon dioxide emission reduction goals – long-term – 80 per cent by 2050 and with interim budgets (as shown in Table 3.13) to ensure progress towards the long-term goal is adequate.

The areas for improvement are presented under six main headings, as follows:

▶ Buildings
▶ Transport
▶ Industry
▶ Secure, low carbon electricity generation
▶ Agriculture, forestry and land management
▶ Waste and resource efficiency.

For further details see the DECC website and The Carbon Plan, available at: http://www.decc.gov.uk/en/content/cms/tackling/carbon_plan/carbon_plan.aspx.

There are many and varied commitments made within the Carbon Plan but a few examples are included below to illustrate the scale of change envisaged.

3.19.4.3 Carbon capture and storage (CCS)

Carbon capture and storage (CCS) is an approach to mitigate global warming by capturing carbon dioxide (CO_2) from large point sources such as fossil fuel power plants and storing it instead of releasing it into the atmosphere. Although CO_2 has been injected into geological formations for various purposes, the long-term storage of CO_2 is a relatively untried concept with significant uncertainty around the net benefit in terms of CO_2 savings. It is thought that CCS applied to a modern conventional power plant could reduce CO_2 emissions to the atmosphere by approximately 80–90 per cent compared to a plant without CCS. However, capturing and compressing CO_2 require a lot of energy and would increase the fuel needs of a coal-fired plant with CCS by about 25 per cent. These estimates apply to purpose-built plants near an appropriate storage location: applying the technology to pre-existing plants and/or plants far from a storage location will be more expensive and require even more energy to transport emissions.

Storage of the CO_2 is envisaged either in deep geological formations, in deep ocean masses, or in the form of mineral carbonates. In the case

Table 3.12 Main greenhouse gases and their sources

Greenhouse gas	Source
Carbon dioxide	Power generation and transport
Ozone	Photo-chemical smog (from transport)
Methane	Agriculture, sewage and landfills
Nitrous oxide	Power generation and transport
Chlorofluorocarbons	Aerosols and refrigerants

Table 3.13 UK national carbon budgets

	First carbon budget (2008–12)	Second carbon budget (2013–17)	Third carbon budget (2018–22)	Fourth carbon budget (2023–27)
Carbon budget level (million tonnes carbon dioxide equivalent MtCO₂e)	3,018	2,782	2,544	1,950
Percentage reduction below base year levels (1990)	23	29	35	50

Source: Department of Energy and Climate Change.

of deep ocean storage, there is a risk of greatly increasing the problem of ocean acidification, a problem that also stems from the excess of carbon dioxide already in the atmosphere and oceans. Geological formations are currently considered the most promising sequestration sites, and these are estimated to have a storage capacity of at least 2000 Gt CO_2 (currently, 30 Gt per year of CO_2 is emitted due to human activities).

As part of the UK climate change programme, the UK government is promoting and funding a £125 million research and development programme which will run to 2015 and includes a pilot plant at Ferrybridge in Yorkshire.

3.19.4.4 The Green Deal

The Low Carbon Plan commits the UK government to a series of funding schemes aimed at facilitating investment in home energy efficiency schemes including such things as 'pay as you save' grants and 'clean energy cash back' schemes to help fund investment in domestic renewable energy schemes. The Green Deal launched in 2012 is a key element of this strategy.

Specific improvements have also been included in the plan, for example, the phase-out and eventual ban on the sale of traditional 'filament' light bulbs. Since September 2012 retailers have only been able to sell existing stock and users can use existing bulbs but no more will be introduced into the marketplace.

3.19.4.5 The Climate Change Levy

A key tool in the drive to increase the efficiency of energy use by commercial users was introduced in 2001 in the form of the Climate Change Levy (introduced under the Climate Change Levy Regulations, 2001). The levy is charged on industrial and commercial use of energy. This covers fuel for lighting, heating and power for appliances by consumers in any of the following sectors of business:

▶ industry (including fuel industries);
▶ commerce;
▶ agriculture;
▶ public administration; and
▶ other services.

The levy is charged by energy suppliers to users, and subsequently collected by Her Majesty's Revenue and Customs. The levy is a ring-fenced tax with a dedicated body (the Carbon Trust) established to invest the funds collected in projects that facilitate the achievement of climate change targets.

The Climate Change Levy may be partially offset by reductions in National Insurance contributions paid by organisations but essentially provides an economic incentive to increase energy efficiency. Some industries via their trade associations have negotiated discounts of up to 90 per cent on the levy payable under so-called Climate Change Agreements. In return, however, individual businesses are required to achieve energy efficiency targets, again negotiated on a sector basis, over an agreed period. Failure to achieve the targets leads to a loss of the discount.

3.19.4.6 Emissions trading

A voluntary emissions trading scheme was established in the UK in April 2002, with participating companies receiving government funds in return for reductions in direct and indirect carbon dioxide emissions over a five-year period. The aim of emissions trading schemes is to encourage the most cost-efficient cuts in emissions by effectively assigning a value to the savings which can be materialised either through government 'rewards' or through trading with other participants of the scheme.

From January 2005 this voluntary programme was expanded to incorporate a mandatory regime for selected (high energy intensity) industrial sectors as part of a European Scheme in line with the 2003 EC Emissions Trading Directive.

Focused on the largest industrial users of energy and direct GHG emitters, the UK participation in the European Union Emissions Trading System (EUETS) is a key driver for those businesses involved to find ways to improve energy efficiency and generate tradable credits which can be sold to others within the scheme.

The EU Emissions Trading Directive (1996 as amended in 2003) has been met in the UK via the Waste and Emissions Trading Act, 2003 and the Greenhouse Gas Emissions Trading Scheme

Regulations, 2003. Essentially the legislation requires the UK government to establish a National Allocation Plan (NAP) over defined periods (currently 2013–2020) that is geared towards achieving the EU emissions reduction targets set out under the Kyoto Protocol.

The NAP is the total GHG emissions from all industries in the scheme (in the UK this relates to nearly 50 per cent of all CO_2 emissions). Although it is set by member states, it is approved by the EU and may be rejected if not ambitious enough in terms of emissions reductions. Under the NAP, an organisation will receive an annual emissions allowance and at the end of the year must be able to demonstrate that they have not exceeded their allowance.

Organisations included within the scheme are allocated a carbon dioxide emissions allowance based on baseline assessments and the agreed reduction plans. They also receive an allocation permit that requires accurate monitoring of carbon dioxide emissions to prove that allocation limits have been met. In the UK, the Environment Agency is responsible for the operation of the Emissions Trading Registry which tracks allocation allowances and performance data and allows 'quota trading' by registered individuals and organisations. The EA, on behalf of DEFRA, also oversees the compliance of individual parties and pools (including the verification of emissions records), while the European Union vets and approves the National Allocation Plans for each member state.

In the current phase of the regime – Phase III – there is a centralised EU-wide cap on emissions set, linked to the EU Climate Change targets. This annual GHG emissions cap will reduce by at least 1.74 per cent per year, so that emissions in 2020 will be at least 21 per cent below their level in 2005. The scheme now covers about 45 per cent of total EU GHG emissions including intra-European aviation.

3.19.4.7 Carbon Reduction Commitment (CRC)

The Carbon Reduction Commitment (CRC) is a UK emissions trading scheme targeting energy use by larger businesses and public sector organisations *not* covered by the EU ETS. This is essentially a roll-out of the existing energy quota and trading scheme to all organisations (private and public) whose total half-hourly metered electricity use is greater than 6,000 megawatt-hours (MWh).

Organisations are considered at group level if appropriate. To minimise administrative overlap, the CRC will cover emissions outside of Climate Change Agreements (CCAs) (mentioned above in relation to the Climate Change Levy) and outside the direct emissions already covered by the EU Emissions Trading Scheme (EU ETS). Around 5,000 public and private sector organisations are included in the scheme, ranging from retail, leisure and manufacturing companies through to Local Authorities, universities and NHS Trusts.

On an annual basis, organisations report all their UK-based CO_2 emissions from all their fixed point energy sources. This includes electricity, gas and other fuel types such as LPG and diesel. However, organisations are not required to report on their transport emissions. At the start of each compliance year, participants purchase allowances to cover their total emissions. They are also able to sell excess on the secondary market and/or buy additional allowances within the secondary market or the 'buy-only' link to the EU-ETS.

At the end of each compliance year, registrants must submit a report to the EA/SEPA demonstrating a balancing of allowances with GHG emissions for which they were responsible over the year. The regulators periodically audit the accounts submitted as evidence in the end-of-year reports.

Phase I of the scheme (2010–2014) has generated significant criticism related to excessive bureaucracy and over-complexity. In 2011, the UK government announced a series of proposals to simplify the scheme while retaining its basic features.

IV EFFLUENT AND WATER

3.20 The issue

Essentially this section is concerned with the protection of water resources from accidental and intentional discharges and from abstraction activities.

3.20.1 Planned and unplanned discharge routes from trade premises

Typical discharge routes from trade premises and industrial sites include:

▶ direct from site/operations to controlled water (e.g. river, stream, groundwater) without pre-treatment by the operator before discharge;
▶ direct from site/operations to controlled water (e.g. river, stream, groundwater) with pre-treatment;
▶ from site/operations to sewer without pre-treatment;
▶ from site/operations to sewer with pre-treatment.

Discharges to sewer can be expected to be treated by the sewerage undertaker at 'municipal' treatment works, and then be discharged into controlled waters – typically a river or the sea.

The Water Resources Act 1991 defines controlled waters. The definition covers virtually all fresh and saline natural waters out to the offshore UK territorial limit, including:

▶ rivers and streams;
▶ canals;
▶ relevant lakes and ponds, and certain reservoirs;
▶ estuaries and coastal waters (up to 3 miles from shore);
▶ groundwaters (underground strata).

A good rule of thumb is if a water body has an inlet and an outlet (even to groundwater), then it generally falls under the definition of controlled waters.

The public sewer is a term used to describe the network of pipes collecting effluent discharges from domestic and commercial premises for transfer to sewage treatment works operated by the regional Water Company (also known in this context as the 'sewerage undertaker'). In the vast majority of cases, sewage treatment works then discharge 'treated effluent' to controlled waters.

Commercial discharges to sewer are often referred to as 'trade effluent' or as 'discharges to foul sewer'. The legal context of these terms is described below.

3.20.2 Water pollutants summary

A number of categories of water pollutant were introduced in Chapter 1. They included:

▶ pathogens
▶ organic wastes
▶ inorganic wastes
▶ inorganic plant nutrients
▶ organic chemicals
▶ sedimentation
▶ thermal pollution
▶ endocrine disruptors.

3.21 International agreements

There are limited international agreements directly relevant to this area of environmental law though some of the key Millennium Goals agreed by the international community at the RIO+10 conference in 2002 related to the protection of, and human access to, freshwater resources. The principal agreements worthy of mention relate to the control of pollution in the marine environment, as detailed here.

3.21.1 MARPOL and UNCLOS (marine agreements and international shipping controls)

The *United Nations Convention on the Law of the Sea (UNCLOS)* was agreed during a series of conferences that took place between 1973 and 1982 but only came into force in 1994 when sufficient countries had ratified the Convention. Aside from its provisions defining ocean boundaries and territorial limits, the Convention establishes general obligations for safeguarding the marine environment and protecting freedom of scientific research on the high seas, and also creates a legal regime for controlling mineral resource exploitation in deep seabed areas beyond national jurisdictions.

The *International Convention for the Prevention of Pollution from Ships (MARPOL)* is the main international convention covering prevention of pollution of the marine environment by ships from operational or accidental causes. The MARPOL Convention was adopted on 2 November 1973 at the International Maritime Organisation. The Protocol of 1978 was adopted in response to a spate of tanker accidents in 1976–77. Requirements include the use of double-hulled tankers for the transport of oil, the regulation or banning of the discharge of listed hazardous substances (including sewage) from ships. The dumping of rubbish was also included in 1988.

In more recent years amendments have set limits on sulphur oxide and nitrogen oxide emissions from ship exhausts and prohibited deliberate emissions of ozone-depleting substances (2005). In addition, in designated emission control areas, more stringent standards for SO_x, NO_x and particulate matter have been set.

In 2011, after extensive work and debate, mandatory technical and operational energy efficiency measures were agreed which are expected to significantly reduce the amount of greenhouse gas emissions from ships. These measures came into force on 1 January 2013.

3.22 European legislation

Directive 2000/60/EC is the *European Union Water Framework Directive*. It commits European Union member states to achieve good qualitative and quantitative status of all water bodies (including marine waters up to one nautical mile from shore) by 2015. It is a framework in the sense that it prescribes steps to reach the common goal rather than adopting the more traditional limit value approach. It does, however, set out criteria for assessing water quality and requires states to take a 'river basin' (also known as catchment management) approach to surface water quality management. In mainland Europe, where many rivers cross national boundaries, this requires international cooperation to ensure that the specified water quality standards are met. River basin management plans are required to be prepared and updated every six years to ensure progress towards the 2015 objectives.

3.23 UK legislation

3.23.1 Legal framework

The Environment Agency, the SEPA and the Northern Ireland Environment and Heritage Service are responsible for the protection of 'controlled waters' from pollution under the *Environmental Permitting (England and Wales) Regulations 2010*, the *Water Resources Act 1991* in England and Wales, the *Control of Pollution Act 1974 (as amended)* in Scotland and the *Water Act 1972* in Northern Ireland.

The principal pieces of legislation dealing with the protection of the water environment and with releases to water from industry and business are:

▶ *The Salmon and Freshwater Fisheries Act 1975*. Under this legislation, it is an offence to discharge any substance to such an extent as to cause the receiving waters to be 'poisonous or injurious to fish, or the spawning grounds, spawn or food of fish'.
▶ *The Environmental Permitting (England and Wales) Regulations 2010*. Since April 2010, this is the main UK legislation regulating discharges to surface and groundwater. Unless an exemption applies, all discharges must be subject to a permit issued by the Environment Agency.
▶ *The Water Resources Act 1991*. Prior to the Environmental Permitting Regulations, 2010, this was the main statute that controlled water pollution. Although the discharge licensing elements of the Act are now dealt with under the environmental permitting regime, the Act is still important as it defines 'controlled waters' and provides for an abstraction licensing regime administered by the Environment Agency. It also provides for the setting up of a system of statutory water quality objectives (SWQOs).
▶ *The Water Industry Act 1991* (as amended 1999). This statute includes the requirement for trade effluent discharge to the public sewer to be authorised by a system of consents issued by the sewerage undertaker/Water Company.

Table 3.14 Summary of legal requirements applicable to effluent discharges in England

Discharge route	England and Wales
To sewer	Consent from Water Service Company/sewerage undertaker
To sewer with 'special category effluent'	Consent from Water Company plus referral to the EA
From regulated facilities to sewer	Environmental Permit from EA (consult with Water Service Company), plus Consent from Water Service Company/sewerage undertaker
To controlled waters	Environmental Permit from EA
From regulated facilities listed in the Environmental Permitting Regulations, 2010 to controlled waters	Environmental Permit from EA or Local Authority

▶ *The Environmental Protection Act 1990 – Part III.* This part of the Act established a regulatory system for dealing with statutory nuisance. Certain water pollution matters can constitute statutory nuisances.

A summary of the regulatory regimes for effluent discharges is shown in Table 3.14. Each area is explained in the sections that follow.

3.23.2 Abstractions from and discharges to controlled waters

The protection of controlled waters (and the regulation of discharges to them) are the responsibility of the Environment Agency or in Scotland, the SEPA. These agencies are the regulatory authorities issuing consents for effluent discharges to controlled waters under powers granted by the Environmental Permitting (England and Wales) Regulations 2010 and the Water Environment (Controlled Activities) (Scotland) Regulations 2005. They also control water company sewer discharges, which in turn can affect discharge conditions of trade effluent to sewer.

3.23.2.1 Abstraction consents from controlled waters

Under the terms of the Water Resources Act 1991, most abstractions from controlled waters that exceed 20 cubic metres per day require a licence. One-off abstraction of up to 5 cubic metres can be made without a licence and up to 20 cubic metres per day with prior agreement from the Environment Agency.

3.23.2.2 Permits to discharge into controlled waters

Under the terms of the Environmental Permitting (England and Wales) Regulations 2010, it is an offence to discharge to controlled waters, either deliberately or accidentally, without possession of an appropriate Environmental Permit issued by the Environment Agency. The only exceptions to this is are:

▶ surface water run-off (see below)
▶ where less than 5 cubic metres per day of appropriately treated domestic sewage (not trade effluent) is discharged to surface water or 2 cubic metres per day to groundwater – as long as defined treatment and monitoring conditions are met.

Environment Agency guidance defines domestic sewage and trade effluent as follows:

▶ domestic sewage effluent – 'domestic sewage' is defined as that arising from normal domestic activities, wherever carried out.
▶ trade effluent – this is defined as effluent discharged from any premises carrying on a trade or industry and is taken to mean an effluent generated by a commercial enterprise where the effluent is different to that which would arise from domestic activities in a normal home.

This revision in interpretation grants small businesses the opportunity, if they qualify, to benefit from exemptions (as above) and Standard Permits under the Environmental Permitting Regulations 2010. A Standard Permit may be appropriate for qualifying:

▶ sewage discharges (5–20 m³/day to surface water or to ground);

▶ discharges or cooling water up to 1,000 cubic metres per day from a cooling circuit or heat exchanger, to inland freshwaters, coastal waters or relevant territorial waters.

Bespoke permits would normally be required for all other discharges.

3.23.3 Discharges of surface water run-off

Discharges of uncontaminated surface water do not need a permit. However, to avoid surface or groundwater contamination, pollution prevention methods should be used (in line with the EA's Pollution Prevention Guidelines (PPGs), especially PPG1 (General Guide), PPG 2 (above ground oil storage tanks) and PPG3 (Use & design of oil separators)).

Discharge of surface water from public roads and parking areas is generally also considered acceptable if it has passed through a well-designed and maintained oil separator or sustainable drainage (SUDS) system. All trade effluents, however, must be kept separate from run-off water. A bespoke permit is typically required to discharge trade effluent to controlled waters.

3.23.4 Prescribed substances

Substances subject to specific consent requirements have been identified at both the European and the UK level. The *EC Dangerous Substance Directive* (1976) was aimed at reducing or eliminating the discharge to the aquatic environment of the most toxic, persistent and/or bio-accumulative substances. The Directive produced two lists:

▶ *List I (Black List) substances* considered being most harmful with discharge standards, generally set at EC level and with priority given to elimination, includes substances such as mercury and cadmium.
▶ *List II (Grey List) substances* are somewhat less harmful and include substances such as copper, zinc and chromium, with discharge standards set at national level but with an emphasis on reduction in discharges.

In the UK, the requirements of this Directive are met through the Environmental Permitting

Regime, and the Water Industry Act, 1991, supported by the Trade Effluent (Prescribed Substances and Processes) Regulations, 1989. Essentially, enforcing authorities must be convinced that 'strict environmental quality standards are being met and maintained in receiving waters', before giving any organisation 'discharge consents' for any of the dangerous substances on these lists.

3.23.5 Discharges to sewer

Under the *Water Industry Act 1991*, in England and Wales all discharges to the foul sewer require authorisation by the sewerage undertaker and may be subject to the terms and conditions of a trade effluent consent.

If an effluent contains substances at quantities listed in the *Trade Effluent [Prescribed Substances and Processes] Regulations 1989 (as amended)*, then prior approval is required from the Environment Agency before the issue of a consent to discharge to the public sewer. These prescribed substances mirror the substances covered in the Environmental Permitting Regulations 2010 and effluents containing them are known as 'special category' effluents. In practice, it is the Water Company (rather than the effluent generator) who seeks EA approval to issue a licence to discharge such effluent to the public sewer.

Under the terms of the Water Industry Act, the Water Companies/sewerage undertakers are also empowered to charge for such discharges in accordance with an agreed format, known as the Mogden formula, which reflects:

▶ the quantity of effluent;
▶ the degree of pollution (by a variety of measures);
▶ the conveyance costs to transfer the effluent from point of generation to the Water Company's sewage treatment works.

3.23.6 Discharge standards

For any consent to discharge effluent, whether to controlled waters or to the public sewer system, discharge standards or consent conditions are applied to ensure that pollution levels fall within acceptable limits. A variety of parameters are used in specifying discharge standards but will normally include a selection of the following:

▶ *Quantity* – expressed as maximum flow rate (e.g. cubic metres per hour), and/or time limited totals (e.g. maximum of 100,000 cubic metres per annum).

▶ *Dissolved Oxygen* (DO) – expressed as mg/l and normally as a minimum limit.

▶ *Biochemical Oxygen Demand* (BOD) – a measure of pollution in a body of water based on the organic material it contains. The organic material provides food for aerobic bacteria, which require oxygen to be able to bring about the biodegradation of such pollutants. The greater the volume of organic material, the greater will be the numbers of bacteria and, therefore, the greater the demand for oxygen. Thus, the BOD value (expressed as mg/l) gives an indication of organic pollution levels in the water. It provides no information on other pollutants such as suspended mineral sediments or heavy metals. If the BOD exceeds the available dissolved oxygen in the water, oxygen depletion occurs, and aquatic organisms will suffer. Fish kills are not uncommon under such circumstances.

▶ *Chemical Oxygen Demand (COD)* – it should be noted that the oxygen demand of effluent is also influenced by the chemical oxygen consumption requirements, i.e. the presence of pollutants that will normally react with any available oxygen. This is not measured by BOD but is provided by an alternative measure, namely: COD, which is an indication of the total pollution level in a water sample determined by boiling the sample in an oxidising agent and measuring the amount of oxygen consumed by the sample over a given period of time. As with BOD, this is typically expressed as mg/l.

▶ *Total Solids (TS)* – the total quantity of solids in an effluent sample, typically found by evaporation. Includes suspended solids, dissolved solids and any settled materials, expressed as mg/l.

▶ *Suspended Solids (SS)* – a subset of the total solids measure that relates only to those materials held in suspension in an effluent

sample (normally found by filtration and again expressed as mg/l).

▶ *pH* – the acidity or alkalinity of the sample, expressed as 0–14 on the pH scale.

▶ *Nitrogen* – normally expressed as mg/l of ammoniacal nitrogen and an important measure of the eutrophication potential of an effluent.

▶ *Temperature* – normally expressed as a variation from incoming water temperature or an increase range above or below receiving water temperature.

▶ *Metals* – either individual concentrations or a group measurement as 'total conductivity' of the effluent.

▶ *Oil and grease* – either as parts per million, mg/l or even as 'no visible sheen'.

A typical permit relating to a water discharge activity is shown in Table 3.15.

3.23.7 Groundwater protection

The *EU Groundwater Directive* (80/68/EEC) and subsequently the EU Water Framework Directive aim to protect groundwater from pollution by controlling discharges and disposal of certain dangerous substances to groundwater. In the UK, the Directive is now implemented through the Environmental Permitting Regulations (EPR) 2010, which revoked the 1998 and 2009 Groundwater Regulations. The Environment Agency protects groundwater under EPR by preventing or limiting the inputs of polluting substances into groundwater. Substances controlled under these regulations fall into two categories:

▶ Hazardous substances are the most toxic and must be prevented from entering groundwater. Substances in this list may be disposed of to the ground, under a permit, but must not reach groundwater. They include pesticides, sheep dip, solvents, hydrocarbons, mercury, cadmium and cyanide.

▶ Non-hazardous pollutants are less dangerous, and can be discharged to groundwater under a permit, but must not cause pollution. Examples include sewage, trade effluent and most wastes.

Further information on groundwater protection measures and policies implemented by the

Table 3.15 Example emission limits included in a water discharge activity permit

Parameter	Emission point		Monitoring frequency
	W1	W2	
Flow rate m³/day	15,000	3,900	Daily
Flow rate m³/hour	625	240	Daily
pH maximum	9.0	9.0	Continuous
pH minimum	6.0	6.0	Continuous
Temperature °C	20	20	Continuous
Suspended solids mg/l	10	35	W1 continuous W2 daily
Total copper mg/l	1.0	1.0	Daily
Soluble copper mg/l	0.5	0.5	Daily
Total nickel mg/l	2.5	2.0	Daily
Soluble nickel mg/l	2.0	1.0	Daily
Total cobalt mg/l	1.0	1.0	Daily
Soluble cobalt mg/l	0.5	0.5	Daily
Fixed and free ammonia as N mg/l	5.0	10.0	Weekly

Environment Agency, including those in response to the European Water Framework Directive, are described in their 2012 guidance note, *Groundwater protection: Principles and practice (GP3)* (DEFRA, 2012c).

3.23.8 Anti-Pollution Works Regulations 1999

These Regulations implement the provisions contained in the Water Resources Act 1991 relating to Anti-Pollution Works Notices. The provisions were introduced by the Environment Act 1995 and provide the Environment Agency with a useful enforcement option.

Anti-Pollution Works Notices are a type of statutory notice that require the person served to carry out specified works or actions to deal with pollution of controlled waters. The recipient of Works Notices can be any person who caused, or knowingly permitted, poisonous or noxious or polluting matter or solid waste matter to be present at a place from which it is likely, in the opinion of the Environment Agency, to enter any controlled waters. A Works Notice can also be served on any person who caused or knowingly permitted the matter to be present in a place from where it might be released.

The clear advantage of an Anti-Pollution Works Notice for the Agency is that it does not have to

do the work itself and go through the process of recovering its costs from the polluter. However, the regulations clearly are only of help to the Environment Agency where the polluter is known.

Powers granted under these regulations are now significantly enhanced by the *Environmental Damage (Prevention and Remediation) Regulations, 2009* described in Section 3.7.

V WASTE DISPOSAL IN THE UK

3.24 The issue

There are two key issues associated with waste which have been discussed in Chapter 1, namely:

▶ resource loss/consumption
▶ disposal-related impacts.

Waste management strategy is based on the premise that we should aim to move away from disposal and towards waste minimisation. The waste hierarchy shown in Figure 3.6 illustrates the perceived preferences for dealing with waste with minimising/reduction being the favoured option wherever possible. The inference with this hierarchy is that in general terms the nearer the top we are, the lower the overall environmental impact.

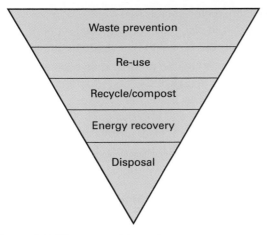

Figure 3.6 The waste hierarchy

3.25 International agreements

Attempts to control the movement of hazardous waste between countries have and continue to be made under the Basle Convention (trans frontier shipment of hazardous waste). The aim of this agreement, more formally known as the 1989 Convention on the Control of Transboundary Movements of Hazardous Wastes and Their Disposal, is to control the export and dumping of hazardous/toxic waste by industrialised nations in developed countries. It essentially demands high levels of pre-notification, waste description and assurance of safe transport and disposal arrangements by producers of hazardous waste prior to export. An amendment that introduces a general ban on the export of hazardous waste from OECD (Organisation for Economic Cooperation and Development) countries to non-OECD countries for disposal or recycling/recovery has failed so far to be ratified.

3.26 European legislation

There is considerable European law and policy relating to waste, the detailed requirements of which are explored in the UK context in Section 3.27. However, an overview of the key requirements is provided in Table 3.16.

3.27 UK legislation

The huge quantities of waste generated in the UK and the issues associated with our current reliance on landfill for disposal were discussed in Chapter 1. In recognition of the environmental issues of such a situation, the UK government has

Table 3.16 Key European waste law

EC Directive	Key requirements/summary
The 1991 European Framework Directive on Waste (Directive 91/156/EEC revised in 2008)	Defines 'waste' and 'holders' of waste and set out the principles for the duty of care to ensure waste minimisation and appropriate control. The 2008 Directive also sets recycling and reuse targets for specified waste streams.
The 1999 Landfill Directive – Directive 99/31/EC	Relates to the management of landfill sites, including the exclusion of certain wastes to landfill and total reduction targets.
The 1991 Hazardous Waste Directive – Directive 91/689/EEC	Sets out definitions of and requirements for the segregation of hazardous from non-hazardous wastes.
The 1994 Directive on Packaging and Packaging Waste – Directive 94/62/EC and updated in 2004	Sets out requirements for minimisation and end of life recovery of packaging waste as well as the elimination of specified hazardous materials from packaging.
The 2000 End of Life Vehicles Directive – Directive 2000/53/EC)	Sets out requirements for the safe disposal of end of life vehicles and establishes recovery and reuse targets. Also specifies the elimination of listed hazardous materials.
The 2003 Waste Electrical and Electronic Equipment (WEEE) Directive and the RoHS Directive (2002/96/EC & 2002/95/EC)	Sets out requirements for manufacturers to recover WEEE and to restrict the use of certain hazardous substances in the manufacture of electrical and electronic equipment.
The 2006 Batteries Directive (Directive 2006/66/EC)	Requires the collection and recycling of all batteries sold within the EU.

produced a number of national Waste Strategies, the latest of which was published in 2007 and was subject to a review in 2011. The strategy is based on the so-called waste hierarchy (as outlined above).

3.27.1 Delivery of the Waste Strategy

The government considers that there should be no major additions to the regulatory framework now in place and that the strategy will (in addition to regulatory controls based on existing legislation), be delivered by such measures as:

▶ market-based measures, e.g. landfill tax;
▶ encouragement of producer responsibility initiatives (aimed at those who would produce products and/or materials that end up as waste);
▶ planning considerations;
▶ promotion of best practice in waste minimisation and alternative waste management options, including information on BPEO for specific waste streams.

One of the key market-based measures which has had a major impact on waste management trends since its introduction in 1996 is the landfill tax. It now represents a significant proportion of the cost of waste disposal to landfill and therefore provides a clear incentive to commerce and industry to find alternative disposal routes. Table 3.17 shows the escalator plans and rates until 2014.

Voluntary action by industry (in response to such market-based measures) is a key element of the government's strategy on waste. However, in

Table 3.17 UK landfill tax rates

Landfill Tax	Introduced from 1 October 1996 for waste going to landfill.
	In 2010 the UK government announced that the landfill tax escalator will continue to increase by £8 per tonne until 2014 at which point it will be £80 per tonne. At the time of writing no information is available as to what will happen after that time.
	A lower rate of £2.50 per tonne for (listed) inactive waste rate exists but may be changed by Chancellor in the annual Budget each year.

Source: Adapted from DEFRA.

addition, a well-established regime of regulatory control is in place to ensure the appropriate management of waste that is generated.

Government-funded programmes such as the Waste and Resources Action Programme (WRAP) and the National Industrial Symbiosis Programme (NISP) are aimed at improving resource efficiency in general but with clear emphasis in relation to waste reduction and recovery. See www.wrap.org.uk and www.nisp.org.uk for further details.

3.27.2 Legal overview

The principal statutes dealing with waste are summarised below.

▶ Environmental Protection Act 1990 – Part II
  ▶ Section 33 – Prohibition on unauthorised or harmful depositing, treatment or disposal of waste
  ▶ Section 34 – Duty of care requirements for 'holders' of waste.
▶ Environmental Permitting (England & Wales) Regulations 2010
  ▶ Permit requirements for waste storage, treatment and disposal operations.
▶ The Waste (England & Wales) Regulations 2011
  ▶ Requires confirmation of use of waste hierarchy
  ▶ Introduces two-tier system of waste carrier registration.
▶ Environment Act 1995 – Part V plus supporting 'waste specific' regulations
  ▶ Producer responsibility provisions.

There also exist a significant number of relevant pieces of secondary legislation relating to the control of waste. It is, however, perhaps most logical to approach such legislation from the perspective of legal requirements relating to different waste classifications and the role of individual parties within the disposal chain, rather than on a statute-by-statute basis.

3.27.3 Definitions of waste

Clearly in relation to waste legislation, clarity of what constitutes waste is important in determining whether licensing/registration is required. In European and hence UK law the key issue is whether a material has been discarded and, if not, whether significant processing is required to enable its on-going use.

Figure 3.7 summarises the definition of waste provided by the *EC Framework Directive on Waste* which was revised in 2008.

Once it is established whether a material is waste, different legal requirements relate to different waste classifications, with the definitions of waste types set as part of the secondary legislation. The majority of waste legislation applies to what is known as 'controlled waste' – namely household, commercial and industrial waste (including agricultural and mining wastes). The key waste activities with correlating legislation are shown in Figure 3.8.

The legal responsibilities under the legislation shown in Figure 3.8 will be considered below.

3.27.4 Duty of Care requirements

Section 34 of the *Environmental Protection Act 1990* defines the duty of care requirements for 'holders' of waste. A waste holder is any person who 'produces, imports, keeps, treats, carries, disposes, or is a waste broker of controlled waste'. Essentially anyone defined as a holder of waste is required to ensure that it is managed properly and that it is recovered or disposed of safely. Details of the duty of care are set out in a Code of Practice, which is regularly updated. It recommends a series of steps, which

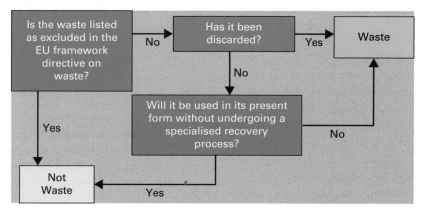

Figure 3.7 Defining waste
Source: Adapted from EC Framework Directive on Waste 2008.

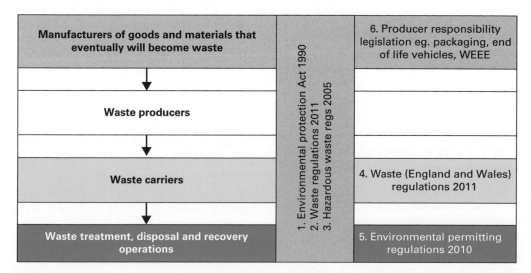

Figure 3.8 The waste disposal chain and applicable legislation

should normally be enough to meet the duty of care. While the Code of Practice is not a legal standard in itself, it is admissible as evidence in any legal proceedings and is a key document for environmental managers. It outlines the requirements of the duty which are for holders of waste as follows:

▶ to contain it securely to avoid escape;
▶ to transfer it only to someone with authority to take it (e.g. registered carrier/licensed waste manager), or be allowed to actually take the waste if receiving it;
▶ to keep appropriate records using transfer notes (including description and information on any potential problems), making sure the documentation is properly completed.

Also, all waste holders should be alert to the non-compliance of others. A breach of the duty of care is an offence with a penalty of an unlimited fine if convicted on indictment.

The same duties are applied to household waste and agricultural waste under the Waste (Household Waste Duty of Care) (England & Wales) Regulations 2005 and Waste Management (England & Wales) Regulations, 2006 respectively.

3.27.4.1 Documentation of waste transfers

Transfers of waste must be accompanied by a transfer note containing specified information. There is no compulsory form for keeping waste transfer records, however, an example is given in the Code of Practice (and included as Figure 3.9). The transfer note must be kept for at least two years and copies must be provided to the Environment Agency or the SEPA if requested. The same requirement exists for hazardous waste and non-hazardous waste. Hazardous waste records must be kept for three years rather than two and are commonly referred to as consignment notes rather than transfer notes.

Waste descriptions have to include the appropriate reference code from the European Waste Catalogue (EWC) which appears in the devolved *List of Waste Regulations 2005*. The codes are six digit numbers that describe standard categories of hazardous and non-

hazardous waste some of which are generic, while many link to particular activities and sources of waste.

There is also a requirement to use a Standard Industry Code (SIC) to identify the type of activity from which the waste is arising. The list of codes is available from the Office of National Statistics (www.ons.gov.uk).

Where there are a series of transfers of the same waste between the same parties within a 12-month period, one transfer note can cover that series of transfers. This is referred to as a 'Season Ticket'.

Finally, the *Waste (England and Wales) Regulations 2011* require organisations to confirm that they have applied the waste management hierarchy (see Section 3.24) when transferring waste, and to include a declaration to this effect on their waste transfer note or consignment note. An example waste transfer note is shown in Figure 3.9.

3.27.4.2 Pre-treatment of non-hazardous waste to landfill

Under the terms of the Landfill Directive, from October 2007, all liquid wastes are banned from landfill and all solid waste must be *treated* before it can be landfilled. The UK government's view is that to meet this requirement, the treatment process must fulfil three criteria, namely:

> It must be a physical, thermal, chemical or biological process including sorting.
> It must change the characteristics of the waste.
> It must do so in order to do the following:
> reduce its volume, or
> reduce its hazardous nature, or
> facilitate its handling, or
> enhance its recovery.

The need to meet these three criteria is referred to as the 'three-point test'. The legal obligation for ensuring that no waste ends up in landfill which has not been pre-treated lies with the landfill operator. However, waste producers, as part of their Duty of Care, are expected to either:

▶ treat their own waste, for example, through sorting recyclables out from the mixed waste stream; or

Duty of care: waste transfer note

Keep this page and copy it for future use. Please write as clearly as possible.

Section A – Description of waste

A1 Description of the waste being transferred

List of Waste Regulations code(s)

A2 How is the waste contained?

Loose ☐   Sacks ☐   Skip ☐   Drum ☐

Other ☐

A3 How much waste? For example, number of sacks, weight

Section B – Current holder of the waste – Transferor

By signing in Section D below I confirm that I have fulfilled my duty to apply the waste hierarchy as required by regulation 12 of the Waste (England and Wales) Regulations 2011   Yes ☐

B1 Full name

Company name and address

Postcode [ ]   SIC code (2007) [ ]

B2 Name of your unitary authority or council

B3 Are you:

The producer of the waste? ☐

The importer of the waste? ☐

The local authority? ☐

The holder of an environmental permit? ☐

Permit number [ ]

Issued by [ ]

Registered waste exemption? ☐

Details, including registration number

A registered waste carrier, broker or dealer? ☐

Registration number [ ]

Details (are you a carrier, broker or dealer?)

Section C – Person collecting the waste – Transferee

C1 Full name

Company name and address

Postcode [ ]

C2 Are you:

The local authority? ☐

C3 Are you:

The holder of an environmental permit? ☐

Permit number [ ]

Issued by [ ]

Registered waste exemption? ☐

Details, including registration number

A registered waste carrier, broker or dealer? ☐

Registration number [ ]

Details (are you a carrier, broker or dealer?)

Section D – The transfer

D1 Address of transfer or collection point

Postcode [ ]

Date of transfer (DD/MM/YYYY) [ ]

D2 Broker or dealer who arranged this transfer (if applicable)

Postcode [ ]

Registration number [ ]

Time(s) [ ]

Transferor's signature [ ]

Name [ ]

Representing [ ]

Transferee's signature [ ]

Name [ ]

Representing [ ]

WMC2A version 3, August 2011

Page 1 of 1

Figure 3.9 An example waste transfer note

Source: Contains Environment Agency information ©Environment Agency and database right.

▶ ensure that their waste management company carry out such pre-treatment on their behalf.

Enforcement of the pre-treatment requirements is the responsibility of the Environment Agency through information provided by the landfill operators. From April 2008, landfill operators have had to obtain written evidence from waste producers that their waste has been treated. The Environment Agency can thus monitor compliance with the rules under its inspection programme and audit of landfills and where necessary, take enforcement action.

3.27.5 Hazardous waste requirements

The *Hazardous Waste Regulations 2005* replaced the 1996 Special Waste Regulations in England and Wales from 16 July 2005. The new regulations define hazardous waste in line with the European Waste Catalogue, now incorporated into the *List of Wastes Regulations, 2005* (devolved versions). A key guidance document produced by the Environment Agency describes the process of classification of hazardous waste (*Technical Guidance WM2 – Interpretation of the definition and classification of hazardous waste*) (Environment Agency, 2013a).

The disposal requirements relating to hazardous waste are similar to, but more stringent than, those for 'general waste' as outlined in the Duty of Care section above. The key additional requirement is that of producer registration with the Environment Agency. Since April 2009, premises that generate less than 500kg of hazardous waste per year are exempt from EA registration as long as registered carriers are used, i.e. the Duty of Care still applies. Any premises producing more than 500kg of hazardous waste per year must be registered with the Environment Agency as a hazardous waste producer. Registered producers are issued with a registration number (the 'premises code') that must be included on all consignment documentation. These codes are valid for 12 months and must, therefore, be renewed annually.

As with the Duty of Care requirements, it is the responsibility of the waste producer to ensure that their waste is appropriately disposed of. However, the registration process gives the EA the opportunity to track back any waste that may be inappropriately classified, stored or disposed. In addition to producer registration, persons receiving hazardous waste for recovery or disposal are required to submit quarterly returns to the EA of all wastes received.

It is also worth noting that the 2004 ban (under the *EC Directive on the Landfill of Waste 1999*) on the co-disposal of 'hazardous' and 'non-hazardous' waste in the same landfill site has led to a drastic reduction in the number of landfill sites that were licensed to accept hazardous waste. This has led to a significant increase in the cost of hazardous waste disposal to landfill and provides a major incentive for producers to minimise hazardous waste generation and find alternatives to landfill disposal.

3.27.6 Registration of waste carriers

The *Waste (England and Wales) Regulations 2011* introduced a two-tier system for waste carrier and broker registration:

▶ *Upper tier registration*: You need to register as an upper tier carrier or broker if you want to carry, broker or deal in other people's controlled waste, unless you are in one of the lower tier categories listed below. You also need to register as an upper tier carrier if you carry your own construction or demolition waste. Upper tier registration lasts for three years, the same as the previous waste carrier or broker registrations.
▶ *Lower tier registration*: You need to register as a lower tier carrier if you only carry, broker or deal in:
  ▶ animal by-products;
  ▶ waste from mines and quarries;
  ▶ waste from agricultural premises.

You also need to register as a lower tier carrier if you carry, broker or deal in other people's waste and are:
  ▶ a waste collection, disposal or regulation authority
  ▶ a charity or voluntary organisation.

From the end of December 2013, it is also a requirement to register as a lower tier carrier if you regularly carry controlled waste produced by your own business, other than construction or demolition waste.

3.27.6.1 Definitions of carriers, brokers and dealers

- A carrier is someone who transports waste produced by other people, e.g. a skip operator or waste disposal company.
- A waste broker is someone who arranges for waste from other businesses or organisations to be transported, disposed of or recovered.
- A waste dealer is someone who buys and sells waste, or uses an agent to do so.

3.27.7 Waste Management Permitting

Section 33 of the Environmental Protection Act 1990 prohibits the unauthorised or harmful depositing, treatment or disposal of waste. Since April 2008, all waste management operations in England and Wales have been required to have a permit under the *Environmental Permitting Regulations, 2010* or a valid exemption. In some cases, e.g. transfer stations, composting operations and many recycling activities, facilities are eligible for a 'standard permit'. Waste management operations with greater pollution potential, e.g. landfill sites and incineration plants are, however, required to apply for a 'bespoke permit'.

3.27.7.1 Exemptions from environmental permitting

Schedules 2 and 3 of the Regulations describe a range of waste activities that may be exempt under environmental permitting, including:

- composting biodegradable waste for cultivating mushrooms;
- crushing, grinding or other size reduction of waste;
- cleaning or coating of waste packaging, containers and textiles;
- storing waste securely;
- treating waste to recover materials;

- burning waste in an exempt incinerator at the place where the waste was produced;
- storing waste electrical and electronic equipment (WEEE) for recovery elsewhere.

Exempt activities may still need to be registered with the Environment Agency or the Local Authority. They must also comply with all the conditions of the exemption and make sure that activities do not cause pollution or harm to human health.

Typically activities carried out by the producer of a waste prior to regular collection are exempt but it should be noted that the Environmental Permitting Regulations have introduced more detailed restrictions in terms of quantities and duration of waste holding in order to qualify for such exemptions.

Some exempt activities, though not required to have an Environmental Permit, are required to make a 'notification' to the Environment Agency. The activities subject to notification (which is renewable annually) are:

- land treatment activities;
- storage and spreading of sludge;
- reclamation or improvement of land;
- recovery operations at water and sewage treatment works;
- storage and use of building waste;
- burning of waste at docks.

It is worth noting that since 2006, agricultural, mine and quarry waste are all subject to the Duty of Care. Since July 2009, the management of mine and quarry waste has required the possession of an Environmental Permit (again with defined exemptions).

3.27.8 Producer Responsibility regime

The Environment Act 1995 (Part V) created a framework whereby those organisations involved in the manufacture of materials/products that subsequently become waste, may be required to share the responsibility for recovering such products at the end of their useful life.

3.27.9 Producer Responsibility: Packaging

3.27.9.1 Producer Responsibility Obligations (Packaging Waste) Regulations 2010

The regulations were originally published in 1997 but revised and replaced in 2007 and again in 2010. They require 'obligated businesses' to recover and recycle specific tonnages of packaging waste based on a calculation incorporating national targets linked to European targets.

Obligated businesses are required to register with the Environment Agency or SEPA, unless in a 'compliance scheme'. The compliance schemes exempt member businesses from individual requirements, but the scheme must meet aggregate obligation.

3.27.9.2 Definition of obligated businesses

In outline, these are businesses which perform any of the following activities:

▶ manufacture of packaging materials;
▶ conversion of materials into packaging;

▶ pack/fill packaging with product;
▶ sell products with packaging;
▶ packaging service provider (e.g. lease pallets to users);
▶ import packaging with a product for sale to an end user;
▶ own the packaging involved;
▶ supply to another stage in the packaging chain or the final user;
▶ handle more than 50 tonnes of obligated packaging or packaging materials a year and have an annual turnover of £2 million or more (calculated at group level). This includes imports, excludes exports, process waste and any re-usable packaging that is being re-used. The thresholds apply at group level.

Packaging categories covered by the regulations are board (including paper), glass, aluminium, steel, plastics and wood.

3.27.9.3 A sample calculation

The simplified example shown in Figure 3.10 explains the basic process required to calculate the actual tonnages of packaging required to be recovered in any particular year:

A manufacturer of widgets buys in cardboard boxes from a UK supplier to use to package his goods which he sells to end users via a retailer. In 2013 he uses 100 tonnes of such packaging.				
Recovery obligation	**Obligated packaging handled (tonnes)**	**Activity obligation (as a packer/filler)**	**UK Recovery target 2013**	
X =	100	X 37%	X 75%	= 27.75 tonnes
Recycling obligation (cardboard)	**Obligated packaging handled (tonnes)**	**Activity obligation (as a packer/filler)**	**UK Recycling target (paper) 2013**	
X =	100	X 37%	X 69.5%	= 25.72 tonnes
His obligations at the end of 2013 are thus to demonstrate recovery of 27.75 tonnes of packaging materials of which 25.72 tonnes must be in the 'recycled' paper category while 2.03 tonnes may comprise non-recycling recovery techniques e.g. waste to energy, composting etc. (and may be other packaging materials than paper).				

Figure 3.10 A simplified example of a calculation for an obligated business under the Producer Responsibility (Packaging Waste) Regulations 2010

Table 3.18 UK packaging recovery targets

Material	2013 (%)	2014 (%)	2015 (%)	2016 (%)	2017 (%)
Paper/card	69.5	69.5	69.5	69.5	69.5
Glass	81	81	81	81	81
Aluminium	43	46	49	52	55
Steel	72	73	74	75	76
Plastic	37	42	47	52	57
Wood	22	22	22	22	22
Total recovery	75	76	77	78	79
Of which recycling	69	69.9	70.8	71.8	72.7

Source: Adapted from DEFRA.

Packaging handled x activity obligation x UK recovery target = recovery obligation
Packaging handled by material x activity obligation x UK recycling target = recycling obligation by material*

* This calculation must be repeated for each packaging material handled and the recycling obligation totals form part of the overall recovery obligation.

The activity obligations essentially divide up responsibility for the end of life recovery between all the parties involved in generating the packaging waste. The key categories with associated percentage obligations (to be used in the calculation) are given as follows:

▶ 6 per cent manufacturing
▶ 9 per cent converting
▶ 37 per cent packing/filling
▶ 48 per cent selling product.

3.27.9.4 Business 'recovery' and 'recycling' targets

The UK government has set 'business' targets which must be met by obligated companies each year to ensure that the UK meets its national targets as set out under the *EU Packaging and Packaging Waste Directive*. The UK business targets are higher than the Directive targets, as under the UK system smaller businesses are excluded from the obligations and so only a proportion of all packaging is obligated whereas the EU Directive targets apply to all packaging waste.

The business targets, which should be used by businesses to calculate their obligations and which are designed to enable the UK to meet the Directive targets, are set out in Table 3.18.

3.27.9.5 Demonstrating compliance

Obligated businesses demonstrate compliance through submission of accounts to the Environment Agency. Recovery and recycling credits are supported by certificates issued by organisations that are registered as recyclers or operators of recovery operations. Such certificates are known as Packaging Recovery Notes or PRNs.

Note an organisation does not have to recover or recycle its own packaging at the point of disposal, merely demonstrate that it has been involved in the recovery of an appropriate quantity of an equivalent material.

The only form of evidence that may be used by obligated businesses to demonstrate that they have met their obligation totals are Packaging Waste Recovery Notes (PRN) and/or Packaging Waste Export Recovery Note (PERN). The only businesses which are entitled to issue PRNs or PERNs are packaging waste reprocessors and exporters respectively which have been accredited by the Environment Agency.

Obligated businesses must obtain sufficient PRNs or PERNs to cover their obligation total either by purchase on the open market (either directly or via a compliance scheme) or by receiving them in return for materials submitted to recovery/recycling organisations.

In general, companies may discharge their obligation in two ways: either join one of the compliance schemes registered with the Environment Agency, which takes on their legal responsibility and must ensure that their obligation is met; or register individually and fulfil the obligation themselves.

3.27.10 Compliance schemes

If you join a compliance scheme, you are effectively transferring your obligation to the compliance scheme. It is the responsibility of the scheme to meet the aggregate obligations of all its members.

While a member of a scheme you cannot be prosecuted for failing to meet your recovery and recycling obligations though you can still be prosecuted for providing false or inaccurate data. There are currently 20 registered compliance schemes in the UK, with some of the largest being Valpak, Wastepack and Biffpack. Membership fees vary considerably depending on:

► the size of the member company and its obligation;
► whether the organisation is an existing or new member;
► the length of membership contract committed to;
► the level of service provided by the scheme (e.g. some schemes provide support on calculating obligations, waste management advice, etc.).

3.27.10.1 Registering individually

This means that the company undertakes direct responsibility for ensuring that its obligation is met. It must therefore consider the following actions:

► registration with the appropriate Agency and payment of the annual fee;
► providing data on packaging handled during the previous year to the appropriate Agency by 7 April each year;
► submitting a compliance plan to the Agency if the company turnover exceeds £5 million;

► ensuring the recovery and recycling obligation is met and obtaining the necessary evidence in the form of PRNs;
► retaining evidence that the obligation has been met for four years;
► submitting a certificate of compliance (a statement signed by an authorised person, plus supporting evidence) by 31 January for packaging activities in the preceding calendar year.

3.27.10.2 Offences under the Regulations

Companies may be prosecuted for:

► not registering;
► failing to meet their obligation;
► failing to provide a certificate of compliance;
► providing false data.

Companies that choose to register individually should recognise that the cost of PRNs may vary widely during the year depending on supply and demand.

3.27.10.3 Choosing a compliance strategy

There are advantages and disadvantages to compliance scheme membership and individual registration. Table 3.19 shows a summary of these pros and cons.

On the whole, the majority of companies have joined compliance schemes (more than 90 per cent). That said, companies should consider the pros and cons carefully before adopting a compliance strategy and indeed should review their decision on a regular basis to ensure that they are not paying an unnecessarily high price for compliance.

3.27.10.4 Achievements to date

The UK system has been successful in increasing the levels of packaging waste recovered and recycled from 27 per cent in 1997 when the regime was first introduced to 67 per cent by 2010. In real terms, the total amount of packaging waste recovered and recycled in 1998 was 3.3 million tonnes; in 2010, it was an estimated 7.24 million tonnes. Table 3.20 shows the breakdown

Table 3.19 Compliance scheme vs individual registration

Compliance scheme	Individual registration
Advantages	*Advantages*
Scheme discharges responsibilities for obtaining PRNs so there is less risk of prosecution	More direct control of costs
	Useful if access exists to relevant types and quantities of 'backdoor' waste
Economies of scale may reduce cost of PRN purchase	
Allows organisation to concentrate on core business activities	Avoids associated fees (joining fee, administration fee, etc.)
Provides 'expert support' related to obligation calculations	
Disadvantages	*Disadvantages*
Compliance scheme costs	Time involved in discharging obligation
	Risk of prosecution

Table 3.20 UK recovery and recycling achievement data, 2010

	Total packaging waste arising	Total tonnage recovered/ recycled	EU target (%)	Recovery/recycling rate (%)
Paper	3,787,560	3,099,941	60	81.9
Paper composting		3,445		
Glass	2,712,860	1,647,917	60	60.7
Aluminium	147,500	60,304		40.9
Steel	652,000	386,621		59.3
Metal		446,925	50	55.9
Plastic	2,478,630	598,252	22.5	24.1
Wood composting		666		
Wood	1,023,939	771,224	15	75.4
Other	22,331			
Total recycling		6,568,370	55	60.7
Energy from waste		721,505		
Total recovery	10,824,820	7,289,875	60.0	67.3

Source: DEFRA.

by material type and with reference to the European targets.

In addition to the recovery obligations on producers of packaging described above, there is also parallel legislation relating to the design of packaging. The *Packaging (Essential Requirements) Regulations, 2003* require that all packaging:

▶ is the minimum necessary to meet the safety and hygiene criteria for the product concerned;

▶ can be reused, recycled or recovered and will have minimal environmental impact if disposed of;

▶ is made with the minimum use of substances which become noxious or hazardous when incinerated or landfilled;

▶ complies with limits for concentrations of lead, mercury, cadmium and hexavalent chromium.

Producers must demonstrate compliance with the regulations through the maintenance of appropriate technical documentation for a four-year period following the release of the packaging on the market. These regulations are enforced by trading standards officers.

3.27.11 Producer Responsibility: End of Life Vehicles

Under the Producer Responsibility banner the *EC Directive on End of Life Vehicles 2000* requires producers to set up systems for the free collection of scrap cars. Manufacturers must design new cars with recycling and reuse in mind, and vehicles must contain an increasing quantity of recycled material. The UK has implemented the terms of the regulations via the *End of Life Vehicles Regulations, 2003* which:

▶ restrict the use of heavy metals in vehicle manufacture;
▶ require certificates of destruction;
▶ require marking of components to aid recycling;
▶ use free take-back of end of life vehicles (ELVs).

A network of authorised ELV treatment facilities have been licensed and, from 1 January 2007, vehicle owners (cars and goods vehicles up to 3.5 tonnes) have been entitled to free take-back. Two service provider schemes have been set up which cover different vehicle manufacturers. For details of manufacturers covered and the location of recycling centres, go to www.rewardingrecycling.co.uk and www.cartakeback.com.

3.27.12 Producer Responsibility: WEEE

The *EC Directive on Waste Electrical and Electronic Equipment 2003* and the *EC Directive on the Restriction of Hazardous Substances, 2003* apply to household appliances, IT, telecommunications and lighting equipment. The Directives include targets for the recovery of waste electrical and electronic equipment and the replacement of hazardous materials in the same equipment.

The so-called ROHS Directive has been implemented in the UK via the *Restriction of the Use of Certain Substances in Electrical and Electronic Equipment Regulations, 2006*. With limited exemptions, new electrical and electronic equipment must not contain lead, mercury, cadmium, hexavalent chromium, polybrominated biphenyls (PBBs) or polybrominated diphenyl ethers (PBDEs).

The WEEE Directive is targeted at two main groups – producers (including importers of electrical goods) and retailers. A producer is any company that manufactures electrical and electronic equipment, resells equipment produced by others under its own brand name or imports equipment into the UK. A retailer is any company that supplies electrical or electronic goods to an end user.

Under the *WEEE Regulations 2006* which were superseded with an updated version incorporating multiple amendments and revised European targets in the *WEEE Regulations 2013*:

▶ retailers are responsible for taking back electrical goods which have reached the end of their working life;
▶ producers are responsible for ensuring products are subject to recovery/recycling.

The rules are complex and vary depending on whether the product has been sold to a business or a consumer, but essentially all producers are required to be members of an approved compliance scheme and, through such membership, provide WEEE collection, treatment and recovery arrangements.

3.27.12.1 What is covered?

There are ten categories of WEEE identified in the regulations:

Large household appliances
Small household appliances
IT and telecommunications equipment
Consumer equipment
Lighting equipment
Electrical and electronic tools
Toys, leisure and sports equipment
Medical devices
Monitoring and control equipment
Automatic dispensers.

The regulations apply to electrical and electronic equipment (EEE) in the above categories with a voltage of up to 1,000 volts AC or up to 1,500 volts DC.

3.27.12.2 Classification of WEEE

WEEE products are also classified depending on when they were placed on the market.

▶ Products placed on the market before 13 August 2005 are called 'historic'.
▶ Products placed on the market after 13 August 2005 are called 'future'.
▶ Different rules apply depending on the classification.

3.27.12.3 Exemptions

Certain types of EEE are exempt from the regulations, including:

▶ equipment that does not need electricity to work;

▶ equipment that is part of another type of equipment which is outside the scope of the WEEE Regulations, for example, aircraft and vehicles;

▶ EEE designed to protect the UK's national security or that is used for a military purpose;

▶ filament light bulbs;

▶ household lighting;

▶ large stationary industrial tools – permanently fixed at a given place in industrial machinery or an industrial location;

▶ medical implants and infected medical equipment.

The implications of the regulations are considered below in relation to four distinct groups:

> producers
> retailers
> household users
> non-household/business users.

3.27.12.4 EEE Producer obligations

A *producer* is any company which manufactures electrical and electronic equipment, resells equipment produced by others under its own brand name or imports equipment into the UK. Their responsibilities under the regulations are as follows:

▶ Producers must be registered with the Environment Agency via membership of a producer compliance scheme.

▶ Producers are required to ensure that all EEE are clearly marked with a crossed-out wheeled bin plus information to assist treatment and reuse, including the location of hazardous materials and the nature of materials and components.

▶ Producers are financially responsible for collecting, treating, recovering and disposing of an equivalent amount of WEEE that is calculated according to the amount of EEE that they produce.

▶ Producers can arrange for their producer compliance scheme to collect, treat and recycle WEEE for them.

▶ Responsibility for non-business historic WEEE (products on the market before 13 August 2005) will be shared out between all producers depending on their market share, and added to their 'equivalent amount'.

▶ All WEEE must be taken only to approved authorised treatment facilities (AATF), where it can be treated safely prior to recycling or disposal. Only such facilities (or approved exporters) can issue *standard evidence notes*, which are used to demonstrate compliance (in the same way as PRNs in the packaging regime) and must be kept for a minimum of four years.

▶ Compliance is demonstrated via a reporting system administered by the producer compliance schemes.

3.27.12.5 Household and non-household WEEE

The regulations differentiate between 'household' and 'non-household' WEEE:

▶ *Household WEEE*: For Producers of EEE for household use, the producer compliance scheme is responsible for household WEEE collected through designated collection facilities (see below). The amount that must be collected depends on the amount of EEE sold and the proportion allocation for 'historic WEEE'.

▶ *Non-household WEEE*. For Producers of EEE for non-household use, the producer compliance scheme is responsible for collecting, treating and recycling:

  ▶ WEEE being replaced by the EEE sold, if it is historic WEEE

  ▶ EEE sold since 13 August 2005, when it is discarded as WEEE.

3.27.12.6 Negotiating obligations for non-household WEEE

Producers can negotiate WEEE responsibilities with non-household EEE users. Parties can agree to transfer obligations for WEEE that is being replaced or for new EEE when it becomes WEEE and is discarded. This agreement can occur as part of normal contract negotiations, and can benefit both parties.

3.27.13 EEE Retailer obligations

A *Retailer* is any company that supplies electrical or electronic goods to an end user. Distance sellers are also included in this category, i.e. those selling via mail order or the internet. Retailer obligations may be summarised as follows:

▶ *Supplier monitoring.* Suppliers are registered as EEE producers, via their producer registration number.
▶ *Customer information.* Retailers must provide information to customers on:
  ▶ the environmental impacts of EEE and WEEE;
  ▶ the reasons for separating WEEE from other waste;
  ▶ the meaning of the crossed-out wheeled bin symbol;
  ▶ how they can safely deposit WEEE for proper treatment and recycling free of charge.
  ▶ Retailers must keep records of this information for four years.

3.27.14 Take-back systems (household WEEE)

Retailers must set up a system that household WEEE customers can use to dispose of WEEE free of charge. There are two types of take-back system, and retailers must provide at least one of them. They are:

▶ *In-store take-back scheme.* In this system the retailer accepts a waste item from customers in store when selling them an equivalent new item. Such schemes must:
  ▶ accept all types of EEE sold by the retailer;
  ▶ record the amount and category of items received, and keep these records for four years;
  ▶ arrange the removal of separately collected WEEE. This must be done through either a producer compliance scheme or via a licensed waste carrier.

▶ *Distributor take-back scheme.* This scheme works through a network of designated collection facilities (DCFs). Retailers must join and pay membership fees to the distributor take-back scheme. Consumers can dispose of WEEE at these facilities free of charge. Retailers must inform customers how and where they can do this.

3.27.14.1 Distance sales

Distance sellers must still provide customers with a free take-back system by doing one of the following:

▶ joining the distributor take-back scheme;
▶ collecting WEEE from customers and delivering it to a WEEE collection point free of charge.

3.27.14.2 Non-household EEE sales

Retailers have no obligation to take back EEE from non-household users. However, they may be asked for information on these sales, such as:

▶ supplying contact information for the EEE producer. The producer's compliance scheme is responsible for the end-of-life handling of EEE.
▶ providing records that will help producers to supply their producer compliance scheme with accurate information, such as numbers of sales of EEE to non-household users.

3.27.14.3 Exemptions from the retailer obligations

Take-back obligations do not apply to retailers selling second-hand or reconditioned EEE. This includes charities or shops selling EEE that has been refurbished by the voluntary and social enterprise sector.

Essentially household users of EEE have the right to free take-back (but not collection) of WEEE of any age via one of two routes:

▶ a retailer selling similar products that is operating a take-back scheme;

▶ a designated collection facility operated under the distributor take-back scheme.

Options will vary locally and retailers must provide customers with information on the appropriate routes.

3.27.14.4 Business users of EEE

Business users of EEE have certain responsibilities:

▶ They must store, collect, treat, recycle and dispose of WEEE separately from other waste.

▶ They must obtain and keep proof that WEEE was given to a waste management company, and was treated and disposed of in an environmentally sound way.

Business users are able to return WEEE free of charge if:

▶ it was sold after 13 August 2005;

▶ it is being replaced with new equivalent EEE.

In these circumstances the producer's compliance scheme is responsible for the WEEE. EEE suppliers must provide information on the take-back system available to customers.

Business users will have to pay for the transfer of WEEE to an approved authorised treatment facility (AATF) if:

▶ they are discarding EEE which was purchased before 13 August 2005, and are not replacing it with equivalent EEE

▶ the producer or their compliance scheme cannot be traced

▶ purchasing new EEE and, through negotiation with the producer, the business has chosen to accept the future costs of treating and disposing of it.

3.27.15 **Producer Responsibility: Batteries**

The key provision of the *EC Batteries Directive, 2006* is that batteries should be collected at the end of their life for appropriate disposal. Additionally hazardous components should be reduced and significant proportions of the recovered batteries should be recycled.

3.27.15.1 Collection targets

The following collection targets were set:

▶ A 25 per cent collection rate for waste portable household batteries which was met in the UK on target in 2012. There is also a 45 per cent collection rate to be met ten years after entry into force (2016).

▶ Prohibition of final disposal of automotive and industrial batteries into landfill and incineration, requiring, therefore, all industrial and automotive batteries to be recycled (indirectly, therefore, this means 100 per cent collection rate).

Following the regulations set out in the *Batteries and Accumulators (Placing on the Market) Regulations 2008,* certain conditions must be met. These regulations began the process of implementation of the EC Directive through the requirement for manufacturers and importers to label portable batteries to indicate appropriate disposal via recycling. They also set limits on the content of lead, cadmium and mercury

According to the *Waste Batteries and Accumulators Regulations 2009,* requirements on collecting, treating and recycling waste batteries and accumulators are made through the Waste Batteries and Accumulators Regulations 2009. Producers (or importers) are required to register with a compliance scheme, to keep records of the total batteries put on the market and, if more than one tonne of batteries are sold, then the producer pays for recovery and recycling according to market share. The compliance scheme actually administers the recovery process and the Environment Agency regulates the producer registrations, and audits/enforces the whole scheme.

Retailers selling 32kg or more of household batteries are required to take back batteries in-store, free of charge, when they become waste.

VI CONTAMINATED LAND

3.28 The issue

The causes and problems associated with contaminated land were discussed in Chapter 1, with examples given of both historic and current sources of contamination. Legislation related to contaminated land largely relates to the assignment of responsibility for and appropriate standards of clean-up/remediation.

3.29 International agreements

There are no international agreements directly relevant to this part of UK environmental law.

3.30 European legislation

There is no European legislation directly relevant to this part of UK environmental law.

3.31 UK policy and legislation

3.31.1 UK policy objectives

These were set out in the paper, 'Framework for Contaminated Land' (DoE, November 1994). In outline, they are:

- ▶ to prevent or minimise new contamination;
- ▶ to act on existing contamination depending on risks and cost effectiveness of action;
- ▶ to improve/remediate contaminated sites to acceptable standards based on their actual or planned usage;
- ▶ to encourage development of contaminated sites (brownfield sites), to minimise avoidable pressures on greenfield sites.

The policy paper concluded that the statutory nuisance powers (Part III of EPA 90) provided a basis for dealing with contaminated land, but that there was a need for a specific contaminated land power which also addressed the scope of controls and potential liabilities. The policy is implemented through the provisions subsequently introduced by the Environment Act 1995.

3.31.2 UK Regulatory Framework

Section 57 of Environment Act 1995 has introduced provisions dealing with the identification and remediation of contaminated land. These were inserted into Environmental Protection Act 1990 (EPA 90) as Part IIA.

The provisions of the Act are implemented by subsequent secondary legislation, including Regulations, and statutory guidance. For England, the *Contaminated Land (England) Regulations 2006* are the key implementing legislation. Similar legislation applies in Scotland and Wales.

In outline, under the regulations:

- ▶ Local Authorities are required to inspect their areas from 'time to time' to identify contaminated land.
- ▶ Remediation Notices are to be issued where appropriate (taking into account – costs, seriousness of the problem, and intended use of the site).
- ▶ Notices are to be served on 'appropriate persons' (those who caused or knowingly permitted the contamination or, where this person is not known, the current occupier owner).
- ▶ Remediation may take the form of investigation and monitoring as well as clean-up.

Statutory guidance related to the regime was updated in 2012 in an attempt to simplify and increase transparency of contamination classification and remediation requirements.

3.31.2.1 Regulators and duties

Local Authorities have a duty to inspect their area, from time to time, to identify contaminated land. They are the principal regulators for this legislation, though for special sites (see below) the Environment Agency takes over the regulatory function.

3.31.2.2 Defining contaminated land

Section 78A of EPA 90 provides the legal definition of contaminated land:

Land which appears to the authority to be in such a condition, by reason of substances in, on or under it that either:

▶ significant harm is being caused or there is a significant possibility of such harm being caused; or

▶ pollution of controlled waters is being caused or is likely to be caused.

The 2012 guidance introduced a new risk-based approach which groups contaminated land into four categories, only the top two of which would normally require remediation.

3.31.2.3 Appropriate persons

Those responsible for remediation of the contamination are termed the appropriate persons.

▶ Class A appropriate persons are those who caused or knowingly permitted the contamination.

▶ Class B persons are generally the current owner/occupier of the contaminated land where no Class A persons can be identified.

▶ Both groups may be liable for remediation costs though liability is limited for Class B appropriate persons to on-site remediation (i.e. excluding water pollution and migration contamination).

3.31.2.4 Remediation notices

If land is 'contaminated land', the Local Authority is required to serve a remediation notice to the appropriate person or persons. This is recorded in the public register.

While remediation can mean clean-up, it can also mean prior investigation or subsequent monitoring. The degree of remediation required is to take account of factors of cost and seriousness of harm (based on the statutory guidance) and is to a standard defined as 'suitable for use'.

3.31.2.5 'Suitable for use'

The concept of remediation that is 'suitable for use' means that the risk assessment concept is central to the contaminated land regime. The 'suitable for use' approach consists of three elements determined on a site-by-site basis:

▶ ensuring that land is suitable for its current use;

▶ ensuring that land is made suitable for any new use as planning permission is given for that new use;

▶ limiting requirements for remediation to the work necessary to prevent unacceptable risks to human health or the environment in relation to the current use or future use of the land for which planning permission is being sought.

3.31.2.6 Special sites

When an area of contaminated land is designated a 'special site', all enforcement responsibilities are taken over by the Environment Agency rather than the Local Authority. The categories of special site are:

▶ sites where the contamination is causing particular kinds of water pollution problems;

▶ sites where a process authorised under an Environmental Permit is being or has been carried out;

▶ sites where various petroleum- or explosives-related activities have been carried out;

▶ nuclear sites;

▶ land contaminated by waste acid tars; and

▶ various kinds of defence or military site.

Land adjoining or adjacent to a 'special site', and which is contaminated land because of substances which have escaped from that 'special site', is also a 'special site'.

3.31.2.7 Offences and fines

Failure to comply with a remediation notice is an offence. There is a maximum fine of £20,000, plus a £2,000 daily fine, which applies to each day the notice is not complied with. This is imposed regardless of the level of the maximum one-off fine of up to £20,000.

3.31.2.8 Registers

Public register provisions were introduced by the Environment Act 1995. These registers are to include information regarding remediation works.

3.31.3 Implications for parties associated with land

The implications of the UK contaminated land regime are far-reaching, affecting anyone with interests in land ownership or use. Different groups may be affected as follows.

3.31.3.1 Land owners

▶ Liability assessment and contractual agreement, as part of pre-acquisition investigations, has become an essential requirement of purchasing.
▶ Control over potential sources of contamination arising from present activities.
▶ Control of contractor operations on site to confirm that any contamination can be traced to Class A appropriate person.

3.31.3.2 Landlords

Tenancy agreements are increasingly being tightened to try to ensure that the landlord is protected from inheriting Class B responsibility for remediation works because the Class A tenant is unclear. This often takes the form of pre- and post-tenancy ground surveys/ inspections. For some landlords (such as owners of large light industrial estates), this is a particular concern due to the large number of tenants, short-term occupancies and the use of ubiquitous contaminants such as diesel and engine oil.

3.31.3.3 Tenants

The opposite of the landlords' concerns: tenants must be able to defend themselves against accusations that they have caused contamination as a result of their activities, to avoid the potential of being held liable as a Class A appropriate person.

3.31.3.4 Contractors

As with tenants, contractors must be able to defend themselves from accusations made by clients that they carry Class A liability for any contamination that may exist. In short, all parties must safeguard against the creation of contamination from current activities and the inheritance of Class B liability arising from contamination caused by others.

3.31.4 Tax relief for the development of contaminated land

Although it is logical to want to clean up the significant areas of contaminated land that exist within the UK, there is, at first glance, a significant problem created by the Contaminated Land Regulations in achieving this. That is, developers are unlikely to want to take on areas of land that have been contaminated in the past because of the additional development costs associated with remediation works. This would seem to run counter to the local and national initiatives that champion the redevelopment of so-called 'brownfield sites' or areas that have been used for many years as industrial or commercial sites. In an attempt to balance this situation, the government has introduced tax incentives for investors and developers of contaminated land.

Companies developing or investing in contaminated land can claim the additional costs of remediation works plus a 50 per cent rebate. For example, if the remediation works cost £100,000, then a developer may claim tax relief (or a tax credit) worth £150,000. Qualifying costs include:

▶ work to prevent, minimise, remedy or mitigate pollution;
▶ preparatory works on assessing the condition of land, provided works are subsequently undertaken;
▶ costs of restoring land or buildings.

3.31.5 Actual vs potential liabilities

Local Authorities have published written strategies for identifying contaminated land within their jurisdiction. Progress to date has been slow but clearly for some authorities in particular this is likely to be a significant long-term task.

For landowners/occupiers the main risks associated with contaminated land are likely to be:

- potential civil liability for damage caused by pollution migrating off-site;
- potential remediation costs pursuant to the service of a statutory notice (remediation costs can include investigation, clean-up and follow-up costs);
- potential criminal liability for breaches of legislation (especially in relation to water pollution) resulting from pollution migrating off-site;
- planning conditions or obligations associated with redevelopment requiring investigation, restoration or aftercare, which act as a constraint on the scope of development or involve expenditure;
- valuation issues, i.e. it may reduce the open-market value of the property. This is an area where surveyors are increasing their expertise and involvement. Property valuations will generally either comment on contamination or make it clear that the valuation assumes no contamination.

VII NUISANCE

3.32 The issue

The causes and problems associated with nuisance were discussed in Chapter 1, with examples given of the sources and consequences often involved. Often nuisance issues become the source of civil disputes between individuals or between private individuals and organisations. Legislation related to nuisance largely relates to statutory powers to prevent it occurring in the first place or to deal with it once it manifests.

3.33 International agreements

There are no international agreements directly relevant to this part of UK environmental law.

3.34 European legislation

The *European Union Directive relating to the Assessment and Management of Environmental Noise* (Directive 2002/49/EC) sets standards in relation to noise assessment and management, including noise mapping and abatement planning. The Directive essentially seeks to limit people's exposure to environmental noise (particularly traffic-related noise), in particular in built-up areas, public parks or other quiet areas, and in noise-sensitive buildings such as schools and hospitals.

3.35 UK policy and legislation

Nuisance claims are often the subject of civil cases with an individual or group bringing a claim against another individual or organisation for some kind of interference in their enjoyment of their property. However, in some instances, the same kinds of nuisance dealt with in the civil courts may be the subject of action under criminal law. When this is the case, they are known as statutory nuisances. The legal process is outlined below.

3.35.1 The Environmental Protection Act 1990, Part III

Part III of the Environmental Protection Act 1990 gives Local Authorities in England and Wales considerable and wide-ranging powers to tackle nuisance problems. Nine types of statutory nuisance are identified by the Act:

- Premises in such a state as to be prejudicial to health or a nuisance.
- Smoke emitted from premises so as to be prejudicial to health or a nuisance.
- Fumes or gases emitted from premises (which are private dwellings) so as to be prejudicial to health or a nuisance.
- Dust, steam, smell or other effluvia arising on industrial, trade or business premises and being prejudicial to health or a nuisance.
- Any accumulation or deposit which is prejudicial to health or a nuisance. It should be noted that under the contaminated land regime this statutory nuisance provision has been repealed.
- Any animal kept in such a place or manner as to be prejudicial to health or a nuisance.
- Noise emitted from premises so as to be prejudicial to health or a nuisance.
- Noise that is prejudicial to health, or a nuisance, and is emitted from, or caused by, a vehicle, machinery or equipment in a street

111

(this provision was inserted into EPA 1990 by the Noise and Statutory Nuisance Act 1993).

▶ Any other matter declared by an enactment to be a statutory nuisance.

This list includes non-business-related nuisances, however, noise, dust and odour from industrial premises and from construction sites are the most frequently occurring nuisances. This is, in part, because of the interference that such effects can have on the neighbouring public and the annoyance which can be created. It is also a reflection of there being fewer mechanisms for prosecuting for noise, dust or odour pollution than there are for impacts such as water pollution, smoke or waste accumulations. The latter three types of nuisance are usually dealt with under more specific legislation.

3.35.1.1 Regulatory roles

The primary controls over nuisance are the statutory nuisance powers of Local Authorities, who can issue abatement notices. Where a Local Authority is satisfied that the noise, dust, odour, etc. from any premises is prejudicial to health or a nuisance, it must serve an abatement notice on the person responsible. This notice may require the abatement of the nuisance or prohibit or restrict its occurrence or recurrence, and may also require the execution of such works and the taking of such steps as are necessary for this purpose.

3.35.1.2 Offences

If an abatement notice is not complied with, Local Authorities may bring proceedings in a magistrates court. Fines of up to £5,000 are available where the nuisance arises on domestic premises, and up to £20,000 where the nuisance arises on industrial, trade or business premises.

Section 82 of the 1990 Act also gives individuals the power to complain direct to a magistrates court about a nuisance problem. Magistrates courts are able to make orders requiring the abatement of the nuisance and specifying whatever measures are necessary for this purpose, and to award costs. A person who, without reasonable excuse, contravenes any

requirement of such an order may be guilty of an offence and can be fined.

3.35.2 The Noise and Statutory Nuisance Act 1993

A limitation of the EPA 1990 Part III provisions relating to statutory nuisance is that the powers afforded to Local Authorities are confined to nuisances arising from a property or premises. The Noise and Statutory Nuisance Act extends those powers to tackle noise caused by vehicles, machinery or equipment in the street where they are satisfied that the noise amounts to a statutory nuisance. It allows Local Authorities to adopt provisions relating to the operation of loudspeakers in streets and the control of noise from audible intruder alarms on premises. It also enables the recovery of expenses incurred in abating statutory nuisances by putting a charge on the premises where it is the owner of those premises who is, or was, responsible for the nuisance.

3.35.3 The Noise Act 1996

This Act deals expressly with noise nuisance (and particularly night-time noise) from domestic premises. Local Authorities have a duty to investigate night-time noise complaints and have powers to issue warning notices and even seize 'noise making' equipment where permitted noise limits are exceeded in the complainant's dwelling. The night-time permitted limits (measured inside the complainant's dwelling) set in relation to the Act are:

▶ 35dB(A) where the background noise level does not exceed 25dB(A), and

▶ 10dB(A) above the background level where this exceeds 25dB(A).

3.35.4 The Control of Pollution Act (Amendment) 1989

Part III of the Control of Pollution Act (Amendment) 1989 was largely repealed in England and Wales by the Environmental Protection Act 1990. Those sections that remain, however, give Local Authorities powers to control noise from construction sites. Construction sites may apply for a Section 61 consent issued by the LA, which specifies conditions relating to the control of noisy operations (e.g. timing of works, notification of residents, etc.). Failure to meet the terms of the consent, or for noisy construction operations where no consent exists, may lead to the issuing of a Section 60 notice by the Local Authority – this is a legally enforceable 'abatement notice' which will specify required noise controls.

The Act also introduced the concept of the Noise Abatement Zone (NAZ), which provides a more sophisticated means of controlling, and, where justified, reducing noise from commercial and industrial premises, particularly in areas of mixed development. Although NAZs have been criticised for their complexity, and few have been designated in recent years, the powers available in such zones (for example, noise reduction notices) remain a potentially useful means of tackling some types of urban noise problem.

3.35.5 Town and Country Planning Act 1990

This legislation is discussed in Section VIII. However, it warrants a mention here because noise nuisance in particular must be considered in applications for planning consent. Consideration of noise within the planning process allows limits to be set for both day-time and night-time exposure to noise as part of the planning consent.

3.35.5.1 Noise limits

There are two basic approaches to setting noise limit levels: either at the site boundary or at the noise-sensitive property. Both have advantages and disadvantages, boundaries are easier for access and monitoring, but do not guarantee control of noise at properties; while at the property, monitoring can be difficult because of low noise levels, problems of access and the difficulty of noise source differentiation.

▶ *Reconciling limits at the boundary with limits at the dwelling*. Current practice in setting limits is apparently rather variable, for example, 57 dB(A) at the façade of dwellings, or 65 dB(A) at the site boundary are typical day-time limits.

▶ *Noise limits at the site boundary*. A more common approach to the control of noise nuisance by planning conditions is for the planning authority to set a limit at the boundary of the site generating the noise. Ideally, site boundary noise limits are set in order to ensure that the site operator works within the noise levels which have been confirmed as acceptable for the local environment.

▶ *Noise limits at noise-sensitive properties*. Since the primary objective is to control the exposure of noise-sensitive properties, it is not surprising to find that, where possible, planning authorities will attempt to define a noise limit at the façade of such properties.

Many authorities base their selection of an appropriate noise limit on British Standard BS 4142, 'Method of rating industrial noise affecting mixed residential and industrial areas'. In simple terms, BS 4142 rates the likelihood of complaints in terms of how far the intruding noise is above or below the background noise. This methodology is explored in Chapter 7.

3.35.6 Noise Emission in the Environment by Equipment for Use Outdoors Regulations, 2001

These Regulations implement Council Directive 2000/14/EC on noise emission in the environment by equipment for use outdoors. The Regulations apply to the equipment for use outdoors as listed and defined in Schedules to the Regulations. The equipment must satisfy the relevant requirements concerning noise emission in the environment and must be assessed and CE (*Communauté européenne*) marked in accordance with the Regulations. Certain exemptions exist and the requirements are not applicable to equipment placed on the market before 3 July 2001.

3.35.7 Environmental Noise Regulations, 2006 (devolved versions, England, Wales and Scotland)

These Regulations implement the European Directive relating to the Assessment and Management of Environmental Noise which set standards in relation to noise assessment and management, including noise mapping and abatement planning. The Regulations relate primarily to public bodies and large-scale noise sources such as roads, railways and airports. Reporting and mapping requirements are now in place with 'strategic noise maps' required to be drawn up by appropriate parties. Where such maps identify 'quiet areas', these areas must be afforded protection by the regulatory bodies. All areas subject to noise mapping are viewable on line at the DEFRA website.

VIII DEVELOPMENT CONTROL

3.36 The issue

There are a range of issues considered under this general heading of development control, but they may be thought to fall broadly into two camps:

▶ planning control
▶ protection of wildlife and cultural heritage.

3.37 International agreements

There are no international agreements directly relevant to planning law, unless we consider agreements relating to the non-development of areas such as Antarctica. However, there are a considerable number of agreements relating to the protection of wildlife and habitats. Perhaps the single most important agreement is the Convention on Biological Diversity.

3.37.1 The Convention on Biological Diversity

The *1992 Convention on Biological Diversity (CBD)*, which was agreed as part of the Earth Summit in Rio, has three main objectives:

▶ the conservation of biological diversity;
▶ the sustainable use of the components of biological diversity;
▶ the fair and equitable sharing of the benefits arising out of the use of genetic resources.

A meeting of the signatories in Japan in 2010 gave rise to a framework action plan and a series of targets for the 2011–2020 decade. The 20 so-called Aichi biodiversity targets were intended to drive and guide national biodiversity action plans in order to achieve the three main objectives of the CBD. National governments are required to report on actions taken, and progress achieved every five years and an international summary of progress is then produced. In essence, the plans are intended to address the key pressures on biodiversity which are stated as being:

▶ habitat change
▶ over-exploitation
▶ pollution
▶ invasive alien species
▶ climate change.

The third edition of the *Global Biodiversity Outlook* published in 2010 confirmed that biodiversity is still in significant decline and was a key driver for the agreement of the Aichi targets.

3.37.2 The CITES Treaty

The CITES Treaty, or the *Convention on the International Trade of Endangered Species* to give it its full title, entered into force in 1975 and has been a driving force for the protection of endangered flora and fauna since then. CITES works by subjecting international trade in specimens of selected species to certain controls imposed by individual countries supporting the Treaty. All imports, exports, re-exports and introduction from the wide number of species covered by the Convention have to be authorised through a licensing system. Each Party to the Convention designates one or more Management Authorities in charge of administering that licensing system and one or more Scientific Authorities to advise them on the effects of trade on the status of the species. The species covered by CITES are listed in three Appendices in the Treaty, according to the degree of protection afforded. National legislation then makes it illegal to trade in listed species without the appropriate licences, and prosecutions may be brought against offending parties.

3.37.3 The Ramsar Convention

More formally, the *Convention on Wetlands of International Importance* (*especially as Waterfowl Habitat*) is an international treaty for the conservation and sustainable use of wetlands. It aims to stem the progressive encroachment on and loss of wetlands now and in the future, recognising the fundamental ecological functions of wetlands and their economic, cultural, scientific and recreational value. It is named after the town of Ramsar in Iran where the Convention was signed in 1971.

3.37.4 The Convention on the Conservation of Migratory Species of Wild Animals

Also known as CMS or the Bonn Convention, this treaty aims to conserve terrestrial, marine and avian migratory species throughout their range. It is an intergovernmental agreement, concluded under the aegis of the United Nations Environment Programme, concerned with the conservation of wildlife and habitats on a global scale. Since the Convention's entry into force, its membership has grown steadily to include over 100 Parties from Africa, Central and South America, Asia, Europe and Oceania. The Convention was signed in 1979 in Bonn (hence the name) and entered into force in 1983.

3.38 European legislation

The key European Directives relating to planning law are the *Environmental Impact Assessment Directive* (the 1985 Directive on the assessment of the effects of certain public and private projects on the environment 85/337/EEC – which was updated in 1999) and the *Strategic Environmental Assessment Directive* (the 2001 Directive on the assessment of certain plans and programmes on the environment 2001/42/EC). Both of these are discussed in the context of UK implementation in the sections below.

As with international agreements, there have also been a number of European Conventions and Directives relating to habitat and species protection. Key examples are described below.

3.38.1 The Berne Convention on the Conservation of European Wildlife and Natural Habitats

Also known as the Berne Convention, this agreement covers the natural heritage in Europe, as well as in some African countries. The Convention was open for signature on 19 September 1979, and came into force on 1 June 1982. It is particularly concerned with protecting natural habitats and endangered species, including migratory species.

3.38.2 The Habitats Directive

Directive 92/43/EEC on the conservation of natural habitats and of wild fauna and flora is a European Union Directive adopted in 1992 as an EU response to the Berne Convention. It aims to protect some 220 habitats and approximately 1,000 species listed in the Directive's Annexes. Annex I covers habitats, Annex II covers species requiring designation of Special Areas of Conservation, Annex IV covers species in need of strict protection, and Annex V covers species whose taking from the wild can be restricted by European law. These are species and habitats which are considered to be of European interest, following criteria given in the Directive.

The Directive led to the setting up of a network of *Special Areas of Conservation*, which together with the existing *Special Protection Areas* established under the Wild Birds Directive (see below) form a network of protected sites across the European Union called *Natura 2000*. Article 17 of the Directive requires EU member states to report on the state of their protected areas every six years. The first complete set of country data was reported in 2007.

The Habitats Directive is one of the EU's two key Directives in relation to wildlife and nature conservation, the other being the *Birds Directive*.

3.38.3 The Birds Directive

More formally known as Directive 2009/147/EC on the conservation of wild birds, this was adopted in 2009. It replaced Council Directive 79/409/EEC of 2 April 1979 on the conservation of wild birds which was modified several times and had become very unclear. It aims to protect all European wild birds and the habitats of listed species, in particular through the designation of *Special Protection Areas* (often known by the acronym SPA).

Additional Directives related to environmental impact assessment and strategic environmental impact assessment are referred to in Section 3.39 under the sections describing UK implementation.

3.39 UK policy and legislation

There is a whole body of UK legislation that directly or indirectly affects the development of land, whether that land is currently occupied or not. As a rule of thumb, any significant change to the operation of an industrial site (and particularly if construction is involved) will require some kind of permission from an appropriate authority. Only the basic legal requirements are outlined below.

3.39.1 Town and Country Planning Act 1990 (as amended)

This Act defines the requirements for regional development planning and the requirements for planning permission. Essentially this Act requires all but a small number of exempt developments to seek planning permission prior to construction. Applications are made to the local planning authority (normally part of the Local Authority), who review the application to decide whether the development meets national and regional guidance and development plans. They also act as coordinators for a number of statutory consultees (e.g. the Environment Agency and national conservation bodies) whose opinion must be sought as part of the planning authority's decision-making process.

The planning authority may deny or grant planning permission and, if granted, may attach conditions which are legally binding.

Some major construction projects (e.g. long-distance pipelines, offshore oil platforms, harbour works) are not dealt with under the Town and Country Planning Act. Under the *Planning Act, 2008* and the *Localism Act, 2011*, a new body known as the Planning Inspectorate became the agency responsible for operating the planning process for what are known as nationally significant infrastructure projects (NSIPs). The 2008 Act sets out the thresholds above which certain types of infrastructure development are considered to be nationally significant and required development consent. Example developments in addition to those mentioned above include:

▶ waste water treatment plants with capacity exceeding a population of more than 500,000;
▶ hazardous waste disposal operation with a capacity exceeding 30,000 tonnes per year.

3.39.2 Legal requirements for environmental impact assessment

Environmental Impact Assessment (EIA) legislation first appeared in the USA in the late 1960s. A European Community Directive in 1985 introduced its application in EC member states and, since its introduction in the UK in 1988, it has been a major growth area for planning practice. EC Directive 85/337 and its 1997 successor have been implemented in the England and Wales via the *Town and Country Planning (Environmental Impact Assessment) Regulations, 2011* (essentially an update of the 1999 regulations but with the same basic structure).

The Directive and the regulations that implement it in UK law apply to both public and private sector projects. The developer of a project covered by the regulations is required to complete an Environmental Impact Assessment (EIA). The resulting Environmental Impact Statement (EIS) must then be submitted, with the relevant planning application, to the appropriate planning authority.

Projects covered by Schedule 1 of the Town and Country Planning (Environmental Impact Assessment) (England & Wales) Regulations, 2011 *must* submit an EIA, e.g. a crude oil refinery.

Schedule 2 projects may require an EIA depending on their size and location, e.g. water management for agricultural projects or energy industry projects, such as a non-nuclear thermal power station. It is up to the relevant planning department to decide whether a full EIA is required for such projects.

For nationally significant infrastructure projects identified under the Planning Act 2008, the *Infrastructure Planning (Environmental Impact Assessment) Regulations 2009* provide for the same EIA requirements as described above but with the Planning Inspectorate as the regulatory body rather than the local planning authority.

In addition to the legislation mentioned above, there is also a range of government guidance relating to EIA.

3.39.3 Environmental Assessment of Plans and Programmes Regulations, 2004

Adopted on 27 June 2001, the *EC Directive on the Assessment of the Effects of Certain Plans and Programmes on the Environment (2001/42/EC)* is often referred to as the *Strategic Environmental Assessment (SEA) Directive*. It applies to 'plans' and 'programmes' adopted under the national town and country planning legislation, including plans and programmes in sectors such as transport, energy, waste management, water resource management, industry, telecommunications and tourism (but only where prepared by public authorities). An EIA is a condition for the adoption of these plans and programmes or for the modification of existing ones. Where an environmental assessment is required, the competent authority will be required to prepare an environmental statement containing an assessment of the likely significant environmental effects of implementing the plan or programme and any alternative ways of achieving the objectives of the plan or programme which have been considered during its preparation. EU member states were required to implement the Directive by 21 July 2004.

In England, the *Environmental Assessment of Plans and Programmes Regulations 2004* implements the requirements of the Directive, requiring SEA for specified types of plans and programmes produced after July 2004. Similar legislation also applies in the devolved administrations. The fact that the SEA requirements only apply to plans and programmes produced by public bodies means that the focus in the UK is on local, regional and national government and also the privatised industries, e.g. railways, water companies and the Environment Agency. For example, each local authority when producing their local development plan must complete an accompanying SEA. The actual process of SEA is further considered in Chapter 6.

3.39.4 Planning (Listed Buildings and Conservation Areas) Act 1990

The Act defines the requirements for the listing of buildings of architectural and/or historical interest. In order to demolish, alter or extend a listed building, a 'listed building consent' is required. Consents are issued by the local planning authority (who may consult with English Heritage, Cadw or Historic Scotland). The Act also provides for the establishment of conservation areas, i.e. cities, towns or villages that have special historical interest. Once designated, Local Authorities must take steps to preserve such areas through consideration in the Local Plan and when determining individual planning applications.

3.39.5 Planning (Hazardous Substances) Act 1990 (as amended)

The *Planning (Hazardous Substances) Act 1990* and the *Planning and Hazardous Substance Regulations 1992* (as amended) set out the requirements for facilities with hazardous substances at or above specified quantities to have a consent from the relevant hazardous substance authority (usually the planning authority). The 1992 Regulations list the substances considered as hazardous and set out the storage thresholds at or above which a consent is required. The list and threshold quantities were updated in the 2009 amendment regulations. Consent procedures are similar to those for obtaining planning permission.

117

3.39.6 The Wildlife and Countryside Act, 1981

This Act (as amended by the *Countryside and Rights of Way Act 2000*) forms the basis of most statutory wildlife protection in the UK. Part I deals with the protection of animals and plants. It is an offence to kill, injure or disturb protected species or their habitat.

Part II is concerned with general nature conservation and habitat/site protection through the designation of Sites of Special Scientific Interest (SSSIs) and National Nature Reserves (NNRs). Development of SSSIs and NNRs requires consultation with Natural England, Natural Resources Wales or Scottish Natural Heritage as part of the planning process.

Under this legislation, nature conservation agencies (as listed above) are required to identify and notify SSSIs which cover a range of habitats and species, as well as those which have geological, physiographical and biological features of importance. The notification of such sites by the conservation agencies includes a description of all activities considered detrimental to the site which may then not be performed without appropriate permission. It becomes an offence to conduct any such operations or land management practices without an appropriate notice of assent from the relevant conservation body.

The *Countryside and Rights of Way Act 2000* strengthened the protection of SSSIs within the 1981 Act. Extensive powers are now available to the regulators to maintain and preserve these sites, including the power to influence land management practices in the vicinity of protected areas that have the potential to cause harm. The 2000 Act also improved public access to the open countryside through a nationwide programme of registration of common land.

3.39.7 National Parks and Access to the Countryside Act 1949

This Act was the original legislation allowing the designation of National Parks and Areas of Outstanding Natural Beauty (AONBs) within which tighter planning conditions are applied. In the case of National Parks, a designated National Park Authority takes over the administration of planning controls from the Local Authorities in which the park is located.

The Act also provided for public access to the countryside through access agreements and through the imposition by Local Authorities of restrictions on the destruction, removal, alteration and stopping-off of any means of access.

The Act also allowed for the rationalisation of the pre-existing law governing footpaths and bridleways in the countryside, and laid the basis for schemes of nature conservation. It has been substantially amended three times by the *Countryside Act 1968*, the *Wildlife and Countryside Act 1981* and by the *Environment Act 1995*.

3.39.8 Protection of Badgers Act 1992

This Act provides for the protection of badgers and their setts by making it an offence to take, injure, kill, be cruel to, sell, possess, mark or ring badgers or to interfere with a badger sett. There are certain exceptions, e.g. to put an injured animal out of its misery and a licensing process (to allow disturbance, etc.) administered by the countryside bodies (Natural England, Natural Resources Wales and Scottish Natural Heritage). The Act also provides enforcement powers and penalty provisions.

3.39.9 Conservation (Natural Habitats, etc.) Regulations 1994 (as amended)

These regulations implement the *EU Habitats Directive* (originally 1992 as amended in 1997) within the UK. The Directive is aimed at ensuring the conservation of habitats throughout the EU through the establishment of special areas of conservation (SACs). The regulations impose tight restrictions on developments within SACs. Additionally they provide for protected species and make it an offence to disturb the habitat of such species.

In addition to the Habitats Directive, these regulations also implement the requirements of the Wild Birds Directive, requiring the classification of special protection areas (SPAs) for areas of high importance to birds. All SPAs in the UK are also SSSIs and subject to protection under the Wildlife and Countryside Act, 1981.

IX HAZARDOUS SUBSTANCES

3.40 The issue

A range of issues are considered under this general heading and in its broadest sense it includes all pollution issues. However, included here are examples of controls that relate either to specific chemicals or to particular activities related to hazardous materials, frequently storage. In most cases, the legislation is aimed at ensuring that where hazardous materials are in use they are contained and managed to prevent escape into the environment. In the case of some of the most hazardous materials, policies and law have been created to force the phase-out of the substances either completely or in the majority of applications.

3.41 International agreements

There are many international agreements that feed into both European and UK law. Several examples have been explored in earlier sections including the Basle Convention (trans-frontier shipment of hazardous waste) and the Montreal Protocol (phase-out of ozone-depleting substances). In addition, the following is worthy of mention.

3.41.1 The Stockholm Convention (on Persistent Organic Pollutants)

The *Stockholm Convention on Persistent Organic Pollutants* was signed in 2001 and became effective from May 2004. It aims to eliminate or restrict the production and use of persistent organic pollutants (POPs). POPs are organic compounds that are resistant to environmental degradation through chemical, biological and photolytic (light degradation) processes. Because of this, they have been observed to persist in the environment, to be capable of long-range transport, of bio-accumulation in human and animal tissue, and to have potential significant impacts on human health and the environment.

Many POPs are currently or were in the past used as pesticides. Others are used in industrial processes and in the production of a range of goods such as solvents, polyvinyl chloride and pharmaceuticals.

Under the Convention initially 12 chemicals were either banned or highly restricted in use, including DDT and several other pesticides plus polychlorinated biphenyls (PCBs). A second group of nine chemicals were added in 2010 and there is potential for further additions as required.

3.42 European legislation

In a sense, all pollutant-specific legislation is relevant in this context. However, this category of legislation is being used to capture that legislation that relates primarily to the storage, supply and use of particular chemicals. From this point of view there are a number of key pieces of European legislation that are worthy of mention.

EC Regulation on Ozone Depleting substances (Regulation (EC) 1005/2009) which implements the latest version of the Montreal Protocol in the European Union. It is covered in more detail in Section III above on air pollution.

Council Directive 96/82/EC of 9 December 1996 on the control of major accident hazards involving dangerous substances (as amended) is a European Union law aimed at improving the safety of sites containing large quantities of dangerous substances. It is also known as the Seveso Directive. It is explained in the context of UK law below.

The REACH Regulation (1907/2006) imposes wide-ranging requirements on the registration, testing and information provisions relating to the supply of hazardous substances within the European Union. It is explained in the context of UK law below.

3.43 UK legislation

As with Development Control, there is a whole body of UK legislation that deals with the manufacture, use and storage of hazardous materials. In this section we will consider only a small selection of legislation under this heading, but including at least one (the Oil Storage Regulations) that affects many organisations.

3.43.1 The Control of Pollution (Oil Storage) (England) Regulations, 2001

These regulations apply to anyone storing more than 200 litres of oil above ground at an industrial, commercial or institutional site, or more than 3500 litres at a domestic site. They cover factories, shops, offices, hotels, schools, public sector buildings and hospitals. The regulations apply only in England. Scotland has its own regulations which are essentially the same but also include oil stores within a building. Wales and Northern Ireland do not have equivalent legislation.

3.43.1.1 Types of oil covered

All types of oil, with the exception of waste oil, are covered by these regulations including petrol, diesel, vegetable, synthetic and mineral oil. Waste oil is covered by the Environmental Permitting Regulations and 'duty of care' requirements of the Environmental Protection Act 1990 Part II. It is important to note that in the case of flammable liquids, such as petrol, additional health and safety requirements may also apply.

The regulations apply to mobile and fixed storage containers. This includes drums greater than 200 litres and mobile bowsers. Many self-bunded bowsers are now available. Those that do not contain integral bunds will need to be kept in a bunded area or a drip tray when in use.

3.43.1.2 Containment standards

- Tanks, drums or other containers must be strong enough to hold the oil without leaking or bursting.
- If possible, the oil container must be positioned away from any vehicle traffic to avoid damage from collision.
- A bund or drip tray must be provided to catch any oil leaking from the container or its ancillary pipework and equipment.

- The bund must be sufficient to contain 110 per cent of the maximum contents of the oil container.
- Where more than one container is stored, the bund should be capable of storing 110 per cent of the largest tank or 25 per cent of the total storage capacity, whichever is the greater.
- The bund base and walls must be impermeable to water and oil and checked regularly for leaks.
- Any valve, filter, sight gauge, vent pipe or other ancillary equipment must be kept within the bund when not in use.
- No drainage valve may be fitted to the bund for the purpose of draining out rainwater.
- Above-ground pipework should be properly supported.
- Underground pipework should be protected from physical damage and have adequate leakage detection. If mechanical joints must be used, they should be readily accessible for inspection.
- A number of other detailed requirements are included in the regulations, such as the positioning of sight gauges, fill points, vent pipes and other ancillary equipment.

3.43.1.3 Enforcement and exemptions

The Environment Agency is responsible for enforcing these regulations throughout England. The Agency may serve a notice requiring that non-compliant facilities be brought up to standard. Failure to comply with such a notice is a criminal offence and may result in prosecution.

The regulations do *not* apply:

- at premises used wholly or mainly as a single private dwelling storing less than 3500 litres;
- at premises used for refining oil or its onward distribution;
- to any oil stored in a building (unless in Scotland) or wholly underground;
- to farms – the storage of agricultural fuel oil on farms is subject to the 1991 Silage, Slurry and Agricultural Fuel Oil Regulations;
- to waste oil.

An example of a secondary containment system for a fixed tank is shown in Figure 3.11.

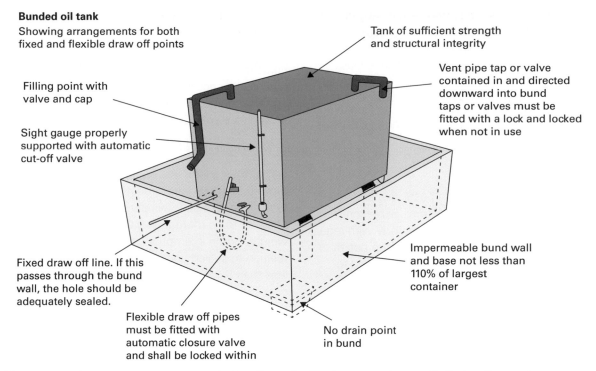

Bunded oil tank
Showing arrangements for both fixed and flexible draw off points

Tank of sufficient strength and structural integrity

Filling point with valve and cap

Vent pipe tap or valve contained in and directed downward into bund taps or valves must be fitted with a lock and locked when not in use

Sight gauge properly supported with automatic cut-off valve

Fixed draw off line. If this passes through the bund wall, the hole should be adequately sealed.

Impermeable bund wall and base not less than 110% of largest container

Flexible draw off pipes must be fitted with automatic closure valve and shall be locked within

No drain point in bund

Figure 3.11 Secondary containment design (from the Environment Agency Pollution Prevention Guidance PPG2)
Source: Contains Environment Agency information ©Environment Agency and database right.

Environment Agency guidance specifies the following additional requirements related to the operation of such a tank:

▶ Any water that accumulates in the tank will reduce its holding capacity. Consider whether it is practicable to roof the bunded area safely to prevent the accumulation of rainwater.

▶ Do NOT put any (open/gravity) drains in the bund.

▶ Remove any water from the bund by bailing or pumping via a manually controlled system.

▶ The water may be contaminated, in which case it must be disposed of as Hazardous Waste. If it is not Hazardous Waste, it must still be disposed of it in accordance with the Duty of Care obligations.

3.43.2 Control of Major Accident Hazards (COMAH) Regulations 1999

These Regulations implement the European Directive on the control of major accident hazards involving dangerous substances. Sites covered by these regulations are required to have emergency plans in place to control accidents. Information on controls must be approved by both the Health and Safety Executive (HSE) and the EA. Sites are also required to prepare a major accident prevention policy with particular emphasis on safety management systems. All COMAH sites are also NIHHS sites (although not vice versa) – both sets of regulations being principally applicable to large-scale installations handling significant quantities of substances with flammable, explosive or toxic properties.

The regulations identify two tiers of sites (Top tier and Lower tier sites) requiring different levels of emergency response planning.

COMAH was updated in July 2005 in line with the European Directive but changes are limited to the expansion of the existing regime to include mining and quarrying activities, plus the inclusion of additional listed substances and the modification of thresholds of some existing ones.

3.43.3 Control of Pesticides Regulations 1986 (as amended)

These Regulations establish a regulatory system for 'approving' pesticides before sale, use, supply, storage or advertisement. The Health and Safety Executive (HSE) and Local Authorities are responsible for enforcing the Regulations. Trading Standards departments enforce commercial aspects.

Sale, supply and storage of pesticides are subject to 'consents' given by Ministers: 'all reasonable precautions shall be taken, particularly with regard to storage and transport, to protect the health of human beings, creatures and plants and to safeguard the environment'.

Anyone involved in the sale, supply or storage of pesticides must be 'competent' to do so. Similar conditions are laid down in the consent to advertise or use pesticides. The requirements for 'competency' are likely to include adequate training in safe storage, handling and disposal procedures. This includes an understanding of the environmental hazards posed by pesticides; in some circumstances a certificate may be required following an approved test.

On commercial sites typical activities where these regulations introduce competency standards for staff and/or contractors include grounds maintenance works and pest control. It is also worth noting that the regulations include the use of pesticides that are freely available for use in the domestic sector.

3.43.4 The Environmental Protection (Disposal of Polychlorinated Biphenyls and Other Dangerous Substances) (England and Wales) Regulations 2000

These Regulations require that any equipment containing polychlorinated biphenyls (PCBs) must be registered with the Environment Agency (EA). Such equipment must be labelled as containing PCBs. From the end of December 2000, it became an offence to hold PCBs, use PCBs or equipment which contains or has contained PCBs and has not been decontaminated.

All sites are required to identify equipment containing specified levels of PCBs. Such equipment should be registered with the EA and arrangements made for decontamination in accordance with EA requirements.

Equipment containing fluids consisting of <0.05 per cent by weight PCBs may be held until the end of their useful life. Equipment containing more than these levels must be decontaminated to achieve less than this percentage and if possible to no more than 0.005 per cent by weight. Some exemptions (particularly in relation to sealed units and those containing less than 5 litres of fluids) are allowable but with consent from the EA. Such equipment must be permanently labelled to ensure correct disposal at end of life. High temperature incineration is typically required to safely dispose of PCBs which are highly persistent and eco-toxic particularly in the aquatic environment.

3.43.5 The REACH programme

As indicated in Section 3.42 above, the *EC REACH Regulation (1907/2006)* imposes wide-ranging requirements on the registration, testing and information provisions relating to the supply of hazardous substances within the European Union. The programme is intended to address the growing concern about the presence of persistent organic pollutants (POPs) in the environment as well as the increasing prevalence of child and adult allergies and conditions such as asthma which have been linked to exposure to chemical sensitisers.

▶ *POPs* – include pesticides such as aldrin, chlordane, DDT, dieldrin and endrin as well as industrial chemicals such as PCBs and hexachlorobenzene and the combustion by-products furans and dioxins. These chemicals can remain in the environment for many years, finding their way into food chains and causing health impacts in humans and animals through bioaccumulation.

▶ *Endocrine disruptors* – are chemicals that mimic the action of sex hormones in humans and other animals. They may impede the appropriate development of reproductive organs and, thereby, affect population stability in affected species. In humans, they may be responsible for reducing sperm counts and causing breast and testicular cancer.

The *European Community Regulation on Chemicals and their Safe Use (EC 1907/2006)* entered into force on 1 June 2007. It is supported in the UK by the *REACH Enforcement Regulations 2008*. The REACH programme gives greater responsibility to industry to manage the risks from chemicals and to provide safety information on the substances. Manufacturers and importers are required to gather information on the properties of their chemical substances, which will allow their safe handling, and to register the information in a central database run by the European Chemicals Agency (ECHA) in Helsinki. The Agency acts as the central point in the REACH system: it manages the databases necessary to operate the system, co-ordinates the in-depth evaluation of suspicious chemicals and runs a public database which consumers and professionals can use to find hazard information.

The Regulation also calls for the progressive substitution of the most dangerous chemicals when suitable alternatives have been identified.

3.43.5.1 Schedule for implementation

The schedule for implementation was divided into two stages:

▶ June–December 2008 – pre-registration of manufacturers and importers of more than 1 tonne of any chemical substance with the European Chemicals Agency;
▶ 2009–2018 – staggered assessment, registration and phase-out process by categories of chemicals set out in the Regulation.

Companies using chemicals are normally classified as 'downstream users' and REACH simply requires them to follow the safety measures provided by the chemical manufacturers/importers. However, care needs to be applied in relation to the 'importer' definition in the Regulation as in some instances registration obligations may exist for 'downstream users'.

3.43.5.2 Exemptions

All chemicals are covered by REACH but there are extensive exemptions from certain parts of the legislation, for example, chemicals in food and medicine are covered by other EU laws. Natural substances are also exempt from registration under REACH, if they are not dangerous and have not been chemically modified.

3.43.6 The Carriage of Dangerous Goods and Use of Transportable Pressure Equipment Regulations 2009

These regulations and the European agreement, on which they are based, lay down strict requirements on those involved in the transport of hazardous materials. Those who wish to move hazardous materials (including wastes) need to be sure that they are fully compliant with all HSE notification requirements (for specified substances) and risk management processes. The risk assessment processes require the appropriate classification, labelling and containment of hazardous materials during transport. This information has clear links to the hazard classifications described in Section 3.45 and the labelling and information requirements imposed on the suppliers of chemicals under the *Chemicals (Hazard Information and Packaging for Supply) Regulations 2002*, as amended (often referred to as the CHIP regulations).

3.44 Hazard classification

With regard to both products and wastes, a standardised waste classification system is now used across the EU under the *EC Regulation No 1272/2008 on the Classification Labelling and Packaging of Chemicals* – the so-called *CLP Regulation*. The EC website provides the following overview of the CLP regulation:

> The CLP regulation sets the rules for classification and labelling of chemicals. It aims to determine whether a substance or mixture displays properties that lead to a classification as hazardous.

CLP itself does not set information requirements (except for determining physical properties). The information requirements laid down in REACH will, however, ensure availability of much data.

Once such properties are identified and the substance or mixture is classified accordingly, manufacturers, importers, downstream users and distributors of substances or mixtures, as well as producers and importers of certain specific articles (explosive articles which are subject to classification according to Part 2 of Annex I to CLP), should communicate the identified hazards of these substances or mixtures to other actors in the supply chain, including to consumers.

The hazard of a substance or mixture is the potential for that substance or mixture to cause harm. It depends on the intrinsic properties of the substance or mixture. In this connection, hazard evaluation is the process by which information about the intrinsic properties of a substance or mixture is assessed to determine its potential to cause harm. In cases where the nature and severity of an identified hazard meet the classification criteria, hazard classification is the assignment of a standardised description of this hazard of a substance or a mixture causing harm to human health or the environment.

Hazard labelling allows for the communication of hazard classification to the user of a substance or mixture, to alert the user to the presence of a hazard and the need to avoid exposures and the resulting risks.

CLP sets general packaging standards, in order to ensure the safe supply of hazardous substances and mixtures. The CLP regulation is referred to in the Environment Agency's technical guidance note relating to the classification of hazardous waste (WM2). The full WM2 guidance note used for the classification of wastes as required under the Duty of Care and the Hazardous Waste Regulations, 2005 is available as a pdf download from the Environment Agency website.

The hazard categories that are used in the WM2 guidance note are shown in Figure 3.12.

Classification of wastes (and products in the context of the CHP Regulation referred to above) is based on comparison of the material under classification with a series of hazardous property

assessments defined in the guidance. These generally require laboratory analysis and testing of the product or waste in accordance with defined protocols.

X CIVIL LIABILITY AND KEY CASE LAW

In Section 3.1 above, we introduced the concept of civil liability in relation to environmental pollution. Civil liability is primarily concerned with the rights and duties of individuals to each other. It relates to actions taken when a 'wrong' or 'tort' is caused by one party to another where a duty of care can be said to exist. Civil actions can only be brought by those with a direct interest in the subject of the action, i.e. the person who has been 'wronged'.

Remedies that can be awarded in relation to civil claims fall under two headings:

▶ *damages*, i.e. sums of money to compensate for the suffering or loss incurred;
▶ *Injunctions,* i.e. a court order requiring those responsible for the wrongdoing to carry out remedial works or refrain from the activity causing the 'damage'.

At the centre of any civil liability case is the requirement to prove:

▶ that a loss or harm has occurred;
▶ a causal link exists between the loss/harm and the 'accused party';
▶ that some kind of liability (responsibility) exists between the accused and the person harmed.

Clearly in the context of environmental impacts, pollution incidents and nuisance events are the most likely occurrences that could give rise to a civil claim.

3.45 Proof of loss or harm

The loss or harm forming the basis of any civil case can be extremely varied and may range from ill health or even death through to property damage or the inability to enjoy one's home or surroundings. Normally, there is a commercial element involved in calculating the loss/harm to enable an appropriate level of compensation to be awarded in the event that the case is found

H1	"Explosive": substances and preparations which may explode under the effect of flame or which are more sensitive to shocks or friction than dinitrobenzene.
H2	"Oxidizing": substances and preparations which exhibit highly exothermic reactions when in contact with other substances, particularly flammable substances.
H3-A	"Highly flammable" - liquid substances and preparations having a flash point below 21°C (including extremely flammable liquids), or - substances and preparations which may become hot and finally catch fire in contact with air at ambient temperature without any application of energy, or - solid substances and preparations which may readily catch fire after brief contact with a source of ignition and Which continue to burn or be consumed after removal of the source of ignition, or - gaseous substances and preparations which are flammable in air at normal pressure, or - Substances and preparations which, in contact with water or damp air, evolve highly flammable gases in dangerous quantities.
H3-B	"Flammable": liquid substances and preparations having a flash point equal to or greater than 21°C and less than or equal to 55°C.
H4	"Irritant": Non-corrosive substances and preparations which, through immediate, prolonged or repeated contact with the skin or mucous membrane, can cause inflammation.
H5	"Harmful": substances and preparations which, if they are inhaled or ingested or if they penetrate the skin, may involve limited health risks.
H6	"Toxic": substances and preparations (Including very toxic substances and preparations) which, if they are inhaled or ingested or if they penetrate the skin, may involve serious, acute or chronic health risks and even death.
H7	"Carcinogenic": substances and preparations which, if they are inhaled or ingested or if they penetrate the skin, may induce cancer or increase its incidence.
H8	"Corrosive": substances and preparations which may destroy living tissue on contact.
H9	"Infectious": substances and preparations containing viable micro-organisms or their toxins which are known or reliably believed to cause disease in man or other living organisms.
H10	"Toxic for reproduction": substances and preparations which, if they are inhaled or ingested or if they penetrate the skin, may induce non-hereditary congenital malformations or increase their incidence.
H11	"Mutagenic": substances and preparations which, if they are inhaled or ingested or if they penetrate the skin, may induce hereditary genetic defects or increase their incidence.
H12	Waste which releases toxic or very toxic gases in contact with water, air or an acid.
H13	"Sensitizing": substances and preparations which, if they are inhaled or if they penetrate the skin, are capable of eliciting a reaction of hypersensitization such that on further exposure to the substance or preparation, characteristic adverse effects are produced. [As far as testing methods are available].
H14	"Ecotoxic": waste which presents or may present immediate or delayed risks for one or more sectors of the environment.
H15	Waste capable by any means, after disposal, of yielding another substance, e.g. a leachate, which possesses any of the characteristics above.

Figure 3.12 Hazard categories as presented in the EA technical guidance note WM2
Source: Contains Environment Agency information ©Environment Agency and database right.

in favour of the plaintiff. Such calculations may consider things like repair/remediation costs, property value impacts, lost earnings and less tangible figures related to disturbance or loss of amenity.

3.46 Proof of causal link

The requirement is to demonstrate that, on the balance of probabilities (a lower level of proof than in criminal cases), the event or activity in question gave rise to the harm or suffering that is the basis of the plaintiff's case. In other words, a causal link exists between the activity of the defendant and the damages experienced by the plaintiff. Without this linkage there is no basis for a civil liability claim.

Although in some cases such a linkage may be extremely obvious in others, where several contributory factors may be involved, proving this link can be a critical part of the plaintiff's and/or the defence case. For example in noise nuisance cases, alternative sources of noise may need to be ruled out.

3.47 Proof of liability

The third requirement of a plaintiff in a civil liability case is to demonstrate that some kind of liability exists. The common law liabilities or 'torts' of trespass, nuisance and negligence, and the strict ruling made in the *Rylands v Fletcher* case have given rise to a body of case law precedents which form the basis of most environmental liability proofs.

3.47.1 The Tort of Trespass

This involves some direct physical interference with property. Like nuisance (explained below) the use of this tort depends upon the person bringing the action having a right in an affected property. It has the advantage over other forms of action that it need not be shown that any actual damage was caused in order to claim an injunction to prevent that interference. Unfortunately, trespass is of very limited use regarding pollution cases because most harm caused by pollution is indirect, that is, mediated by either water, air or transported in

some other indirect way. Illegally dumped waste on another's land would, however, be grounds for a trespass claim. Allowing the same waste to be transmitted by a watercourse, or burning it to cause noxious fumes could not be a trespass (as there has been no unauthorised access to a property by a person), but may be considered as grounds for nuisance.

3.47.2 The Tort of Negligence

Negligence has the advantage of being actionable without reliance upon property rights, i.e. you do not need to own or be a leaseholder of a property affected by pollution or nuisance in order to bring a claim of negligence. Its focus is upon compensation to victims for the damage caused to them by the careless acts of others. An action for water pollution in negligence requires proof, first, that the polluter owed the affected person a duty of care; second, that he or she carelessly breached that duty; and, finally, that this breach caused foreseeable harm. Failure to comply with statutory requirements is a key factor in determining whether negligence has occurred.

3.47.3 The Tort of Nuisance

In contrast to negligence, 'nuisance' requires that the pollution causes some actual harm, in the form of an unreasonable interference with the enjoyment of property rights. Here, the interference may be indirect, making the claim more suitable for those seeking damages or an injunction to prevent pollution or nuisance. Its basis is, however, not the minimisation of pollution, but the protection of property rights and legitimate interests in the use of land or water. It is limited, in that it provides an action against a polluting activity only in so far as that activity interferes to an unreasonable extent with another property owner's (or tenant's) rights such as the abstraction or use of water, or the ability to sleep in a residential property without undue disturbance, etc.

3.47.3.1 Factors influencing nuisance

Case law has identified six important factors which must be considered when assessing an alleged nuisance. These are:

- character of neighbourhood;
- standard of comfort;
- time and duration;
- motive;
- interference;
- reasonableness.

- *Character of neighbourhood*: When considering an alleged nuisance, local standards apply. This means that the test of nuisance allows for the nature and type or class of area. For instance, what may be acceptable in a mixed commercial and residential area may not be acceptable in a purely residential or rural area (and in some cases vice versa).

- *Standard of comfort*: The concept of nuisance does not allow for over-sensitivity, and normal standards of comfort (and hence protection) have to apply. For example, with regard to noise, consider the hypothetical case of a shift worker who sleeps during the day while neighbours go about their normal daily activities. In this situation there is an obvious potential for the shift worker to be disturbed by noise from the use of domestic appliances, radios, cars, etc. Could this disturbance be considered a nuisance? In the case where the neighbours could be shown to be acting reasonably and without malice (see below), it is unlikely that a nuisance could be proved despite the obvious interference. The worker could be said to be overly sensitive to noise because of atypical sleep patterns. Similar arguments are also applicable to situations where a person has a physiological (or psychological) over-sensitivity to noise. This can manifest itself as an over-sensitivity to a particular noise (fairly common) or to noise in general. Whilst the disturbance to such an individual may be severe from the individual's point of view, it may not necessarily be considered a nuisance by a civil court.

- *Time and duration*: The time of day and duration of the alleged nuisance are also relevant. The common law concept of nuisance does not encompass events which could be considered trivial. It may be that an event which is very limited in duration or infrequent would not be considered a nuisance.

- *Motive*: The person against whom legal action would be taken in the case of nuisance is usually referred to as the author of the nuisance. Any indication of malice or similar motive in the conduct of the person alleged responsible for the nuisance is likely to be considered evidence of unreasonable behaviour and hence nuisance.

- *Interference*: Interference is crucial to showing the existence of a nuisance. Material interference must be proved but this need not be exclusively physical loss or damage. This is especially relevant to noise nuisance where it is unlikely that physical loss or damage would result from the interference. What is much more likely is that the interference will manifest itself by such effects as having to modify normal patterns of behaviour to compensate for the nuisance. Obvious examples would be the disruption of sleep, loss of privacy and disturbance of normal activities. In some cases medical evidence may be relevant though in itself medical evidence may not prove the interference and must be treated warily.

- *Reasonableness*: Much of the preceding discussion is based on the argument of reasonableness. Any evidence of unreasonableness, on either side of the argument, is likely to prove crucial to deciding whether or not a nuisance exists. In addition, and in some respect at variance with the ideas of time and duration, no one should suffer a reasonably avoidable nuisance, no matter how slight.

This last point illustrates how the six factors outlined above are interrelated and interdependent. Even if it were it possible, it is not advisable to make or argue any assessment of nuisance without regard to all the factors. The assessment of nuisance in any particular case is made up from the assessment of the whole. This explains why nuisance is not amenable to mechanistic approaches. Superficially the concept of nuisance is simple. However, the reality is that the accurate, consistent and reliable assessment of nuisance is a complex and skilled process.

3.47.4 Defences against charges of nuisance

There are a series of standard defences against action for nuisances as a civil wrong (i.e. in common law) which are as follows:

The nuisance was an act of God.

The nuisance was an act of a trespasser; this assumes, therefore, that the act was committed without the knowledge or control of the alleged author of the nuisance.

The action taken was with plaintiff's consent; this assumes the plaintiff consented to the precise act.

The author was ignorant of causing nuisance. The defence of 'prescription' is available only in the case of private nuisance and can only exist against another piece of land (known as the servient tenement) which suffers from the alleged nuisance. This defence has to show that the acts complained of have been undertaken openly and to the knowledge of the owner of the servient tenement. After 20 years' continuous existence, the operations will achieve legality and cannot be a nuisance with respect to the servient tenement.

3.47.5 Public vs private nuisance

This is a somewhat confusing area of common law due to the lack of clarity over the boundary between statutory and civil jurisdiction. However, the key points are as follows:

▶ Private nuisance involves an infringement of an individual's right to enjoy their own property; public nuisance involves the causing of damage at large to a sector of society and is not necessarily linked to property rights.

▶ Private nuisance forms the basis of a civil claim by an aggrieved individual leading to injunction and compensation if successful; public nuisance proceedings are initiated by the Crown Prosecution Service as criminal proceedings leading to the award of fine and/ or custodial sentence.

A single event could be actioned under both private nuisance proceedings and public nuisance proceedings. The process of a private nuisance claim has been explained above, i.e. a claimant (plaintiff) would have to prove that the defendant was the cause of tangible loss that met the description of the tort of nuisance. If the event leading to a private nuisance claim affected multiple parties, then the Crown Prosecution Service may pursue a criminal prosecution in the criminal courts under this 'common law' definition of public nuisance:

A person is guilty of a public nuisance (also known as common nuisance), who (a) does an act not warranted by law, or (b) omits to discharge a legal duty, if the effect of the act or omission is to endanger the life, health, property, morals, or comfort of the public, or to obstruct the public in the exercise or enjoyment of rights common to all Her Majesty's subjects.

It is an oddity of UK law that this generic 'criminal act' does not appear as a specific offence in any piece of 'statute law' but dates back to the twelfth century and the legal rights of the Crown (hence it is referred to as 'common law'). Clarifications as part of a House of Lords appeal process in 2005 found that in many cases, situations that might otherwise have been criminal public nuisances had now been covered by statutes. Thus, for example, the *Protection from Harassment Act 1997* would now be used in cases involving multiple telephone calls, and the *Criminal Justice and Public Order Act 1994* confers powers on the police to remove persons attending or preparing for a rave

at which amplified music is played during the night (with or without intermissions) and is such as, by reason of its loudness and duration and the time at which it is played, is likely to cause serious distress to the inhabitants of the locality.

These and other similar statutes (including Part III of the Environmental Protection Act 1990) had, in effect, made the common law offence of 'public nuisance' largely redundant and it was argued in the appeal that it should no longer be considered an offence in English law. The Lords agreed that, as a matter of practice, all alleged offences falling within the remit of statutes would now be charged under those statutes.

The 2005 review also accepted, however, that this left a very small scope for the application of the common law offence. But, just as the courts had no power to create new offences and could not widen existing offences so as to retrospectively criminalise conduct, it equally had no power to abolish existing offences and so 'public nuisance' proceedings remain a viable option for dealing with 'nuisances' affecting a wide range of people that do not constitute an offence under UK legislation.

Examples of scenarios that could be subject to public nuisance proceedings include the following:

▶ widespread fallout of dust over a large number of properties;
▶ blasting noise and flying rocks/dust from a quarry;
▶ an offensive smell affecting a town centre area.

3.48 Key cases

The torts described above have been developed through case law into a range of common law rulings relevant to environmental pollution. Some of the key cases that environmental practitioners need to be aware of are as follows.

3.48.1 Rylands v Fletcher (1867)

Even where there are no grounds for negligence, this case establishes a tort known as 'strict liability' imposing liability on those who create situations of special danger or hazard. Under 'strict liability' a person does not have to be shown to be negligent to carry the responsibility for any losses incurred as a result of their actions or omissions. The ruling applies to those who, for their own purposes, keep anything on their property which is likely 'to do mischief' if it escapes where such operations involve what can be considered abnormal risk. In this historic test case, a release of water from a reservoir owned by Mr Fletcher flooded mine shafts owned by Mr Rylands. The court found in favour of the plaintiff with the following ruling:

> If a person brings or accumulates on his land anything, which, if it should escape, may cause damage to his neighbour, he does so

at his own peril. If it does escape and causes damage, he is responsible, however careful he may have been, and whatever precautions he may have taken to prevent the damage.

Use of care and skill and public benefit are not defences. This is the precursor of producer responsibility or the polluter pays principle as we know it today in both civil and criminal law. Its interpretation has been modified by subsequent case law, notably by the *Cambridge Water v Eastern Counties Leather* case in 1994 (see below), where the House of Lords ruled that a key test in determining liability under the tort of negligence, nuisance and the *Rylands v Fletcher* ruling, is the foreseeability of harm or injury.

3.48.2 The Cambridge Water Company v Eastern Counties Leather (1994)

This case illustrates how case law is always developing. The case was originally argued on the basis of the rule in *Rylands v Fletcher*.

Cambridge Water Company (CWC) owned a borehole at Sawston from which it pumped water into the public supply. CWC bought the land in 1976 for the purposes of constructing the borehole to supply water to its customers. The pumping house was commissioned in 1979.

As part of its operations in the 'industrial village' of Sawston, Eastern Counties Leather plc (ECL) used organochlorine compounds for degreasing skins. Until about 1976, the chemical was delivered to the site in drums, kept in storage there and, when needed, tipped into the reservoir supplying the degreasing machine. This transfer led inevitably to spillages which over the years filtered through the ground into the underlying aquifer. The degreaser is volatile and easily evaporated, and ECL might have considered that it evaporated before it could pass through the ground.

In 1980, the *EC Drinking Water Directive* was issued and became effective in 1985. This Directive included maximum permissible concentrations of organochlorine compounds. CWC found that concentrations in its borehole were in excess of those limits and as a result CWC had to stop pumping and find an alternative supply. This cost something in excess of £900,000. The material accidentally spilled by ECL was eventually identified as the cause of the excessive levels in the water, and the water company claimed the costs of relocation from ECL.

Round 1. In the first instance, CWC founded its action on nuisance, negligence and the rule in *Rylands v Fletcher* which imposes strict liability for the natural consequences of the escape of a substance brought onto or accumulated on land when the use of the land is 'non-natural'. The case was decided against the water company on the basis of the rule in *Rylands v Fletcher*, holding that the use of organochlorines by a tannery was not 'non-natural use' of land. Also, because the consequence of these spillages was not considered reasonably foreseeable, the actions under nuisance and negligence also failed. CWC appealed pursuing its case on the basis of *Rylands v Fletcher*.

Round 2. In the Court of Appeal, a judgment was given on 19 November 1992, which reversed the previous decision and held that ECL had indeed interfered with CWC's natural right to pump water from beneath its own land and that the spillage of chemicals was an actionable nuisance and the liability a strict one. The appeal court also held that there was no need to show that a defendant had foreseen the consequences of his or her action. No importance was attached to the fact that CWC had suffered damage only when quality standards had been raised about three years after it had begun to abstract the water and long after ECL had stopped the spillage of chemicals.

It was held that the rule in *Rylands v Fletcher* did not apply because the CWC case involved the tannery's liability for its spillages of chemicals rather than liability for the escape of the chemicals from ECL's property. An example of leakage from a negligently constructed tank was a case in which the rule in *Rylands v Fletcher* might apply, but the court expressly refrained from an 'examination of the conditions for liability under the rule'.

Round 3. ECL appealed to the House of Lords,

who delivered their judgment on 9 December 1993. The Lords overturned the decision made by the Court of Appeal, finding that ECL was again not liable for the £2 million damages which had been ordered. Lord Goff explained that although the fact that a defendant has taken all reasonable care will not exonerate him or her from liability in nuisance, 'it by no means follows *that the defendant should be held liable for damage of a type which he could not reasonably foresee'*. This means that strict liability would only apply if damage could be reasonably foreseen. Lord Goff also commented that the storage of large amounts of chemicals should be considered a non-natural use of land under the rule of *Rylands v Fletcher*. This comment removes the possibility of a defendant claiming that pollution occurring on an industrial site is in fact a 'natural land use' for that kind of site.

It may be useful to summarise these two cases using key words:

Rylands v Fletcher	Strict liability
CWC v ECL	Foreseeability and clarification of non-natural land use

3.48.3 Hancock and Margerson v JW Roberts Ltd (1996)

In *Hancock v J W Roberts Limited* heard in Leeds High Court, the claim related to ill-health suffered as a consequence of asbestos dust released from the factory. June Hancock was diagnosed as suffering mesothelioma in 1994. Mesothelioma is an incurable asbestos-related cancer. Mrs Hancock was exposed to asbestos between 1938 and 1951 while living and playing as a child near the J W Roberts factory in Armley, Leeds. The company is a subsidiary of T&N plc, once the world's largest manufacturer of asbestos products.

Mrs Hancock argued that J W Roberts knew of the dangers of asbestos during the period she was exposed to it and that it should therefore have taken steps to protect its workers and people in the local community. Constructing the case involved studying over 50,000 documents, some dating back to the nineteenth century.

Counsel for the defence denied that the firm had been aware that the Roberts factory posed a risk to nearby residents. Company minutes, however, from 1933, were produced as evidence during the six-week hearing, which suggested that the dust risks were realised.

Mrs Hancock's daughter sued the company after watching her mother die from mesothelioma contracted from the Roberts factory which closed in 1958. It can take 10–50 years for mesothelioma, only caused by exposure to asbestos dust, to show. Some 200 people from Leeds have died of mesothelioma.

The ruling was that J W Roberts knew of the harmful effects of asbestos as early as 1933, and so should have protected people who lived in the Armley area from that date. T&N were ordered to pay £65,000 in compensation to Mrs Hancock. (Foreseeability re Rylands and Fletcher, Negligence and Nuisance are all relevant to this ruling.)

In a parallel claim Evelyn Margerson was awarded £50,000 compensation for the death of her husband from mesothelioma in 1991. He had also lived near the factory as a child.

3.48.4 Hunter & Others v Canary Wharf Ltd and Hunter & Others v London Docklands Corporation (1997)

The plaintiffs in this case claimed damages in respect of interference with the television reception at their homes on the Isle of Dogs. They claimed the interference was caused by the construction of the Canary Wharf Tower which was built on land developed by the defendants. The tower obstructed the television transmissions from the BBC transmitter at Crystal Palace because of its size and the metal in its surface. The interference began in 1989, and a relay transmitter was built to overcome the problem. A satisfactory reception was achieved by 1992 but the plaintiffs claimed damages in respect of the interference with their television reception during the intervening period. Their claim was framed in nuisance and in negligence, but the claim in negligence was abandoned before the case reached the House of Lords.

The House of Lords upheld the decision by the Court of Appeal that interference with television reception by a tall building cannot amount to a nuisance. They agreed that as a general rule, a person is entitled to build on his or her own land, even if the presence of that building may interfere with his or her neighbour's enjoyment of his or her land. The defendants were, therefore, allowed to build on their land, even if this caused interference to television receptions on their neighbour's land. This highlights that an action in private nuisance generally arises from something emanating from the defendant's land, such as noise, dirt, fumes and noxious smells, rather than a structure on the site.

The Lords then considered the plaintiffs' right to sue in private nuisance in view of the fact that the plaintiffs were not restricted to householders who had exclusive right to possess the places where they lived, whether as freeholders or tenants, or even as licensees. They included people with whom the householders shared their homes, for example, spouses, children or other relatives. All of these people were claiming damages in private nuisance by reason of interference with their television viewing or by reason of excessive dust caused by the construction of the Limehouse Link Road. Lord Goff explained that the tort of nuisance is a tort directed against the plaintiff's enjoyment of his or her land, therefore an action in private nuisance will usually be brought by the person in actual possession of land affected. The loss suffered is the infringement of property rights not the actual damage per se suffered by the individuals. This ruling reaffirmed that private nuisance remained essentially an action concerned with interference with property interests.

3.48.5 Dennis v Ministry of Defence (2003)

Mr and Mrs Dennis lived at Wilmot Hall, a large estate near RAF Wittering near Peterborough. The RAF base was home to a squadron of Harrier jump jets and the Dennis couple brought the claim against the MOD based on the noise caused by the training and circuit flying of these jets. The MOD had clearly acknowledged the intrusion caused to local residents as it had provided a voluntary scheme of grants for double glazing and offered to buy out properties especially affected. Wilmot Hall was a listed building for which double glazing was not practicable and in any case this would not have affected the noise experienced outside of the building. Experts on both sides agreed that the noise levels caused disturbance and interference but disagreed as to their seriousness.

The court initially looked favourably on the plaintiff's claim of nuisance not just on the basis of disturbance but also in terms of lost business opportunities (in terms of use of the Hall) and overall value of the estate.

The court also rejected the MOD's arguments that the training of pilots was a 'legitimate use of land' – stating that any activity that generates extreme noise and other pollution should be considered extraordinary uses of land. The key issue thus became whether the MOD could argue that the operations at the airfield were in the public interest and thereby immune from nuisance claims. The court recognised a problem with the public interest defence with regards to the Human Rights Act under which the claimants had submitted separate claims. Basically the argument ran that even if an activity is in the public interest, it could be considered a breach of an individual's rights to suffer 'damage' for the wider benefit of society. Although the court had the option to disregard the nuisance claim and uphold the claim under the Human Rights Act, Mr Justice Buckley chose to uphold the nuisance claim stating that his principle of one person not bearing the damages for many should be recognised within the tort of nuisance.

The final solution was to allow the MOD operations to continue, but to require compensation to be paid to the claimants, for both past and future damage, including the diminution in value of their property. The MOD provided evidence suggesting that the Harriers would be phased out by 2012 and damages were therefore calculated for the period from 1984 (when Mr and Mrs Dennis arrived) to 2012. Damages of £950,000 were awarded. The Harriers were eventually retired from service in 2010.

This case is important for three reasons:

▶ the issue of immunity for operations that may be considered in the public interest has been overturned in nuisance claims, but:
▶ the automatic right of injunction for a successful nuisance claim is removed. The 'normal' outcome of a successful nuisance claim is the award of both damages and an injunction to prevent recurrence of the nuisance. It is this right to an injunction which gives common law nuisance its particular strength as a legal remedy in tackling pollution and environmental problems. The Dennis case suggests that in future a more flexible approach may be employed towards awarding compensation for continuing nuisances, with courts examining closely the social benefits of a defendant's activities in deciding whether or not to grant an injunction. It may be foreseen, for example, that a defendant that is a private company may argue that its activities are in the public interest by, say, creating local employment or creating export markets.
▶ It reinforces the argument that any significantly polluting activity cannot be considered a 'natural use of land'.

3.48.6 Empress Car Company (Abertillery) Limited v. National Rivers Authority – House of Lords ruling, 1998

This is not a civil case but a House of Lords ruling on a criminal prosecution made under the Water Resources Act, 1991. It has wide-reaching implications, however, in both the civil and criminal courts and is therefore included here.

Empress Car Company maintained a diesel storage tank on its premises, which drained directly into a river. Although there was a bund around the tank, the company had overridden this by fixing an extension pipe to the tank outlet which connected to a drum outside the bund. The tank outlet was governed by a tap which had no lock.

In 1995, someone opened the tap, allowing the full contents of the tank to run into the drum, which overflowed into the yard and drained into the river. The person who opened the tap was never identified. It could have been an employee or, since there had been local opposition to the company's business, an act of sabotage may have been involved.

Nevertheless, the company was charged with causing the entry of polluting matter into the river. It was convicted by local magistrates and lost an appeal to the Crown Court on the grounds that it had brought the diesel onto the site and had failed to take adequate preventative measures, such as fitting a lock on the tap and ensuring the integrity of the bund.

Empress appealed but the Court of Appeal upheld the decision and in 1998 a House of Lords ruling laid down guidance to the lower criminal courts as to the issues surrounding whether a company has 'caused' water pollution contrary to the WRA 1991, s. 85(1) (this offence is now covered by the Environmental Permitting Regulations 2010).

The House of Lords stated that courts should not be sidetracked by trying to determine which person or event it was that caused the pollution. The real question is whether it was the defendant who had done something which led to the cause of the pollution. For this purpose, it was necessary that what the defendant had done was

a positive act, but that act did not need to be the immediate cause of the escape. It was found that the act of the defendant maintaining a tank full of diesel fuel in an unsecured bund was something which had caused the problem.

In practical terms, the decision makes it clear that companies should address tampering and vandalism as a risk when devising environmental management systems to prevent pollution occurring from their premises. An escape, even when caused by the act of a third person, may well be a breach of duty. The question is in general whether the intervention should be regarded as a 'fact of life', rather than whether the particular intervention ought to have been foreseen by the defendant. The House of Lords said that, regrettably, there was nothing unusual about 'ordinary vandalism' (as opposed to a terrorist attack).

XI FURTHER RESOURCES

Staying up to date with constantly evolving environmental legislation and policy is a challenge for any environmental professional. Fortunately there are a wide range of public and private resources available to help. Table 3.21 presents some resources for further study.

Table 3.21 Further resources

	Further information	Web links (if relevant)
The Environment Agency	The Environment Agency publishes detailed guidance and technical information regarding many aspects of pollution control legislation falling under their remit. Much of this information can be found on their website.	www.gov.uk/ environment-agency
www.gov.uk	This website is a UK government information hub service and the business pages contain information and links relating to a variety of environmental and in particular waste related information. It also links to source legislation held on the National Archives website at www.legisation.gov.uk. In Scotland and Northern Ireland the Netregs service (www.netregs.org.uk) provides accessible summaries of legal requirements organised by issue and industrial sector – an excellent resource.	
WRAP (Waste and Resources Action Programme)	Government-funded, best practice programme providing extensive best practice and case study information	www.wrap.org.uk
European Environmental Agency	European Directives and policy	www.eea.eu.int
The Department of Environment, Food and	Maintains an excellent website covering legislation, policy and consultation documents. DEFRA also plans to launch a new	www.gov.uk/defra

Table 3.21 (continued)

	Further information	Web links (if relevant)
Rural Affairs (DEFRA)	environmental legislation hub which will be known as DEFRA-LEX but at the time of writing no definite timescales could be confirmed.	
Institute of Environmental Management and Assessment	In particular, the legal brief section of *The Environmentalist* magazine	www.iema.net
The Joint Nature Conservation Committee	Information on UK Special Protection Areas (SPAs) and Special Areas of Conservation	jnccdefra.gov.uk
Magic website	This is an interactive website maintained by Natural England that provides information on all sites of conservation interest including SSSIs across the UK.	www.natureonthemap. naturalengland.org.uk
Commercial organisations	There are a number of commercial organisations which will provide information about current environmental issues for a subscription fee. Examples of the more popular bulletins include: the Ends report; McCormick information bulletins; Technical Indices; Barbour Index; Croner's Environmental Management and Case Law; GEE's *Environmental Compliance Manual.*	

CHAPTER 4

Understanding environmental management and sustainable development in a business context

Chapter summary

This chapter looks at the pressures and opportunities created in relation to organisations in response to environmental problems and a growing global awareness of the issue of sustainability. The general principles of risk management and opportunity assessment using a value chain approach are introduced, and case studies are presented of organisations that have adopted proactive sustainability strategies. The importance of resource efficiency is discussed in general terms, again with case studies illustrating the scale of improvements reported by innovators in this area.

Links to later chapters are also presented, notably:

- Chapter 6 – risk management tools and control techniques
- Chapter 8 – creating a business case for change or improvement
- Chapter 9 – company communications and reporting, including corporate social responsibility and sustainability reporting.

I HOW DO ORGANISATIONS IMPACT THE ENVIRONMENT?

Often our first impression of an organisation's impact on the environment is based on issues such as chemical spills, stack emissions or effluent discharged to rivers. In most cases, however, such 'direct' impacts are only part of the picture. In order to understand the overall impact arising from an organisation's activities we must consider direct impacts, supply chain impacts and product/service impacts (Figure 4.1).

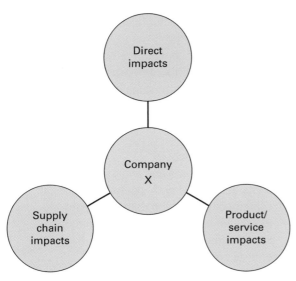

Figure 4.1 How do organisations impact the environment

4.1 Direct impacts

These are environmental issues that occur as a result of activities performed by the organisation itself. The group includes issues such as:

▶ waste-related impacts such as landfill gas generation;
▶ effluent generation and associated water pollution impacts;
▶ emissions to air and associated air quality and climate change impacts;
▶ land use impacts associated with amenity use and/or habitat loss;
▶ raw material usage and impacts related to the consumption of renewable and non-renewable resources;

▶ transport issues including consumption of fuels and the generation of air pollution and nuisance.

4.2 Supply chain impacts

These are environmental issues that are caused by third parties who are carrying out activities linked to or supporting the organisation's activities (i.e. suppliers and contractors) and might include:

▶ raw material production, extraction and consumption;
▶ production impacts (including energy generation) which mirror the organisation's own direct impacts;
▶ transport impacts associated with the supply of products and services.

4.3 Product/service impacts

These are impacts that are caused by customers or others using the product produced or as a result of the service supplied by the organisation. Examples include:

▶ packaging waste disposal by customers;
▶ paper waste disposal by customers;
▶ end-of-life product waste disposal by customers;
▶ energy consumption by customers in using the product.

II HOW DOES THE ENVIRONMENT IMPACT ORGANISATIONS?

In many respects, business and commerce may be considered the prime source of the pollution and resource depletion environmental impacts facing us as a wider society. However, the same organisations that are the source of these issues that aggregate to produce the global environmental threats are also affected by them both directly and as a result of society's response to them. The business pressures created are often presented under the following four categories:

▶ legal
▶ financial

▶ market
▶ social.

4.4 Legal pressures

As the international community comes to grips with the issue of sustainability and as public awareness grows at a more local level, there has been a rapid expansion of new and demanding environmental legislation that increasingly affects all organisations. It includes international agreements, European and national law. Over 400 pieces of legislation have come from Europe alone, covering a range of issues from packaging, waste, discharges and eco-labelling of products.

Poor management and preparation in relation to new legislation may result in a business losing its licence to operate, or at the least becoming involved in expensive and reputation-damaging legal proceedings. Even if significant legal non-compliance issues are avoided, poor reputation with regulators can lead to costly delays and objections to planning applications, permit renewals, etc.

4.5 Financial pressures

The ways in which environmental issues are becoming manifest in organisations as financial issues are continuing to grow each year. Some examples are given below.

4.5.1 Pressure from lenders and insurers

Environmental considerations are increasingly being taken into account in the lending process and are becoming a standard part of loan screening. A requirement to show evidence of effective environmental management systems is becoming as important as sound cash flow management.

Similarly, the satisfaction of environmental performance criteria is increasingly part of assessment by insurance companies. Successful management of the environmental issues in a business is a growing requirement of insurers. Legal compliance audits may be mandatory as part of the agreement to provide cover.

Liability implications for both insurers and lenders are what is driving the trends described above.

4.5.2 Operational costs

Environmental matters can directly affect the profit and loss accounts and balance sheets of any organisation. Simple measures such as reducing energy consumption, minimising waste and improving transport operations bring environmental benefits but also increasingly reduce overheads and save money. Examples of operational costs that also reflect environmental priorities include:

▶ fines and clean-up costs associated with pollution incidents;
▶ pollution control costs, e.g. emissions abatement or effluent treatment techniques, containment of hazardous material storage areas;
▶ civil claims arising from nuisance or pollution incidents;
▶ waste disposal costs;
▶ raw material costs and availability – escalating costs associated with increasing material scarcity or production costs, plus the legislative ban or restricted use of certain substances;
▶ energy costs;
▶ product-related costs associated with design criteria changes (e.g. the ban on inclusion of hazardous substances) or the recovery of product at end of life (e.g. vehicles, batteries, etc.).

Increasingly governments are seeking to use economic measures rather than legislation to improve environmental performance, for example, the cost of landfill or energy taxes will considerably push up manufacturing costs unless offset by effective waste minimisation or energy-efficiency measures.

4.5.3 Shareholder pressures

The financial markets are becoming increasingly attuned to both positive and negative indicators related to environmental credibility. Organisations that can demonstrate high standards of commitment to sustainability tend to manage resources more effectively and have better long-term business development strategies in place. This is clearly of interest to individual and institutional shareholders. Organisations that are perceived as poor environmental performers might, however, be considered to be at greater risk of prosecution (and perhaps business disruption), customer boycott or operational inefficiency. For these reasons, it is no longer just the ethical investor or fund that is considering the reputations of organisations from a 'corporate responsibility' viewpoint. Profit-focused fund managers may also have clear reasons for considering environmental credibility.

4.6 Market pressures

The quality expectations of both high street and business customers are increasing. Environmental criteria are simply another aspect of this growing quality expectation and are being increasingly used to distinguish between brands and/or companies. Good reputation may lead to market gains against competitors or the identification of new market opportunities. Bad reputation may lead to an inability to tender for business contracts or to consumers actively discriminating against the organisation and its products.

This may become the strongest single pressure driving environmental performance improvement in organisations in the future. It has certainly been a key factor in the first decade of the twenty-first century in rolling out the use of environmental management systems such as ISO 14001 throughout business supply chains.

4.7 Social pressures

4.7.1 Staff interests

In line with the rise in the environmental awareness of the general public, employees are becoming increasingly sensitive to the quality of the environment in which they work and the impacts associated with the organisation's activities. In a world of ever increasing demands in terms of productivity, employee satisfaction, attendance and morale are crucial and produce measurable benefits to an organisation.

4.7.2 Community pressures

Society at both a local and wider level is becoming increasingly intolerant of poor environmental performance. Effective environmental management and increasing demands for openness and information are becoming a fundamental part of an organisation's licence to operate.

At a local community level, poor reputation may translate into frequent complaints and objections to planning applications, permit renewals, etc. At a wider community level, press coverage and attention from non-governmental organisations (NGOs), such as Greenpeace, can translate into market pressures as sections of society express their disapproval through their spending powers.

III MANAGING RISK AND ENHANCING OPPORTUNITIES

The legal, financial, market and social pressures described above represent risks or opportunities to organisations depending on their response to and engagement with their environmental impacts. The first step in ensuring that risks are minimised and opportunities embraced is developing an understanding of where the organisation's sustainability priorities lie. In Chapter 1 we introduced the idea of 'aspects and impacts' as an approach developed in ISO 14001 that has become central to the way many organisations approach the assessment and prioritisation of their environmental issues. Chapter 6 deals with risk assessment tools and techniques in some detail but here we will elaborate on the longer-term issues associated with sustainability strategy as well as the general principles of risk management.

4.8 Sustainability threats and opportunities

The long-term implications of sustainability are profound for all organisations. This is the case whether we consider sustainability from the perspective of a national/global goal, or simply as an expression of planetary limits to resource availability and pollution assimilation. An oil company or car manufacturer, for example, cannot expect to be in business in 50 years' time unless it radically alters its current activities. The same principle applies to all organisations that have, to date, used energy from fossil fuels, road transport, raw materials without consideration of renewable status or supply chain impacts – in short, pretty much all of us!

As always, pressure for change represents both threats and opportunities to individual organisations depending on their response.

In addition to the general environmental pressures discussed earlier under the headings of financial, legal, market and social effects, there are a number of additional issues that are likely to become relevant to organisations as the international community grapples with the problem of limited resources and the move towards sustainable development.

4.8.1 Sustainability threats

▶ Raw material availability and increasing cost are likely to become an escalating issue for companies, either driven by non-renewable resources becoming scarcer, e.g. fossil fuel-based materials, or through regulation of supplies to ensure the most socially effective use of non-renewables. In some instances, increased standards of management of renewable resources may also lead to an end of 'cheap goods' where prices do not currently reflect the environmental cost of the materials, e.g. palm oil, low grade plastics, etc.

▶ Escalating energy and transport costs and availability – as with raw materials, energy cost is likely to escalate in the short term, in particular through a combination of decreasing availability of fossil fuels and government-imposed financial incentives to increase efficiency and drive the shift to more renewable sources of energy. Which of these pressures is most important in driving up costs depends on the speed and commitment of response by individual governments and the international community, but it seems likely that one or the other (and potentially both) will be increasingly relevant in the coming decades.

▶ Increased legal constraints are likely to be part of the education-taxation-regulation package of measures used by government to drive the shift to sustainability. Whether related to eco-efficiency, product design, supply chain management, material resourcing or consumer information, we can expect to see a continued flow of legal pressures pushing all organisations and especially stragglers in the move towards sustainability to keep pace with the best in class innovators.

▶ Market share threats is another looming issue with consumer awareness showing a steady (if somewhat patchy) increase over the past 20 years. Organisations that fail to identify and respond to shifting priorities or sustainability concerns among their customers may end up losing ground to competitors.

4.8.2 Sustainability opportunities

In the face of the kinds of pressures on organisations described above, there will be those that turn them to their advantage by responding more quickly or effectively than their competitors. As many of the pressures around sustainability relate to inefficiencies in material and energy usage, organisations that respond well may not simply avoid the disadvantages of escalating costs but can make savings through becoming increasingly efficient and streamlined throughout their direct and indirect operations.

In addition, if consumer pressure continues to move in the way we have seen over the past couple of decades, organisations that position themselves and their products and service well in terms of key reputation measures are likely to be able to turn such trends to their advantage.

In more general terms, the advantage of the long-term perspective demanded by sustainability planning may provide organisations with the vision and wherewithal to adapt and secure their long-term viability in a way that more short-sighted organisations may struggle to achieve. This is particularly true given the fundamental changes required to move us collectively to a sustainable society.

4.9 A value chain and long-term change perspective

The value chain perspective was developed by Michael Porter in the 1980s and has become a common way of thinking about business management. Each part of a business (including the supply chain, distribution and sales elements) is seen to have a role to play that contributes quantifiable value to the delivery of the goods or services on which the business is based. Understanding the value added at each stage allows business managers to identify opportunities for efficiency enhancement or waste reduction that enhance the profitability of the business overall. The value chain is most easily thought of in relation to a manufacturing organisation but can be applied to any business. It is often presented as shown in Figure 4.2.

Consideration of sustainability criteria (for example, those issues discussed under threats and opportunities above) within this model is a relatively simple step. Doing so begins to integrate environmental decision-making into the general business management approach, rather than it being a separate issue that is considered as some kind of moral imperative, but only if and when the organisation has the luxury to consider it or when stakeholder voices can no longer be ignored.

Moreover it is unlikely to be wise to simply 'rebrand' existing environmental management programmes as 'sustainability systems'. This is not because ISO 14001 could not consider sustainability issues, it is simply that in many organisations it is the case that environmental programmes are still at the stage of pollution control. In this context, it may be useful to consider four evolving stages of involvement in environmental management and sustainable development:

> Stage 1 – being reactive to issues
>
> Stage 2 – developing management systems
>
> Stage 3 – adopting an integrated strategic approach
>
> Stage 4 – organisational transformation leading to the sustainable enterprise.

Not so long ago, most companies were at Stage 1, employing 'end of pipe' control techniques such as scrubbers, filters, effluent treatment systems, etc. with the objective simply to stay within legally acceptable limits of pollution.

There has been a growing trend, however, in response to ethical, reputation and financial drivers, towards a more pro-active approach that focuses on a more generic understanding of the environmental impacts associated with an organisation's activities, product and

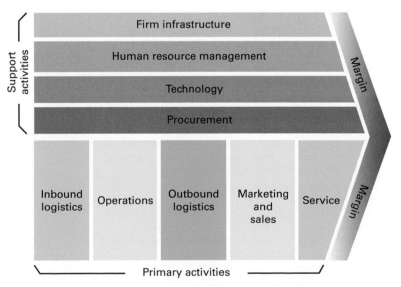

Figure 4.2 Porter's value chain
Source: Image by permission of DP Singh via Wikimedia Commons.

process design rather than simply control of harmful emissions. The increasing popularity of environmental management systems is indicative of a shift towards this second stage in the environmental management spectrum.

The key difference in the transition into Stage 3 is that innovative and often fundamental change to the way an activity is performed is involved. There are numerous tools and techniques aimed at providing the information required to identify opportunities and facilitate the implementation of change. These will be examined in Chapter 6 and include:

▶ life cycle analysis
▶ environmental impact assessment
▶ strategic environmental assessment
▶ eco-efficiency and sustainability indicators.

In Stage 4, these tools and the strategic changes that arise from them form a central part of the organisation's mid to long-term thinking and become embedded in the value chain management approach. Companies at this level will inevitably work closely with government and other stakeholders as change from this point on is unlikely to proceed in isolation. Beyond this level

of engagement we are in uncharted waters but the sustainability objective dictates that, at some point along this path, we will arrive at the elusive goal of a truly sustainable organisation.

Organisations have taken action on environmental issues and moved through the stages described for many different reasons, some provoked by external pressure but increasingly (especially in the later stages) as a result of their own strategic decisions.

Examples of drivers for transition through the stages are shown in Table 4.1. As organisations progress through these stages, it is to be expected that there will be different parts of the organisation at different levels. However, by stage 4 these variations should be decreasing and much greater consensus and involvement will be present across the whole organisation.

An increasing number of large and small organisations are making public commitments to sustainability as a key corporate goal. Often presented under the banner of corporate social responsibility (CSR), the most credible examples involve a definitive shift in business thinking from a short-term 'profit only' focus to a long-term

Table 4.1 Stages of sustainability in organisations

	Characteristics	Drivers
Stage 1 – reactive	A fire fighting approach – dealing with issues as they are brought to our attention by third parties, e.g. complaints, legal breaches or demands.	External stakeholders' interests or the threat of prosecution
Stage 2 – developing environmental management systems (EMS)	A more coordinated and voluntary approach to control and improvement of environmental issues. Often involving ISO 14001 certification, etc.	Reactive – customer specification, regulator requirement Proactive – reputation enhancement, resource efficiency objectives
Stage 3 – adopting a strategic approach	Established EMS in place and focusing on impact reduction not simply control measures. Some redesign of process and product/service to reduce impacts. Developing a longer-term view of improvements that cover infrastructure, product and supply chain change rather than simply 'tweaking' what is already present.	Reactive – customer specification, regulator requirement Proactive – reputation enhancement, resource efficiency objectives
Stage 4 – sustainable enterprise	A major shift in perspective with full life cycle thinking built into corporate decision making at all levels – short, medium and long term. Organisations will not be wholly sustainable today but they will have a clear view of how they intend to move towards this goal and will be communicating with all key stakeholders to develop understanding and cooperation. In some instances companies will be considering fundamental changes to the nature of their products, services, business structure in order to facilitate such change.	Proactive – a high level recognition that sustainability means short-term efficiency, mid-term market gain and long-term business survival

Case study 4.1 M&S's Plan A

Marks and Spencer (M&S) launched Plan A in January 2007: 'committing to change 100 things over five years, because we've only got one world and time is running out'. This came from a realisation that acceptance of the status quo was not working, and that the world was changing. It has since been extended to 180 commitments to be achieved by 2015 across areas including:

▶ Customer involvement
▶ Climate change
▶ Waste
▶ Resource usage
▶ Fair partnerships and community benefits
▶ Health and well-being in relation to supply chains, employees and customers.

In the 2013 performance report, the company recorded the following progress against the 180 commitments: '139 Achieved, 31 On plan, 5 Behind plan, 4 Not achieved and 1 Cancelled'.

The Plan A process began with an in-depth stakeholder engagement process to figure out what the issues were, determine what the solutions might be and provide support along the journey. M&S decided that it wanted to lead the market – to become 'the world's most sustainable major retailer' by 2015. Its business is predicated on leadership, and while it cannot lead on price, it can lead in other areas perceived as important to customers and shareholders. In this case, M&S decided that sustainable development offered a competitive advantage and that it is the only way to do business in the long term.

What was then needed was a plan and hence the statement – Plan A as there is no Plan B. The aim was and is to go beyond simple compliance with legal requirements and tackle intangibles such as bringing together traditional silos within the organisation in order to provide integrated solutions which benefit the business and the larger world around it, and seeking constant measures of performance. The benefits to the business in the short term are presented as a lower cost base through greater efficiencies and high resilience to commodity shocks. However, in the longer term the benefit is seen in terms of a greater trust base with its customers who value a more considered and ethical way of doing business.

Through Plan A, M&S is working with its customers and its suppliers to combat climate change, reduce waste, use sustainable raw materials, trade ethically and help its customers to lead healthier lifestyles. To understand more of what this means in practice, explore M&S's Plan A at http://plana.marksandspencer.com.

---

successful enterprise model that demands profit but without excessive resource consumption or environmental/social cost – in other words: in a way that will allow profitability to be sustained indefinitely.

For an inspiring presentation on this topic by the CEO of a large American corporation, go to: www.ted.com/talks/ray_anderson_on_the_business_logic_of_sustainability.html. To complete this section we'll consider a couple of examples of organisations that may be considered to be well into Stage 3 and perhaps even moving into Stage 4.

4.10 General principles of risk management

The language of risk assessment had permeated modern life and in recent years the areas of environmental management and sustainability have been no exception. In essence, the management of risk is intended to reduce the likelihood and consequences of undesirable situations. The approach as tailored toward environmental issues is dealt with in some detail in Chapter 6, however, the general principles will be introduced here.

Case study 4.2 Unilever's Sustainable Living Plan

In 2010, Unilever, the consumer goods group, launched its Sustainable Living Plan which set out ambitious aims to double the size of its business 'in a way which helps improve people's health and wellbeing, reducing environmental impact and enhances livelihoods'.

There are a number of key elements in the plan that point towards a fundamental shift into the 'sustainable business' arena. The clear statement of intent to grow the business but *'decouple growth from environmental impact'* is a bold commitment to change from 'business as usual'. Second, the long-term targets (see below) and their focus on profitability with social and environmental benefit are again a shift from 'doing less harm while we make a profit' to 'doing more good and still making a profit'. Of course, the proof will be in whether or not the plan succeeds but Unilever already has an impressive track record to back up these commitments so their credibility is strong.

The plan is focused on the following key commitments supported by more than 50 targets to be achieved by 2020:

▶ Help more than a billion people take action to improve their health and well-being.
▶ Halve the environmental footprint of the making and use of products as the business grows, i.e. on a 'per consumer use' basis.
▶ Enhance the livelihoods of hundreds of thousands of people as the business grows.

For the full Sustainable Living Plan, visit: http://www.unilever/sustainable-living. For an example of an innovative social education programme launched under the plan, see also the Unilever Project Sunlight campaign.

4.10.1 Definitions

▶ *Hazard* – a condition or situation with the potential for an undesirable consequence (akin to an environmental aspect in ISO 14001 terminology).
▶ *Harm* – the adverse consequences resulting from the realisation of a hazard (akin to the environmental impact in ISO 14001 terminology).
▶ *Risk* – the potential for the realisation of undesirable consequences, i.e. a combination of the likelihood and consequences of a specific outcome.

To help illustrate these definitions in an environmental context, here are two examples. The first is based on a pollution impact and the second on a resource consumption impact. The risk assessment approach lends itself more easily to pollution issues but can quite easily be expanded to consider resource consumption issues as well.

▶ A pollution example: a drum of oil (or other hazardous substance) represents a *hazard* with the potential to cause water pollution (the *harm*). The *risk* of the water pollution occurring is based on a range of likelihood and consequence criteria including the condition of the drum, any bunding present, the proximity to a storm drain, whether there is a site interceptor in the drainage system and whether or not it is well maintained, the proximity of the water course, the sensitivity and/or conservation value of the water course, the location of downstream users, etc.
▶ A resource consumption example: electricity consumption represents a *hazard* with the potential to deplete non-renewable resources and generate indirect air pollution, waste, etc. (the *harm*). The *risk* of depletion of non-renewable resources is based on the likelihood of usage of electricity (in most organisations a 100 per cent guaranteed likelihood) combined with the consequence of the electricity generation (if power generation is based on fossil fuel consumption, then the consequence will certainly be depletion of non-renewable resources, the indirect pollution impacts will vary by fuel type).

4.10.2 **A risk management approach**

Risk management is based on the understanding that hazards create risks which may lead to harm. The source-pathway-receptor model introduced in Chapter 1 is clearly relevant here in developing management strategies to minimise risk through eliminating the hazard at source (e.g. replacing hazardous materials with less harmful alternatives), breaking the pathway (e.g. placing barriers between a landfill and groundwater) or reducing the likelihood of receptor impacts through separation or isolation (e.g. banning development in ecologically sensitive locations).

The language of risk management can get very involved and complicated but it closely parallels the systematic approach to environmental management described in ISO 14001. In essence, any robust risk management approach will seek to do the following:

▶ *Identify and prioritise risks.* In the ISO 14001 model, it is expressly stated that, when assessing environmental risk, both pollution and resource consumption issues should be considered.

▶ *Minimise and mitigate risks.* Risk management controls are covered in some detail in Chapter 7 but typically encompass a range of measures that cut across the following categories. These controls may aim to eliminate or reduce the nature of the hazard and/or reduce the risk by reducing the likelihood of the hazard being realised.

    ▶ technological/mechanical controls related to plant, equipment and process/product design, etc.;

    ▶ operational controls related to working methods, procedures, operating hours, location, etc.;

    ▶ strategic controls related to decisions around supply chain sourcing, scheduling of operations, location of operations, product development and marketing, etc.;

    ▶ human controls related to staff awareness, competencies, motivation and engagement, etc.

▶ *Monitor and provide assurance that hazards are adequately controlled.* Typically some kind of monitoring programme will be in place, particularly in relation to those risks deemed most significant. Monitoring, supervision inspections and audits all constitute assurance mechanisms aimed at providing evidence that any risks are being managed within limits deemed acceptable to the organisation.

For further details on risk assessment and risk management, refer to Chapters 6 and 7 respectively.

IV THE BUSINESS CASE FOR CHANGE

Many companies are still at Stages 1 or 2 as described in Section 4.9 and for most the shift further along the spectrum is likely to involve changes in the three key areas described below: eco-efficiency, supply chain management and stakeholder information. In general terms, however, whether making the case for a single investment or change within an organisation, or petitioning for a more widespread shift in product or service planning, we will have to present the case with reference to the following:

▶ *Cost implications.* As with standard cost-benefit analysis calculations, any proposal should consider the initial investment, on-going costs and/or savings and include adjustment factors not just for standard elements like the rate of inflation, but also predictions related to material or energy price changes based on best available information.

▶ *Sales implications.* Wherever possible, a business case for a sustainability initiative should attempt to quantify the potential impact on sales and profitability. This can be somewhat tricky to quantify in terms of predicted sales (or lack thereof) based on perceived customer preference. However, where initiatives relate to eco-efficiency savings that can be converted into a profitability per unit of sales, this can be a very powerful statistic if the organisation is focused on the 'value chain management' principles outline above.

▶ *Indirect benefits.* Even more difficult to quantify but, in some cases at least, important business drivers for

145

sustainability initiatives, are those relating to relationships with key stakeholders. Easier to provide benchmark data on are the benefits of improved reputation with investors and insurers. Insurance premium negotiations or cover restrictions may provide an immediate reference point in support of an environmental improvement action. There is also an increasing body of evidence supporting the idea that businesses with active sustainability strategies are seen by corporate investors as being better performers. This is aside from the marketing appeal to ethically minded customers who may be interested in buying into funds operated by such investors.

A 2012 Harvard Business School study compared 'high sustainability' companies with 'low sustainability' companies and found that the former outperformed the latter over the long term in both stock market and actual accounting terms. In their words:

> The outperformance is stronger in sectors where the customers are individual consumers, companies compete on the basis of brands and reputation, and in sectors where companies' products significantly depend upon extracting large amounts of natural resources.
> (Eccles *et al.*, 2013)

It is less easy to quantify the benefits of improved reputation with the likes of regulators, neighbouring communities, non-governmental organisations and even employees. However, many organisations have reported improvements in these third party relationships as being advantageous outcomes associated with successful environmental programmes that, in general terms, just make doing business easier.

Driving long-term change in organisations is further considered in Chapter 10, while business case development, improvement programmes and stakeholder communications are dealt with in Chapters 8 and 9. An overview of eco-efficiency, supply chain management and stakeholder information is, however, provided below.

4.11 Moving towards resource efficiency

In the 2005 UK government policy 'Securing the Future', commitments to 'sustainable production and consumption patterns' point towards greater resource efficiency or 'eco-efficiency' in organisations and society as a whole as a key element in the move towards sustainability. Eco-efficiency has been defined as:

> the delivery of competitively priced goods and services that satisfy human needs and bring quality of life, while progressively reducing ecological impacts and resource intensity throughout the life cycle to a level at least in line with the earth's estimated carrying capacity.
> (World Business Council for Sustainable Development)

Proctor & Gamble define it rather more succinctly in their company goal: '[to] do more with less'.

There is a clear link here to the UK priority of sustainable production and consumption, with the emphasis being on reducing the energy and raw material inputs to a production process. Although described in terms of product manufacture, these same principles apply to the service sector.

These are challenging principles and perhaps best illustrated with an old but excellent example from the World Business Council for Sustainable Development publication *Sustainable Production and Consumption: A Business Perspective* (Falkman, 1997) (see case study 4.3).

4.12 Other resource efficiency case studies

The World Business Council for Sustainable Development in its 2000 publication, *Eco-efficiency: Creating More Value* produced the image shown in Figure 4.3 as a way of summarising where an organisation may find eco-efficiency improvement opportunities.

There are many examples, often involving very simple changes in product design or manufacturing process to produce significant improvements in terms of resource efficiency with corresponding reductions in cost. Case

Case study 4.3 Compact laundry detergents: eco-efficiency in action

Proctor & Gamble first introduced compact powdered detergents, also known as Ultra formulations, in 1989. Ultras generally require consumer usage of half the volume or less of traditional laundry powders due to a combination of a denser product and 30 per cent less product raw materials. Consequently, the smaller volume requires 30 per cent less packaging, and less energy is used to ship and distribute enough powder for a given number of wash-loads. In fact, P&G estimates that trucking needs have decreased by 40 per cent worldwide for compact detergents vs traditional detergents.

Since their introduction, further improvements in the manufacture and packaging of Ultra detergents have resulted in additional gains in resource efficiency. Switching from a wet to a dry manufacturing process reduced by half the amount of energy needed to produce the detergent, as well as reducing water consumption. Plastic refill bags, now available for powdered laundry detergents in the US and Europe, utilise 80 per cent less material than traditional cartons, and require less energy to make and ship the product/package. They also contain 25 per cent recycled content. In other parts of the world, detergent is sold only in plastic bags.

P&G uses life cycle analysis to understand the amount of energy and waste produced through the life cycle of its products – beginning with raw material extraction to final disposal of the product and packaging. P&G's compact detergents offer multiple environmental benefits throughout their life cycle, some of which are not immediately apparent. For example, if less raw materials and packaging are used per wash-load, less raw materials are mined, processed and transported. This requires less energy and produces less solid waste.

A life cycle study of laundry detergents conducted for P&G under US conditions indicates that Ultras save approximately 700 mega joules of energy for every 1,000 wash-loads and a total of 500 tonnes of solid waste per year compared with traditional powders. If all wash-loads in the US were done with Ultra detergents, this would be an equivalent of saving over 500,000 litres of petrol per year and the amount of solid waste generated in a year by 765,000 people.

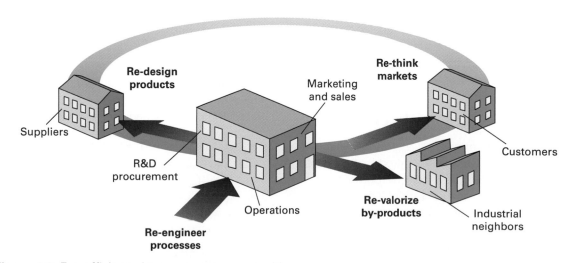

Figure 4.3 Eco-efficiency improvement opportunities
Source: (WBCSD, 2000).

studies across a variety of business sectors collected under the UK Envirowise (now WRAP) programme suggested that, on average, waste costs in an organisation amount to about 4.5 per cent of turnover. The following mini case studies serve as further examples of the financial benefits associated with resource efficiency initiatives:

▶ Sony – nearly £3.5 million per annum in packaging costs saved through the resale of good packaging to supplier (NB savings arising from one UK plant only).
▶ Xerox – returnable packaging programme reduced waste by 10,000 tonnes and saved up to $15 million dollars annually.
▶ Johnson & Johnson – packaging for surgical gloves – folding lengthwise reduced package size by 40 per cent with $1.3 million dollar saving in packaging material costs

These dramatic examples illustrate the step change approach which marks a change of strategy from 'tweaking' existing processes to a complete re-think of product and manufacturing in order to increase eco-efficiency. Clearly there is a significant cost saving to a business in achieving such improvements.

The high financial incentives associated with increasing eco-efficiency suggest that the job of initiating such changes in most businesses would be an easy one. Unfortunately rarely is a change of approach easy to accomplish and, in this case, there is a fairly fundamental mind shift required from focusing on labour costs and per unit cost of goods or services produced to a focus on per unit profit of raw materials consumed, however, once benefits become apparent, momentum builds in most organisations.

4.13 Managing supply chain issues

The concept of eco-efficiency introduces the need to begin to consider the wider environmental implications of the manufacture of a product or the provision of a service. To use the ISO 14001 terminology, organisations must consider their indirect environmental aspects as well as the direct. This means incorporating within the scope of consideration the environmental implications

of organisations that supply materials, energy and services. Indeed, the 2004 revision of ISO 14001 increased the emphasis on supply chain management by changes to wording relating to those areas of an organisation that must be included within a credible EMS.

For many organisations, the difficulty lies in translating a consideration of impacts into positive action. To date, there have been a number of strategies employed including:

▶ Supplier specification, e.g. the statement made by the Ford Motor Company that no company without certification to ISO 14001 will retain or achieve preferred supplier status after 2003.
▶ Partnership approach, e.g. the approach employed by B&Q among others where, while blanket standards are not prerequisites for involvement, suppliers and contractors are recognised for high environmental standards and encouraged to develop their approach through provision of information, training and transparent supplier rating methodologies.
▶ Selection approach – perhaps the simplest approach, which depends on incorporating a general assessment of environmental performance as part of the supplier/contractor selection process. Approaches vary from simple pre-qualification questionnaires to company audits.

The supply chain strategy employed by an organisation is a function of two things. The following two elements combine to form what has been described as the organisation's 'sphere of influence':

▶ its perception of acceptable sphere of influence, i.e. how far up and down the supply chain the company and its stakeholders consider would be reasonable for them to consider;
▶ the company's degree of influence over other elements within the supply chain – this clearly varies significantly depending on the size and purchasing power of the organisation involved.

The following organisations have all produced information for internal or wider use that relates to sustainable procurement:

- the Environment Agency
- the Institute of Environmental Management and Assessment
- the United Nations sustainable procurement group
- the UK government in the procurement section of the sustainable development website.

This is an emerging part of the environmental management field and is dealt with in more detail in Chapter 6.

4.14 Improving stakeholder communications

The final area of change for an organisation developing a sustainability strategy involves stakeholder liaison. This includes the following elements:

- the provision of product-/service-related environmental information;
- the provision of performance information related to company activities;
- consultation with key stakeholders as part of organisational decision-making.

The provision of product- and activity-related environmental performance information means that such issues can become part of how stakeholders come to decisions related to purchasing company products, investing in shares, supporting or objecting to new developments, etc. Such informed choice is described in the UK sustainability strategy as a prerequisite for success in meeting the priority of sustainable production and consumption.

The development of consumer awareness is driven by a number of different mechanisms including:

- the activities of lobby groups and non-governmental organisations;
- media coverage;
- public education programmes;
- company consumer information.

It is likely that as public awareness of environmental and sustainability issues grows, governments and consumer groups alike will bring increasing pressure to bear on business to provide sufficient information, in an accurate and appropriate manner, that enables consideration of environmental performance as part of purchasing and investment decisions. This is likely to mean increased pressure in relation to:

- company reporting;
- provision of product-related information and the associated issue of eco-labelling.

Organisations may see this as an opportunity or a threat. However, the growth in the 'green market' in recent years has led some companies to examine whether they can gain a niche in the market by manufacturing products which are less harmful to the environment in production and use. This has sometimes led to attempts to exploit the demand from the consumer for greener products by making misleading and erroneous claims. In many cases such attempts have proved to be risky and opportunist. There are, of course, other companies, such as Interface, Ecotricity, Bodyshop International, etc., which have made ethical and environmentally sound methods central to their whole business ethos, and which have proved highly successful.

Businesses of all scales and interests should recognise the opportunities that exist in the commitment of governments and international institutions to pursuing sustainable development. Since the 1960s, we have seen a phenomenal change in the way institutions and the general public perceive environmental issues. The relatively recent commitment to sustainable development confirms that this trend is only likely to continue with increasing incorporation of environmental considerations into economic decisions.

For further details on corporate reporting and corporate social responsibility (CSR) see Chapter 9.

V FURTHER RESOURCES

Table 4.2 presents an overview of further resources.

Table 4.2 Further resources

Topic area	Further information sources	Web links (if relevant)
Risk management	IEMA practitioner guide, 2006. Risk management for the environmental practitioner. IEMA	www.iema.net
Eco-efficiency	World Business Council for Sustainable Development	www.wbcsd.org
Sustainability strategy	Information on a company transformed by commitments to sustainability	www.interfaceglobal.com/sustainability

CHAPTER 5

Collecting, analysing and reporting on environmental information and data

Chapter summary

This chapter is about the data and information we need to obtain, and in some cases monitor, in order to effectively manage the environmental impacts of an organisation. It begins by considering the wide variety of sources, types and ways of presenting data and refers to relevant guidance found in ISO 14031 and AA1000. The way we convert raw data into indicators can be crucial in developing tools that can provide effective assurance and performance information over the longer term.

The chapter then considers a number of sources of guidance relating to the assurance and verification of data with particular reference to its application to various forms of reporting. Related to this we also consider a selection of the most established corporate benchmarking schemes, which use such data and reporting to gauge the environmental, and/or sustainability engagement of organisations.

An overview is then provided of data collection methods in the context of air, water and land pollutants, plus noise, dust and odour nuisance.

The chapter concludes with some summary statements regarding interpretation and communication of data and information and makes links to other chapters where further detail is provided on these aspects.

5 Analysing and reporting environmental data

I   WHAT DO WE NEED TO KNOW?

The old adage of 'if you can't measure it you can't manage it' is as true for environmental performance as for anything else. In effect, if you do not have good data on key environmental issues, it is not possible to make reliable statements about the scale of impacts/risks or to track performance improvements over time. In basic terms, we cannot reliably say what impact we're having and whether we are getting better or worse.

The aim of this section is to consider:

    The types and sources of data and information that might be useful/appropriate.
    Methods for ensuring the accuracy and relevance of data and information.
    Verification and assurance methods.
    Benchmarking performance – third party assessment schemes.
    Data collection – pollution and nuisance monitoring.
    Methods for the interpretation of data and information.
    Methods for the communication of data and information.

II   IDENTIFYING TYPES AND SOURCES OF DATA AND INFORMATION

We often use the analysis of data to provide us with information. An environmental example might be as follows:

Box 5.1 Definitions

▶ *Data*: a collection of relevant facts and statistics used to assess a particular issue.
▶ *Information*: that which is communicated or understood by assessment of a collection of data.

Measurements made at an effluent discharge point provide *data* in the form of readings in mg/l of specified pollutants. Collation of the data into peak readings, daily averages and calculated monthly emission totals provides us with *information* that we can compare with limits set by the regulator to determine our compliance status with permit conditions.

In an organisation embarking on an environmental management improvement programme there might be considered to be two reasons for sourcing and monitoring a wide variety of data and information:

▶ to inform decisions about environmental priorities, i.e. to tell us what our issues are;
▶ to track performance over a period of time, i.e. to tell us whether our efforts are getting the results we hope for.

5.1 Information and data to help set priorities

The sorts of information consulted during a preliminary environmental review can be very varied, depending on the nature of the organisation and its environmental aspects. Table 5.1 gives a sense of the types of information and data that might be consulted or collected and some example reasons for doing so.

5.2 Information and data to help track performance

Once an organisation has an idea of its environmental priorities, there is a need to generate a clear set of indicators and references to determine assurance benchmarks and also to monitor improvement progress. Not infrequently the same indicators may be used for both purposes. For example, emissions data from a chimney stack monitoring station may provide continuous readings of pollutant releases. Such monitoring may actually be specified within an Environmental Permit and be used to check compliance with emission limits also set with the Permit. This data is thus being used primarily for assurance monitoring purposes, i.e. assuring us and the regulators that the abatement techniques are effective and that pollutant loads are therefore within acceptable limits. However, if we are also involved in an improvement programme to reduce the amount of hazardous material used within the process generating the emissions, it may be that the same data becomes a key performance indicator in tracking our efforts. In both cases, the way the data is presented may be crucial in ensuring on-going relevance and accuracy.

5.2.1 ISO 14031 Environmental Performance Evaluation (EPE)

ISO 14031 is part of the ISO 14000 series and introduces a performance evaluation process that can be used as part of an EMS or that may be used as a feedback loop and performance improvement mechanism in its own right. The approach focuses on the use of performance indicators relevant to the organisation's activities and the context in which it operates. The Standard describes two general categories of indicators:

▶ environmental performance indicators (EPIs);
▶ environmental condition indicators (ECIs).

There are two types of EPI:

▶ Management performance indicators (MPIs) are a type of EPI that provide information about management efforts to influence the environmental performance of the organisation's operations. These might also be considered as 'active monitoring' criteria. An example might be 'number of audits conducted per month'.
▶ Operational performance indicators (OPIs) are a type of EPI that provide information about the environmental performance of the organisation's operations. These might also be considered as 'reactive monitoring' criteria. An

Table 5.1 Types of data consulted during a preliminary environmental review

Aspect/Issue	Potential sources of data/ information	Data/information examples	Purpose for reference/collation
Material inputs	• Substance Safety Data Sheets • COSHH records • Purchase records	• Quantities of materials used • Environmental toxicity/hazard • Waste disposal requirements/ classification	• Identify priority material usage by quantity or hazard • Identify appropriate controls relating to storage use and disposal of materials
Energy inputs	• Supplier invoices • Meter reading records • Fleet vehicle refuelling records • Production or other activity based data (e.g. turnover, man-hours worked, etc.) • GHG reporting guidelines – conversion factors to enable impact comparison across energy use types	• Fuel breakdown and consumption rates/totals • Production or activity information to allow internal benchmarking	• Identify key users of energy and hence priorities for efficiency optimisation • Identify opportunities for improvement
Processes	• Process flow diagrams • Maintenance records • Environmental permits and guidance notes • Solvent inventory data • Relevant operational instructions • COSHH risk assessments • Environmental Permitting Regulations (schedule 1 to determine whether any of the 'regulated activities' fit)	• Activity breakdown of the organisation • Aspect identification • Audit reports/results • Compliance criteria including key emissions limits • Operational standards used to establish 'required behaviour' • Relevant and realistic abnormal and emergency scenarios	• Understand the range of activities and organisational structure of a business • Understand environmental priorities and key areas of risk • Establish and monitor compliance with external and internal standards • Understand efficiency of control methods
Materials handling and storage activities	• Site plan showing facilities • Emergency procedures and plans • Site drainage plans • Incident reports • Environmental Permitting Regulations	• Material storage quantities and location • Pollution pathway descriptions • Incident history and response arrangements	• Identify risk of uncontrolled releases • Identify hazardous waste streams • Identification of cumulative contaminated land potential
Atmospheric emissions	• Environmental permits • Monitoring records	• Emissions data at monitoring locations	• Identification of regulated and non-regulated emission routes • Evaluation of legal compliance risk
Effluent	• Discharge content to sewer • Environmental permits • Site drainage plans	• Emissions data at monitoring locations	• Identification of regulated and non-regulated emission routes • Evaluation of legal compliance risk
Waste	• Duty of care transfer notes • Hazardous waste consignment notes • Environmental permits • Certificates of technical competence	• Waste stream quantity and disposal route data • Disposal cost data	• Identify waste streams and quantities to provide baseline performance data • Identify legal compliance status

Table 5.1 (continued)

Aspect/Issue	Potential sources of data/information	Data/information examples	Purpose for reference/collation
	• Contracts with waste carrier/disposer • Certificate of registration of waste carriers		• Identify existing control standards and legally defined minimums • Quantify cost to company and begin to set up business case for improvement programmes
Nuisance	• Complaint records • Abatement notices	• Specific occurrences/data relating to nuisance occurrences	• Identify nuisance potential and key receptors • Identify complaints management process
Product	• Output/production records • Any life cycle studies • Product specification information	• Output data • Input – output data and ratios from life cycle stages • Design specifications e.g. weight of glass per component	• Identify variability of operations and provide a benchmarking reference for other data gathered • Understand the relative importance of life cycle stages • Identify design criteria priorities for existing products
General Management Issues	• Environmental policy statement • Organisation chart • Site plan • Map of surrounding area • Training records • Correspondence with regulators and other stakeholders • Interviews and questionnaires	• Number of employees • Number of training events conducted in a defined time period • Data based on answers to consultation actions eg number of people aware of waste target • Distance to and location of nearest receptors	• Identify existing management strengths and weaknesses • Identify administrative structures and potential ways of implementing an EMS or environmental improvement programme • Identify staff competency and awareness and also mechanisms for improvement • Identify key receptors • Identify key stakeholders

example might be 'tonnes of waste generated per year'.

ECIs provide information about the condition of the environment. This information can help an organisation to better understand the actual impact or potential impact of its environmental aspects, and thus assist in the environmental management process. Development and application of ECIs are frequently the function of local, regional, national or international government agencies, non-governmental organisations, and scientific and research institutions rather than the function of an individual organisation. For purposes such as scientific investigations, development of environmental standards and regulations, or communication to the public, these agencies, organisations and institutions may collect data and information on:

▶ the properties and quality of major bodies of water;
▶ regional air quality;
▶ endangered species;
▶ resource quantities or quality;
▶ ocean temperatures;
▶ concentration of contaminants in tissue of living organisms;
▶ ozone depletion;
▶ global climate change;
▶ and many other parameters.

Some of this information may be in the form of ECIs, e.g. mg/l SO_2 in ambient air at the site boundary. These could be useful to an organisation in managing its environmental aspects or indicating specific issues that an organisation should consider in its environmental management programme. Organisations that can identify a clear relationship between their activities and the condition of some component of the local environment may choose to develop their own ECIs as an aid in evaluating their environmental performance as appropriate to their capabilities, interests and needs.

Examples of the types of indicators that might be included in both the environmental performance indicator and environmental condition indicator categories are provided in Table 5.2 which represents a selection of the examples provided in the appendices of ISO 14031.

In most cases, it will be appropriate to select a group of condition and performance indicators that will reflect the organisation's key areas of influence in terms of environmental impact. The indicators may then be used to track performance trends over time and provide feedback as to the positive or negative implications of the organisation's activities.

In the regulatory context, these two different kinds of performance indicators are used in different circumstances, as each is considered to have different strengths and weaknesses as summarised in Table 5.3.

III PRINCIPLES FOR ENSURING THE RELEVANCE OF DATA AND INFORMATION

A glimpse into the techniques and principles of environmental monitoring and analysis is provided in Section VI. In this section, the emphasis is not on the technical processes of collation, sampling, analysis and interpretation but, rather, on the general assurance principles that are used to ensure the accuracy and relevance of data and information. We will consider two areas:

▶ data categories and applications – different data types and their uses;
▶ data selection principles – definitions and guidance on materiality, responsiveness and completeness.

5.3 Data categories and applications

Section II highlights the wide variety of data and information that might be used by an organisation to assess priorities and monitor performance. It is perhaps already clear that such a variety of data is likely to vary in its accuracy and relevance. We have already implied that care should be taken to ensure that appropriate information is used and presented in a manner relevant to both the organisation and the environmental aspect under consideration. It may be useful here to highlight three areas where decisions will need to be made about the type of data to be used and the way it should be managed, interpreted and reported.

5.3.1 Qualitative vs quantitative data

Qualitative data may be defined as information based on opinion, judgment or interpretation. An example in the environmental management context might be perceived nuisance potential based on detectable levels of noise or odour, for example, at a site boundary.

Quantitative data may be defined as numerically measured criteria that can be subject to statistical analysis and/or benchmarking. An example in the environmental management context might be air emissions measured at a discharge point in milligrams per cubic metre (mg/m^3).

Generally speaking, it is preferable to use quantitative data as it allows for clearer referencing to acceptable standards (including legally permitted levels) and/or previous measurements. It is also likely to be subject to less variability based on the subjective experiences, preferences and competencies of individuals. For that reason the majority of data used in legal standards and operational performance monitoring will be quantitative. There are, however, occasions when it may be appropriate to ignore this rule of thumb. They exist mainly (but not exclusively) in relation to nuisance potential. Consider the examples included in boxes 5.2–5.4.

Table 5.2 Examples of indicators adapted from ISO 14031

Indicator category	Issue areas	Indicator examples
Management performance indicators (MPIs)	Implementation of policies and programmes	• number of achieved objectives and targets • number of organisational units achieving environmental objectives and targets • degree of implementation of specified codes of management or operating practice • number of prevention of pollution initiatives implemented
	Conformance	• degree of compliance with regulations • number and frequency of specific activities (e.g. audits) • number of audits completed versus planned • number of audit findings per period • frequency of review of operating procedures
	Financial performance	• costs (operational and capital) that are associated with a product's or process' environmental aspects • return on investment for environmental improvement projects • savings achieved through reductions in resource usage, prevention of pollution or waste recycling
	Community relations	• number of inquiries or comments about environmentally related matters • number of press reports on the organisation's environmental performance • number of environmental educational programmes or materials provided for the community
Operational Performance Indicators	Materials	• quantity of materials used per unit of product • quantity of processed, recycled or reused materials used • quantity of packaging materials discarded or reused per unit of product
	Energy	• quantity of energy used per year or per unit of product • quantity of energy used per service or customer • quantity of each type of energy used
	Services supporting the organisation's operations	• amount of hazardous materials used by contracted service providers • amount of cleaning agents used by contracted service providers • amount of recyclable and reusable materials used by contracted service providers
	Physical facilities and equipment	• number of pieces of equipment with parts designed for easy disassembly, recycling and reuse • number of hours per year a specific piece of equipment is in operation • number of emergency events (e.g. explosions) or non-routine operations (e.g. shut-downs) per year • total land area used for production purposes
	Supply and delivery	• average fuel consumption of vehicle fleet • number of freight deliveries by mode of transportation per day • number of vehicles in fleet with pollution-abatement technology
	Products	• number of products introduced in the market with reduced hazardous properties • number of products which can be reused or recycled • percentage of a product's content that can be reused or recycled

Table 5.2 (continued)

Indicator category	Issue areas	Indicator examples
		• rate of defective products
	Services provided by the organisation	• amount of cleaning agent used per square metre (for a cleaning services organisation) • amount of fuel consumption (for an organisation whose service is transportation).
	Wastes	• total waste for disposal • quantity of waste stored on site • quantity of waste controlled by permits • quantity of waste converted to reusable material per year
	Emissions	• quantity of specific emissions per year • quantity of specific emissions per unit of product • quantity of waste energy released to air
	Effluent	• quantity of specific material discharged per year • quantity of specific material discharged to water per unit of product • quantity of waste energy released to water
Environmental condition indicators	Regional, national or global ECIs	• thickness of the ozone layer • average global temperature • the size of fish population in oceans
	Local or regional ECIs – air	• concentration of a specific contaminant in ambient air at selected monitoring locations • ambient temperature at locations within a specific distance of the organisation's facility
	Local or regional ECIs – water	• concentration of a specific contaminant in groundwater or surface water • turbidity measured in a stream adjacent to its facility upstream and downstream of a wastewater discharge point • dissolved oxygen in receiving waters
	Local or regional ECIs – land	• concentration of a specific contaminant in surface soils at selected locations in the area surrounding the organisation's facility • concentration of selected nutrients in soils adjacent to the organisation's facility • area rehabilitated in a defined local area • area dedicated to landfill, tourism or wetlands in a defined local area
	Local or regional ECIs – flora	• crop yield over time from fields in the surrounding area • population of a particular plant species within a defined distance of the organisation's facility • number of total flora species in a defined local area

5.3.2 Pollutant vs resource usage data

In most organisations there are likely to be two major categories of data collated in relation to its environmental aspects. From a pollution control perspective, data relating to the presence and emission of pollutants is critical. Some of this data may be sourced from records such as inventory quantities, delivery notes and perhaps production specifications. However, it is also likely that key data will be generated from pollutant monitoring programmes, e.g. effluent analysis, stack emissions monitoring, etc. Further details on the principles and methodologies involved in pollutant sampling and analysis are included in Section VI.

159

Table 5.3 Environmental performance indicators vs environmental condition indicators

	Advantages	Disadvantages
Environmental performance indicators/ standards	• they are easier to administer and compliance is easier to monitor than measurements made in the receiving environment (the regulator does not have to analyse the receiving medium) • they are easier for all parties to understand • it is easier to quantify compliance costs • it is easier to measure emissions at point of discharge • all operators in the same sector are treated the same regardless of location • they facilitate controlling environmental quality in respect of trans-boundary pollution • they can be used to force technological development and pollution abatement techniques • the operator (polluter) carries out much of the compliance monitoring at its expense (polluter pays): the regulator only needs to verify data	• they do not take account of the quality of the receiving medium • they may be set at the level of the lowest common denominator, i.e. the level at which agreement can be reached in negotiations, rather than what is justified in environmental terms • they may lead to unjustified expenditure on the part of the process operator • changes in legislation which contain such limits may be difficult or protracted when knowledge or technology improves
Environmental condition indicators/ standards	• control can be at the most appropriate level for the receiving medium, in that authorisation conditions can be tightest where the environment is the poorest and justifies the level of intervention • standards are not applied blindly but take account of the quality and nature of the receiving medium • they can avoid wasteful or unnecessary controls or costly abatement measures on the part of operators • there is an assessment of the quality of the receiving medium which takes account of all sources of pollution, and allows a more comprehensive control strategy • they can reflect the use of the medium, particularly water, more effectively • they can set the objective at a level which is most appropriate to protect public health	• different operators in the same industrial sector can have different levels and costs of control leading to competitive disadvantage • they do not force operators to employ the best available techniques to minimise polluting discharges and emissions • there may be emissions at a level known to be harmful but which are diluted in the environmental medium into which they are discharged • if objectives are set nationally, there can be a deterioration in those areas initially of higher quality: a levelling down • if different objectives are set for different uses of the medium which reflect the current condition or use, e.g. for water, progress on securing improvements may be slow • more extensive sampling and analysis of the medium is required, which can be expensive for the regulator, but those responsible for emissions will still have to undertake their own discharge monitoring

Resource usage data is often generated more from existing record-keeping sources than from separate sampling regimes, for example, purchasing records, invoices, product specifications, inventory listings, production output data. This make the collation and assessment of resource usage more focused on identifying sources of information and analysing existing data with perhaps a different objective.

However, where existing data has gaps or is presented in a format that is not useful, it may be necessary to conduct sampling and assessment activities not dissimilar to those used in pollutant monitoring. Some examples of such activities might include:

▶ the installation of temporary sub-metering to enable localised allocation of energy consumption data;

Box 5.2 Example 1 Detectable odour

An organisation operates a process that is particularly malodorous, e.g. a slaughter house/meat packing operation. It is a reasonable assumption that if people can detect the odour coming from the works, then it will constitute a nuisance and may lead to complaints. It would therefore seem appropriate to consider the qualitative measurement of 'detectable odour' at the site boundary as a key performance indicator for the site. The advantage of this strategy is that it gives a quick and easy method for determining nuisance potential or efficacy of control measures. The disadvantage might be that, if the measurement is carried out by someone working on the site, they may be less able to detect what to them is a familiar odour, than a neighbouring resident who is not exposed to the odour on a daily basis.

Box 5.3 Example 2 The Ringlemann scale

The Ringlemann scale is a reference method widely used by Local Authority environmental health officers (EHO) in determining compliance with the Clean Air Act, 1993. It involves the EHO making a visual (and therefore qualitative) comparison of the colour of smoke emissions from a chimney with a (semi-) standard scale of reference. There is no 'official' standard reference chart, rather an agreed gradation of four steps between 0 (white) and 5 (black). Figure 5.1 provides an example chart.

If the EHO considers the smoke emissions to be darker than the level set as the limit by the Local Authority, they may raise an abatement notice requiring the polluter to take action to reduce the levels of particulates in the emission. Although qualitative and potentially subject to some discrepancies in allocation of reference number, the system works as a quick and easy

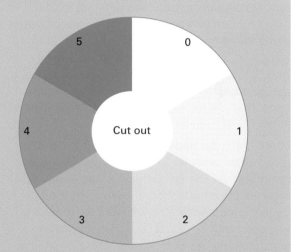

Figure 5.1 A Ringlemann scale reference chart

reference guide for operators and regulators alike in determining the pollution and/or nuisance potential associated with smoke emissions.

▶ workplace or ambient air monitoring of volatile organic compounds to provide an estimate of percentage fugitive emissions when assessing usage patterns;

▶ analysis of waste streams to provide clearer understanding of the usage patterns and losses of a particular material resource.

Methods used to generate data and evaluate resource usage are further considered in Chapter 6 on resource efficiency surveys.

5.3.3 Normalised vs absolute data

Quantitative data may be measured and presented in either normalised or absolute format.

Absolute data is the actual amount of something measured, e.g. total tonnes of waste generated by an organisation in a day/month/year. Absolute data gives the truest picture of the scale of impact created by the organisation, however, it can be misleading in judging levels of control or improvement over time if there are multiple variables involved.

Box 5.4 Example 3 Hydrocarbon sheen

Due to the high visibility of hydrocarbon sheen on water, many permits relating to discharge of effluent that is at risk of hydrocarbon contamination will specify a limit of 'no visible hydrocarbon sheen'. This becomes a quick and easy monitoring criterion for operators and regulators alike to assess legal compliance and pollution potential.

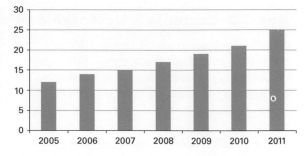

Figure 5.2 Absolute waste generation data

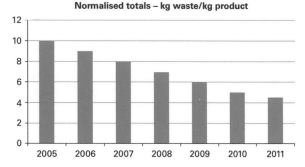

Figure 5.3 Normalised waste generation data

Normalised data is based on absolute data but is subject to some kind of referencing process to enable comparison between different locations or over time. An example might be tonnes of waste generated per unit of product produced. Although often a much more useful way of presenting data from the point of view of tracking performance, it can be challenging to find appropriate 'normalising units' that will remain relevant to an organisation in the medium to long term.

Often there is a trade-off to be considered when deciding on whether to report data in absolute or normalised format. Generally normalising data involves the introduction of assumptions and hence some inaccuracy, whereas absolute data remains free from such corruption, however, absolute data can also be misleading in terms of interpretation. Consider the examples provided above in the following context.

A manufacturing organisation operates a production schedule that is considerably seasonal in output. It is also experiencing a period of significant growth in overall business – a trend that has been on-going over a period of years.

If the organisation reported absolute waste generation data on a monthly and then annual basis, we may well expect there to be a picture of total waste tonnage that follows (at least in overall trends) the production output (see Figure 5.2).

If, however, we use 'normalised data' (see Figure 5.3), we get a very different and arguably fairer picture of the organisation's efforts and environmental performance over an extended time period. In practice, very often we would use both data formats – the first to give a clear

idea of total impact and the second to track progress and change in relation to the issue under consideration.

5.3.4 Trends and benchmarking

The final area where we are commonly required to make a choice about what we want the data to do for us and therefore how it should be collected and presented relates to trending and benchmarking. This may be determined for us in terms of regulator-specified monitoring or information requirements or it may be something we need to decide on in relation to our own performance improvement programmes.

Often benchmarking data is best presented in a quantitative and absolute format to enable clearest comparison with whatever benchmark we are relating to – a legal standard, a competitor's performance, etc. The main exception to this is where the benchmark itself is presented in a normalised format, e.g. mg of pollutant per cubic metre of effluent discharge (a legal standard) or litres of water per 12 kg wash load (a product performance standard to enable consumer benchmarking with competitor products).

When we are interested in generating data to track trends, we more commonly work with normalised data as this tends to be important in filtering out variations caused by other variables (such as in the production schedule variability example given above). It may even be more appropriate in trend analysis to consider qualitative data such as stakeholder feedback, as we are often more interested in the underlying message conveyed, than in the complete accuracy of any individual piece of data.

5.4 Data selection and reporting principles

AA1000AS (2008) is the latest version of an assurance and reporting standard produced by the AccountAbility organisation. Its stated purpose is 'to provide organisations with an internationally accepted, freely available set of principles to frame and structure the way in which they understand, govern, administer, implement, evaluate and communicate their accountability'. It has become widely recognised and referenced by those involved in sustainability reporting. It is further discussed in Section IV in terms of assurance processes, but here its relevance is in the clarification of data selection principles. It sets out three basic principles that should be applied when selecting performance monitoring data within an environmental or sustainability programme: (1) inclusivity; (2) materiality; and (3) responsiveness.

5.4.1 Inclusivity

AA1000 explains inclusivity as a principle that demands an organisation accept its accountability to all key stakeholders and therefore understands the main concerns or interests of those stakeholders. By inference, performance data and reporting should relate to those interests. This principle leads essentially to the following two supporting principles: materiality and responsiveness.

5.4.2 Materiality

Materiality is a concept from accounting that relates to the importance, significance and relevance of information, including the degree of accuracy of the data involved. At its simplest it might simply mean monitoring and reporting data relevant to the organisation's most significant environmental aspects or stakeholder concerns. It may also mean presenting the data in such a way as to be a true representation of impact, avoiding misleading or 'glossy interpretations' that present information in a way biased towards the organisation.

Associated with the principle of materiality is the concept of 'completeness', i.e. that all material issues are covered and reported, again rather than omissions that perhaps reflect favourably on the organisation's performance or reputation.

To put it succinctly, we should identify, monitor and report on all key performance data relating to issues of concern to our stakeholders.

5.4.3 Responsiveness

Responsiveness means demonstrating reactivity to stakeholder concerns and meeting their information needs.

So in essence, AA1000 is saying any organisation when using data and information as part of its environmental programme should do the following:

▶ consider who and what will be impacted by its activities;
▶ monitor those things relevant to such impacts and process information in a complete and unbiased way;
▶ report such information in an appropriate way to all key stakeholders.

IV VERIFICATION AND ASSURANCE METHODS

Any data or information management process should be subject to some kind of periodic verification and assurance process. Assurance is defined as a formal guarantee – a positive declaration that a thing is true. Verification may be defined as the process of establishing the truth or accuracy of data. So verification processes provide assurance.

How this is done will depend on the information and data being processed and the underlying goal for doing so. However, whether we are considering the management of mandatory reporting data as part of an environmental permit or performance data for corporate reporting, the following issues may be relevant.

5.5 Calibration and corroboration of data sources

Wherever equipment is used to collect data, there should be consideration of the accuracy of that equipment. This is true across a wide range of data and information collection methods, e.g. noise meters, in-line analysers, electricity meters and even qualitative data such as site boundary nuisance assessments like detectable noise or odour.

Critical monitoring equipment, for example, at the discharge point from an effluent treatment plant, should be inspected, maintained and calibrated on a regular basis in order to ensure a high level of assurance that the data generated is accurate. In high risk circumstances, it may even be appropriate to have cross-check monitoring or data collection to ensure on-going accuracy. For example, on many permitted air emissions streams there would be continuous automatic monitoring subject to regular calibration checks, however, there would also be periodic manual monitoring between calibrations to help identify discrepancies or gaps in data collection.

Less accuracy-critical data should also be checked periodically, e.g. supervisory checks to provide corroboration that routine subjective assessment of nuisance risk is accurate.

Where data collection is being conducted not as part of an on-going programme but as single point assessment, it is even more important perhaps to ensure that data is corroborated to ensure that decisions are not based on inaccurate information.

5.6 Internal and independent auditing/validation

Whether data and information are generated as part of an on-going environmental programme or management system, it is generally seen to be good practice to undertake periodic auditing of the collection and reporting process. Internal audits may be carried out as part of wider assessments of operational controls or may be set up specifically to deal with data collection as a stand-alone assessment. Checks are typically made against the organisation's own procedures for data collection and reporting.

5.6.1 AA1000AS (2008)

External or independent auditing and validation of data may be particularly important where a high degree of credibility or reliability is required. The AccountAbility standard AA1000AS (2008) refers to the need for assurance processes that assess an organisation's accountability against the three principles set out in Section 5.4. The guidance document in support of the standard suggests that an independent assurance assessor be commissioned to evaluate compliance with the principles by asking such questions as the following in relation to materiality:

Is there a process in place to determine what is material?
Does the process include an evaluation of relevance?
Does the process include an evaluation of importance?
Does the process fairly represent the views and importance of stakeholders?
Are the criteria for evaluation clear and understandable?
Is there a process for resolving conflicts or dilemmas between different expectations regarding materiality?
Have the processes been systematically applied?
Is the determination of materiality consistent with stakeholder views?
In your professional judgement, are there any material omissions or misrepresentations?

5.6.2 Regulator assurance processes

Regulators will frequently carry out their own corroborative sampling and analysis of pollutant

Table 5.4 Excerpt from the Environment Agency OMA scoring methodology

Element	Qualification for OMA scoring	OMA Score
Air OMA 1A Management structure	There is a poorly defined management structure for monitoring issues. Posts are not clearly identified as having responsibility for monitoring issues. There are inadequate resources available for monitoring.	1
	There is an acceptable management structure for monitoring issues. Monitoring is the responsibility of defined personnel. This is not documented in detail. Sufficient resources are normally available for monitoring.	2
	There is a well-defined and formally documented management structure for monitoring issues. Posts are clearly and formally identified as having responsibility for monitoring issues. Sufficient resources are always available for monitoring.	3

Source: Contains Environment Agency information © Environment Agency and database right.

data to ensure that controls and compliance arrangements are being met, for example, sampling by Water Companies of trade effluent discharges into the public sewer.

In England, the Environment Agency also uses and encourages operators to use an assessment process related to pollution monitoring programmes that it calls *Operator Monitoring Assessment (OMA)*. Full details are available in the guidance note provided by the Environment Agency but essentially all data collection and monitoring activities are assessed against the following criteria:

▶ Management, training and competence of personnel
▶ Fitness for purpose of monitoring methods
▶ Maintenance and calibration of monitoring equipment
▶ Quality assurance of monitoring.

Each criterion is presented with a series of sub-elements for which descriptions are provided with associated scores. So, for example, for the monitoring of air emissions from a site regulated under an environmental permit, under the first criterion in the list above there are five sub-elements:

Management structure
Schedules
Use of results
Understanding of requirements
Competence of personnel.

Under each of these a score is allocated to the operator-based defined criteria. An excerpt from

the OMA is included for illustrative purposes in Table 5.4.

The EA assessment then combines scores to give an overall 'reliability result' that helps the Agency plan its regulatory inspections. This approach could be used by any organisation with on-going data or information management requirements (whether subject to environmental permit or not) to provide the basis for an assessment and development of internal data management arrangements.

5.6.3 EMAS and DEFRA guidance

Some organisations commission independent verification of data and information that is used in corporate environmental or sustainability reports. This may be done as a completely voluntary measure to enhance the credibility of any performance data and claims related to improvement. It may also be done as a requirement under the EMAS verification process (see Chapter 6).

Guidance is available within the EMAS process and from DEFRA as to the appropriate standards for external verification of environmental reporting. At its most rigorous, however, assurance auditors will evaluate everything from calibration of equipment through competency of personnel, sampling of data sets produced at ground level and management and calculation processes used to generate corporate performance figures.

V BENCHMARKING PERFORMANCE: THIRD PARTY ASSESSMENT SCHEMES

As environmental and sustainability programmes have become more common, there has been increasing interest in being able to compare the relative performance of organisations in the same or even in radically different sectors. This interest has arisen from the organisations themselves to get an idea of 'how well are we doing?' and also from customers, investors and lobby groups interested in being able to compare or choose between organisations. While not as quantitative or reliable as the other data and information methodologies discussed above, there are a number of benchmarking schemes that are becoming increasingly popular because of their ease of use as performance league tables. Three examples appear to be leading the pack at present – the Dow Jones Sustainability Index (DJSI), the Business in the Community (BITC) Corporate Responsibility Index, and the FTSE4good index. All are based on independent reviews of questionnaires completed and data provided by participating companies.

5.7 The Dow Jones Sustainability Index (DJSI)

The corporate sustainability performance of each of the 2,500 companies (2013 figures) in the DJSI investment stocks group is evaluated by the DJSI corporate sustainability assessment based on voluntary questionnaires and publicly available information. Each company assessed is assigned a corporate sustainability performance score. The top 10 per cent are then included in a listing known as the Dow Jones Sustainability World Index.

The methodology is based on the application of specific criteria to assess the opportunities and risks deriving from economic, environmental and social dimensions of each of the companies. These consist of both general criteria applicable to all industries and criteria applicable to companies in a specific industry group.

For each company, the input sources of information for the assessment consist of the responses to the corporate sustainability assessment questionnaire, submitted documentation, policies and reports, publicly available information and personal contact with companies.

Monitoring media and stakeholder information assesses a company's on-going involvement in critical social, economic and environmental issues and its management of these situations.

Companies that score poorly in the on-going Corporate Sustainability Monitoring are excluded from the annual Corporate Sustainability Assessment. Companies that successfully pass the on-going monitoring process and annual assessment process qualify for the DJSI component selection and, if scores are high enough, inclusion in the World Index.

To ensure quality and objectivity, external audit and internal quality assurance procedures, such as cross-checking of information sources, are used to monitor and maintain the accuracy of the input data, assessment procedures and results.

The overall outcome of the annual assessment process is a list of organisations that fall within set performance criteria that can also be ranked as a whole or by sector. This enables investors to assess companies on the basis of sustainability as well as considering pure financial performance. Research from financial and academic institutions in various parts of the world is continuing to produce an ever expanding body of proof that in the longer term, organisations with strong sustainability credentials are outperforming competitors in financial terms.

5.8 BitC Corporate Responsibility Index

In the UK, Business in the Community (BitC) takes a similar approach to the Dow Jones Sustainability Index in considering the results of questionnaires completed by participating companies to produce a ranking in terms of environmental and social engagement. Consideration of environmental issues are made alongside community (e.g. investment in community programmes), marketplace (e.g. product safety) and workplace (e.g. occupational health and safety) measures. Companies are measured in relation to the degree of implementation of corporate strategy in each of the areas listed. The placing of a company in the index is based on the overall score achieved across all the components and also on the level of assurance provided by their submission. Questions look for consideration of key environmental management issues, as well as improvement plans and demonstrable change in performance.

Media and government interest in the annually published index is growing and companies within the FTSE 350 are coming under increasing pressure not only to participate in the voluntary scheme but to perform well. The scheme has been running since 1996. In 2013, 126 companies participated, each with global revenue in excess of £250 million and collectively representing more than 4 million employees.

As well as acting as a ranking/performance measurement tool that allows participating companies and other interested parties to rate organisations within or between sectors, the annual survey also produces 'industry status'- type information by looking at results across all participant companies.

5.9 FTSE4Good

Produced by the FTSE, one of the world's leading global index providers, FTSE4Good is an index aimed at facilitating socially responsible investment. The FTSE4Good selection criteria cover the following three areas:

▶ Working towards environmental sustainability.
▶ Developing positive relationships with stakeholders.
▶ Upholding and supporting universal human rights.

As with their financial indices, companies scoring highest in all three areas appear at the top of the FTSE4Good index.

VI POLLUTION AND NUISANCE MONITORING

5.10 Introduction

The monitoring of pollutants and nuisance at one level or another is an integral part of any environmental programme. It is a specialist discipline in its own right requiring good understanding of chemistry and laboratory techniques, however, a general understanding of principles and available techniques is beneficial to those acting in a general environmental management role.

Box 5.5 Definition of terms

▶ *Measurement* – the quantification of pollutants achieved by some kind of gauging.
▶ *Monitoring* – the collection and interpretation of a number of measurements or estimates over a period of time. Often monitoring involves some kind of comparison with a reference standard.

Guidance for operators on appropriate standards for equipment, personnel competency and sampling and analysis protocols are provided by the Environment Agency under the Environmental Permitting horizontal and technical guidance notes, the Monitoring Certification Scheme (MCERTS) and the Operator Monitoring Assessment (OMA) guidance note. The MCERTS scheme is based on international standards and provides for the product certification of instruments, the competency certification of personnel and the accreditation of laboratories. MCERTS is progressively being extended to cover all regulatory monitoring activities. The relevant guidance note/MCERTS document is referenced within each of the following sections for those requiring additional detail. All MCERTS guidance and Environmental Permitting guidance documents are available for download from the EA website (www.environment-agency.gov.uk).

This section is intended to be an introductory overview accessible to all and requiring minimal prior knowledge of chemistry or physics. Those seeking further information or detail in relation to the topics covered may find a range of information on the EA website and in Brady's (2011) *Environmental Management in Organizations: The IEMA Handbook* . Roger Reeve's (2002) book, *An Introduction to Environmental Analysis*, provides more detailed information on analytical techniques in particular, while wikipedia.com has some excellent explanations of sampling and analysis techniques with some useful sketches.

5.11 Analytical techniques for environmental pollutants

It is beyond the scope of this book to consider in detail the equipment, methods or scientific basis of the analytical techniques used in environmental monitoring, however, Table 5.5 provides an indication of the types of techniques used and their application to different pollutants.

The technique chosen depends on a variety of criteria, including emission type and complexity, sampling methods, reliability requirements and the nature of the pollutants. In most cases, specialist equipment requiring calibration and competent operation is involved. Also in most

Table 5.5 An overview of analytical techniques used in pollution monitoring

Type of analysis	Basic principle	Examples of analytical techniques/pollutants
Chromatography	Separates pollutants in a mixture allowing them to be quantified individually by another technique	Ion chromatography, gas chromatography, liquid chromatography. Used for the separation of a wide variety of pollutants including pesticides, chlorinated solvents, polychlorinated biphenyls, dioxins and endocrine disruptors, VOCs, etc.
Electrochemical	Measures electrical properties related to chemical composition	Ion selective electrodes (including pH meters) and conductivity meters are used for individual pollutants, e.g. metals and total salts (ie combined pollutant load).
Gravimetric analysis	Measures the mass of pollutant present	Filtration techniques, e.g. total solids and suspended solids in water; PM_{10} and total suspended particles in air
Optical	Assess optical properties of a sample either qualitatively (e.g. colour) or quantitatively (e.g. via light transfer)	Colour, turbidity (of water samples), obscuration, opacity (for particulates samples) Visible oil and grease
Spectrometry	Techniques involve energy from different parts of the spectrum. In all cases, from radiation absorbed or emitted, information is obtained on the composition of the sample and the amounts of constituent pollutants	Atomic absorption spectrometry (used for heavy metals) Chemiluminescence analysis (used for oxides of nitrogen in air) Infrared spectrometry (used for SO_2, total hydrocarbons) Mass spectrometry (often coupled with gas chromatography and used for many organic pollutants, e.g. PCBs, solvents etc.)
Volumetric	Measures the volume of one (known) substance reacting at a fixed ratio enabling the amount of the other chemical (unknown) to be inferred	Titrations for pH – generic chemical properties – exact pollutant unspecified Sorbent tubes and diffusion tubes – e.g. Draeger tubes (a particular brand name). These glass tubes contain a pollutant specific reagent adsorbed onto an inert solid. A fixed volume of gas is drawn through the tube using a hand pump. The sample time is a few seconds, during which (if the pollutant is present) a colour develops from the sampling end of the tube. At the end of the sampling period, the colour should extend along a fraction of the length of the tube. The tubes are pre-calibrated with a concentration scale, so that the distance the colour has travelled can be directly related to the gas concentration. This technique can be applied to a wide variety of pollutants by varying the reagents in the tubes.

Source: Adapted from Brady *et al.* (2011)

cases, both field and laboratory equipment is available to suit particular requirements.

5.12 Monitoring strategies: general principles

Monitoring may be carried out for a variety of reasons including legal requirements, process control, to reduce the likelihood of complaints, etc. The purpose will largely determine whether it is most appropriate to monitor pollutants as they emerge from the source (*source monitoring*) or once they have been dispersed into the receiving medium (*ambient monitoring*). In some instances, both will be appropriate, e.g. to confirm that control systems or abatement equipment are achieving the expected results at or near key receptors.

5.12.1 Monitoring programme types

Typically monitoring falls under one of three classifications:

▶ *Continuous monitoring* – measurements made continuously with few or no gaps in data collection.
▶ *Periodic monitoring* – measurements made at defined intervals or under defined operating conditions.
▶ *Surrogate monitoring* – the pollutant itself is not measured but is estimated from another parameter, e.g. fugitive emissions of VOCs estimated from consumption totals of cleaning solvents.

Both continuous and periodic monitoring may be conducted manually or automatically. Automatic monitoring uses equipment that normally provides a real-time data read-out. Manual techniques involve the collection of a sample which is subsequently analysed on site or in a laboratory. In both cases, the analytical techniques used must be appropriate to the pollutants being monitored and the sampling methods adopted must ensure that the monitoring programme represents an accurate picture of pollutant loading (see below).

5.12.2 Sampling

Sampling itself is a specialised subject but the golden rule is that sampling must be representative both spatially and temporally. In all cases a relatively small amount of collected material must provide a reasonable estimate of the overall character of the material (whether ambient air, an effluent, etc.). The number and location of samples required depend on the variability of the 'bulk material' in both spatial and temporal terms.

Effluent concentrations and air emissions may vary over time depending on the exact operational conditions. Concentrations in soil can be different even in adjacent samples. Samples taken from watercourses or the air can vary widely depending on flow rates, dispersion/dilution factors, etc. Any comprehensive sampling strategy must consider such factors and take a number of samples at different times and locations in order to take account of this variability. Different strategies may need to be employed for different pollutants depending on their individual characteristics. The chemical characteristics of pollutants may also affect the choice of collection point, abstraction speed, materials used in collection vessels, temperature and light exposure of collection lines and vessels, storage times before analysis, etc.

This variability of pollutant concentrations means that at any sampling point we typically require the average concentration over a period of time as well as an instantaneous measurement. These are known as the *time-weighted average concentrations*. There are two approaches to this, depending on the sampling and analytical method used as well as whether the monitoring programme is continuous or periodic:

▶ Sample over an extended period and calculate concentrations based on known volumes.
▶ Take many instantaneous samples (grab samples) and use computerised data control and storage to calculate averages.

The principles outlined above apply to any monitoring programme in any receiving medium, however, there is some variation in detail depending on whether the monitoring programme is related to emissions to air, land

169

or water. There are also specific issues relevant to monitoring actual or potential sources of nuisance, e.g. noise, odour and dust. Each of these six areas will now be considered individually.

5.13 Monitoring air pollution

As indicated above, pollutants may be measured at source or in the receiving medium – ambient air in the case of air pollution. In terms of industrial pollutants, air emissions are either controlled or fugitive releases. Table 5.6 illustrates the categories (with examples) of emissions that may be the subject of a monitoring programme.

5.13.1 Source-based monitoring

Depending on the standards to be achieved and the factors discussed in Section 5.12, monitoring may be continuous or periodic. For monitoring programmes conducted as part of an Environmental Permit, the preferred approach of the Environment Agency is for day-to-day monitoring using continuous emissions monitoring systems (by necessity these are usually automated), augmented by occasional checks using periodic monitoring based on manual or automatic techniques. The analytical techniques may also vary between the continuous and periodic monitoring systems, e.g. for a programme measuring particulate emissions from a chimney stack, the continuous monitoring equipment may use optical techniques measuring opacity, while the periodic monitoring may be based on gravimetric analysis.

5.13.2 Ambient air monitoring

Ambient air monitoring techniques may also be divided into continuous automatic measurement and periodic sampling requiring subsequent analysis. Continuous analysers are widely used by local authorities in the UK to measure CO, NO_x, SO_2, PM_{10} and ozone as part of the national air quality strategy.

Where manual sampling takes place, the process may be active – where a known volume of air is collected using pumps – or passive – where a collection vessel or substrate is simply left exposed for a defined period.

An example of active sampling is when a known volume of air is drawn through a filter which is subsequently weighed to provide particulates mass (filters of different sizes may be used to give specific particulate group readings, e.g. PM_{10}).

An example of passive sampling is the use of sticky plates to collect deposited particulates over a defined time. Calculations are then used to provide an average ambient air concentration of the particulates over the reference period.

5.13.3 MCERTS documents relating to air emissions monitoring

The following MCERTS monitor air emissions:

▶ continuous monitoring of industrial chimneys, stacks and flues;
▶ emissions monitoring from chimney stacks – using accredited laboratories and certified staff;
▶ monitoring ambient air quality;
▶ portable equipment for emissions monitoring;
▶ monitoring with isokinetic samplers.

Table 5.6 Categories of air emissions

	Point source	Line source	Area source
Controlled release	Emissions from fixed plant, often released via a stack, vent or duct	Exhaust emissions from vehicles (which are mobile)	Open process tanks
Fugitive release	Intermittently leaking valve	Dust re-suspended in a vehicle's wake; wind whipping of dusty material on an open conveyor belt	Wind whipping of a stockpile of dusty material

Source: Adapted from Brady et al. (2011)

Other important reference material is contained in the Environment Agency's Monitoring Technical Guidance Notes, especially:

M1 – sampling requirements for stack emissions monitoring

M2 – monitoring of stack emissions to air

M8 – monitoring ambient air.

5.14 Monitoring water pollution

As with air pollution, water pollution may be measured at source (e.g. the effluent discharge pipe) or in the receiving water body.

5.14.1 Source-based monitoring

Effluent discharge variation occurs largely as a result of changes in the operation or the treatment plant giving rise to the discharge. Diurnal variation, abnormal operations and even accidental discharges may change the standard parameters or flow rates. For that reason many effluent discharge routes are subject to both continuous monitoring of key parameters and periodic sampling of the full range of potential pollutants. Discharge consents often specify the sampling regime that must be followed.

5.14.2 Monitoring in the receiving water body

When sampling in the receiving medium there is even more scope for variation. Variation may be both periodic and spatial. Consider a river receiving effluent from a factory. Periodic water quality variations may occur due to:

▶ seasonal variations in natural processes (e.g. periods of fish spawning, high/low plant growth), or pollutant use (e.g. pesticides and fertilisers), or significant variations in flow levels in the river (affecting dilution rates);

▶ weekly variation, e.g. because a factory only discharges Monday to Friday;

▶ daily variation again perhaps because of discharge variability but also because some pollutants may be changed by biological processes requiring sunlight (e.g. heavy organic load). In addition, dissolved oxygen levels will rise and fall on a diurnal basis

because of the contribution played by photosynthesis during daylight hours only.

Spatial variation within the river is also a very real issue in deciding where to take samples in order to get a representative water picture of pollution. Although ease of access is a crucial consideration, if we only take samples from the surface of a slow moving eddy at the side of the river, we will undoubtedly get inaccurate results. This is because mixing and sedimentation rates in such areas will be very different from the main flow of the river. Typically when sampling a river, the following guidance is given:

▶ Samples should be taken far enough downstream to ensure complete mixing (unless very localised impacts are being assessed, in which case sampling takes place close to the discharge point).

▶ Placid areas away from the main flow should be avoided.

▶ Samples should, if possible, be taken in a sub-surface profile across the river to ensure that they are representative of the main flow.

▶ Enough samples should be taken to be statistically valid.

▶ Care should be taken to avoid the introduction of additional contaminants during the sampling procedure, e.g. fuel from boat engines, additional sediment stirred up by sampler's feet when taking manual samples.

▶ Samples should be stored appropriately to minimise pollutant change prior to analysis, e.g. in non-reactive containers, with a storage temperature (below 4°C to reduce biological degradation), in the dark (to avoid photochemical decomposition), with nil storage time (for BOD measurements).

5.14.3 Effluent/water monitoring parameters

Below is a summary of common parameters used in the analysis of water pollution both at source and in the receiving water body.

5.14.3.1 Dissolved Oxygen (DO)

Oxygen is partially soluble in water and at about 10°C the concentration at saturation is about 10 mg/l; as the temperature of the water increases, the dissolved oxygen saturation falls. The minimum desirable DO for a balanced population of aquatic life is about 5 mg/l. The natural bacterial oxidation of organic matter requires oxygen, which is taken from the water, hence lowering the DO. Dissolved oxygen was previously assessed using titration methods but is now most commonly measured using ion-specific electrodes.

5.14.3.2 Biochemical Oxygen Demand (BOD)

This is the concentration of oxygen, measured in mg/l, dissolved in the water sample which micro-organisms require to break down the organic matter present. It is a measure of the polluting strength of the sample. The original test takes 5 days and the BOD5 test, as it is designated, is a classic test. The DO level of a fully aerated sample is first measured (best practice states that this should be done as soon as possible after sampling). The sample is then kept in the dark in a completely full container under standard conditions designed to be ideal to promote microbial activity (20°C, pH adjusted to 6.5–8.5, with trace nutrient addition if necessary). At the end of 5 days, the DO content is again measured and the BOD is then calculated as:

BOD = (initial oxygen concentration – final oxygen concentration) mg/l

Some difficult wastes may be incubated over 7 rather than 5 days and the test is then referred to as a BOD7. Care has to be taken with difficult wastes to ensure that any pollutants that might lower microbial activity, e.g. chlorine, are removed prior to testing. Equipment has recently been developed, based on microbial probes, which can give a BOD result in 30 minutes. The type of test used must be indicated when quoting results.

BOD results are comparative over time and between locations because of the standardised nature of the test and for that reason are commonly used in the long-term monitoring of natural water. Table 5.7 shows some typical BOD results.

Table 5.7 Typical BOD values

Sample	BOD (mg/l)
Good quality river water	<3
Well-treated sewage effluent	<20
Raw sewage	300

Source: Adapted from UK Royal Commission standards for treated sewage and the Scottish Environmental Protection Agency Prevention of Environmental Pollution from Agricultural Activity (PEPFAA Code).

5.14.3.3 Chemical Oxygen Demand (COD)

COD is the amount of oxygen required, expressed as mg/l, to chemically oxidise available pollutants in the waste water. This is a standard volumetric test, which involves reacting the sample with an excess of oxidising agent (i.e. a material that releases oxygen during its chemical reaction). After a fixed period (e.g. 2 hours in the standard potassium dichromate test), the concentration of unreacted oxidising agent can be measured and the oxygen equivalent then calculated. COD tests are normally much faster than the BOD tests and are thus commonly used as a rapid analysis method for heavily polluted effluent. They represent an indicator, produced by artificial chemical means, of the amount of oxygen that the effluent will consume naturally from dissolved oxygen in the receiving water body. They do not match BOD values, however, because they measure the oxygen demand of both the organic and inorganic fractions of the sample. However, for effluents generated from fairly consistent processes, the BOD:COD ratio is often relatively constant so that if the COD is measured, the BOD may also be inferred/estimated. Some typical values of BOD and COD are shown in Table 5.8.

Table 5.8 Typical values of oxygen demand for various wastes

Type of waste	BOD	COD
Landfill leachate	8,650	11,000
Food canning effluent	2,500	4,600
Soft drink bottling effluent	7,000	10,000
Enzyme manufacture	3,800	5,600
Fermentation processes	17,000	24,000
Domestic wastewater	220	500

5.14.3.4 Total Solids (TS)

This is the amount of dry solid material present in a sample of the effluent, either as dissolved or suspended matter. It is measured by drying the sample at 105°C to evaporate the water and the result is usually expressed as a percentage of the total.

5.14.3.5 Suspended Solids (SS)

Suspended solids are fine solid particles suspended in effluent or water. These can cause harmful effects if, for example, light penetration is inhibited leading to a reduction in photosynthesis of aquatic plants. Fish may be harmed or killed through damage to their gills which affects respiration. The settling out of solids may also result in the smothering of organisms on the river bed.

Suspended solids are measured using filtration and weighing. The Royal Commission Standards for suspended solids in effluent discharges from sewage treatment plants is 30 mg/l.

5.14.3.6 pH

The strength of an acid or an alkali is indicated by the value of pH, a logarithmic scale, ranging from 0–14. The pH of pure water is neutral at 7. The value of pH can be measured by a universal indicator – a standard mixture of dyes which turns from red for strong acids through to blue for strong alkalis. Alternatively, pH electrodes may be used.

5.14.3.7 Nitrogen

Nitrogen is essential for living things, including plants, but they obtain it not directly from the air but from nitrates in the soil or, in the case of animals, by eating the plants. Nitrogen is converted to nitrates naturally by lightning and by certain bacteria (nitrogen fixing bacteria) and also is man-made in the form of fertilisers such as potassium nitrate or ammonium nitrate. Some nitrates get converted back into nitrogen by bacteria (denitrifying bacteria).

The presence of nitrogen in a waste is determined by measuring the ammonia, nitrite and nitrate and thus total nitrogen. Different methods are used for each compound, e.g. nitrates are often measured using ion chromatography, while ammonia is measured using spectrometric techniques.

5.14.3.8 Electrical conductivity

Electrical conductivity is a measure of the total inorganic salt content in a sample (e.g. iron oxide, sodium hydroxide, sodium chloride, etc.). Conductivity cells are used to measure the total conductivity of the sample in micro siemens per centimetre. The cells are calibrated against solutions of known conductivity at a standard temperature (typically 25°C). The readings give an overall indication of salt content (though no breakdown of any different salt types present).

5.14.4 Classification of inland waters

The Environment Agency is responsible for monitoring the quality of inland waterways in England, as are the SEPA in Scotland and Natural Resources Wales in Wales. In all cases, inland waters (i.e. rivers and canals) are assessed both in terms of their physical characteristics (physico-chemical conditions) and on their ability to sustain aquatic life (compared with reference conditions for the type of water coursebeing assessed). Since 2009, an ecological classification methodology introduced under the EU Water Framework Directive (WFD) has been used. It is summarised in Figure 5.4.

Under the terms of the EU WFD, national regulators are responsible for monitoring water quality as part of the River Basin Management Plans. In the UK, for each River Basin Planning cycle, the Environment Agency/SEPA/Natural Resources Wales defines environmental status objectives for each water body. Objectives may be to 'achieve good status' or to 'maintain high status' within a specified time period (with the status definitions based on EU classifications).

The Directive recognises that in some water bodies it may be impossible to get to a near natural (high) condition because of useful changes, such as to protect people from floods,

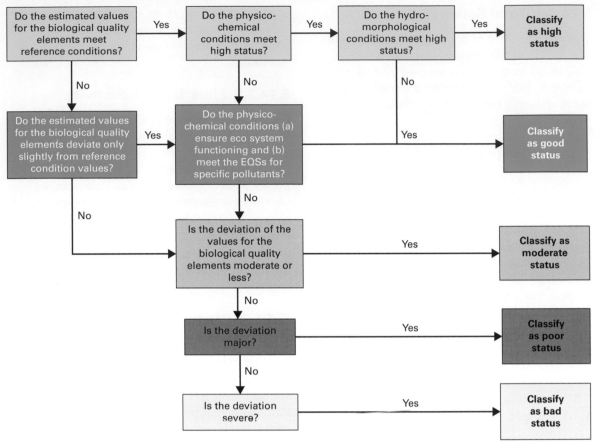

Figure 5.4 Ecological classification of inland waters
Source: Environment Agency. Contains Environment Agency information ©Environment Agency and database right.

to allow navigation, or to hold back water for abstraction or power generation. For these water bodies, the regulators set a target of 'good' ecological potential.

The classification systems are used to assess the state of the environment. They show the regulator where the quality of the environment is good, and where it may need improvement. Environmental standards are the values for water quality, quantity and habitat structure, which will ensure the right environmental conditions are created to achieve the 'ecological potential status' objectives. Again, they are calculated by the regulator based on monitoring data and modelling of 'optimum' conditions for the river basin concerned.

5.14.5 MCERTS documents related to effluent/water monitoring

MCERTS related to effluent/water monitoring are:

▶ equipment for monitoring discharges to rivers, smaller watercourses and the sea;
▶ self-monitoring of effluent flow;
▶ direct toxicity assessment of effluents.

5.15 Monitoring/assessing land pollution

Unlike emissions to air and water, emissions to land are rarely measured at source but rather once they have been deposited and perhaps dispersed. Contamination may not be limited to soil but may involve pollution of groundwater or the build-up of gases. In addition, frequently land contamination evaluations are carried out as one-off assessments rather than on-going monitoring programmes.

There are a variety of reasons why a monitoring programme/assessment of land contamination may be carried out:

▶ in response to a request from a regulator under the Contaminated Land Regulations;
▶ as a requirement under an Environmental Permit;
▶ as part of a due diligence assessment prior to land or company acquisition;
▶ as part of a land development project where the site history indicates the potential for contamination.

Generally contaminated land assessments are carried out in two stages. It is important to note that the presence of contamination in itself does not imply a requirement for remediation. The presence of actual or potential harm is the crucial factor and therefore the identification of pathways and receptors is an important part of both stages of the assessment process. This is also the case where an on-going monitoring programme (rather than a single point assessment) is being undertaken. In addition, as with air and water monitoring, monitoring programmes relating to land contamination may involve the use of continuous or periodic techniques, with analysis being carried out either in situ or at off-site laboratories for extractive sampling. Frequently even on-going monitoring programmes follow the same initial stages as a single point assessment, namely:

Stage 1 – is aimed at identifying the potential for and likelihood of contamination, as well as providing an indication of possible consequences.

Stage 2 – is aimed at 'proving' the risk assessment carried out in Stage 1 through intrusive survey and detailed evaluation of source, pathway and receptors. Stage 2 will normally also provide some recommendation as to appropriate containment or remediation requirements and/or on-going monitoring requirements in terms of location, frequency and parameters.

5.15.1 Stage 1 assessments

These are further discussed in the environmental audit section in Chapter 6. A summary of the audit format is shown in Table 5.9.

5.15.2 Stage 2 assessments

Stage 2 assessments involve intrusive sampling and analytical techniques to determine:

▶ whether contamination exists;
▶ whether a risk is posed to human health and/or the environment;
▶ whether there is a need for clean-up to mitigate such impacts;
▶ what is the nature of on-going monitoring programmes, if required.

A variety of methods may be employed, depending on ground conditions, the presence of groundwater, aquifers and the likely contaminants. Whichever techniques are used, great care must be taken to minimise the risk of spreading any contamination during the survey, e.g. inappropriately reinstated boreholes may act as a migratory pathway through an impermeable layer that previously prevented the migration of pollutants.

Ground investigations should be targeted on potential sources and on contaminants of concern highlighted by the Stage 1 assessment. The investigations should also be targeted on providing an understanding of the geological and hydro-geological conditions, so that a view can be taken on whether any contamination that is present has the potential to migrate to sensitive receptors.

Such ground investigations include soil and groundwater sampling:

▶ to delineate the extent of impacts;
▶ to confirm groundwater flow directions and also groundwater hydraulics testing (to provide such data as migration rates, etc.);
▶ to confirm the potential for the off-site movement of contaminants.

The aim of any such intrusive investigation is to confirm whether or not a contamination problem exists; the scale of the contamination (vertically and horizontally); whose responsibility it is; and

Table 5.9 Overview of a Stage 1 contaminated land assessment

Scope	Generally open ended evaluations of the likelihood of contamination at a defined site. A desk and visual inspection study aimed at identifying the potential for and likelihood of contamination.
Data review	Typically some or all of the following sources of information will be consulted: • current and historic topographic and land use maps • current and historic site plans • regulatory registers • geological and hydrogeological surveys • any available borehole quality data associated with aquifers potentially connected to the site • current registers of hazardous material usage and any associated procedures relating to delivery, storage, use and disposal
Site investigation	As a minimum, an observation and discussion-based visit looking for sources of current pollution (chemical/fuel stores, underground tanks, poor site drainage controls, evidence of leaks or spills, etc.) and anecdotal evidence of past sources of contaminants (long-term employees can be a valuable source of information where standards have improved considerably over recent years of operation).
Assessment of findings	Essentially a risk assessment based on the identification of contaminant sources, likelihood of release and pollution potential through consideration of pathways and receptors. The outcome at this stage can only be an educated 'guess' based on current standards and available historical evidence. Some indication of the level of confidence of assessment may be given based largely on the amount of risk associated with historic contamination.
Report compilation	By their nature these assessments require detailed reports that clearly lay out findings and indicate the basis of any conclusions, as well as highlighting assumptions made and any gaps in the historic record. Recommendations may also be included that relate either to phase 2 assessment requirements or to remediation work or to actions required to prevent contamination from current or future activities.

whether there is current or potential off-site pollution.

Samples may be taken using abstractive or in situ methods which include:

▶ surface samples;
▶ trenches;
▶ boreholes, where samples may comprise soil, water or gas depending on the expected pollutant type.

Analytical techniques vary widely depending on pollutants and their state in the ground (liquid, solid or gaseous), e.g. pH (electrochemical meters), metals (atomic absorption spectrometry).

The risks posed by the identified pollutant concentrations and distribution can then be evaluated by risk assessment methodologies, which range from a simple qualitative assessments to complete quantitative risk assessment, evaluating every hazard, exposure, pathway and risk. In the UK, a standard Environment Agency methodology is required where evaluations are conducted for permitting or planning purpose (see Chapter 7 for details).

5.15.3 On-going monitoring programmes

Many of the principles introduced above are applicable in on-going monitoring of contaminated land, however, typically sampling points will be fixed (e.g. boreholes created during a phase 2 survey may be utilised for on-going monitoring) and which pollutants to be assessed will be defined. Monitoring programmes may be aimed at:

▶ detecting contamination, e.g. as part of a pollution prevention programme at a major oil storage facility, or,
▶ monitoring the spread of existing or known contamination, e.g. methane monitoring programmes around the perimeter of a landfill site.

5.15.4 MCERTS documents relevant to land

There are also several Technical Guidance Notes (TGNs) related to the management and monitoring of landfill gas and leachate, groundwater and surface water.

5.16 Monitoring nuisance: noise

5.16.1 Noise measurement terms

Noise may be defined as unwanted sound or sound that causes sufficient disturbance or annoyance that it has social and/or medical implications. Sound is a wave motion transferred through the air from the source to the receiver. It may consist of variable characteristics such as bangs, clanks and whines or be relatively uniform with no special distinguishing features.

The loudness of sound is measured in decibels (dB), with values being calculated by comparing the power of a specific sound with a reference level. From a health, safety and environment perspective, the reference level is the range of human hearing and the measurement scale (which is adjusted to that range) is referred to as dB(A). The scale is logarithmic, which means that an increase of 10dB(A) represents a ten-fold increase in the sound intensity, though typically such an increase would only seem twice as loud to most people. Table 5.10 gives some examples of noise sources, dB(A) readings and perceived loudness.

As a rule of thumb, if you are having to raise your voice to make yourself heard over background noise, then that noise is likely to be around 75–80 dB(A). In many cases, noise levels vary over time and a calculation is used to provide what is called the equivalent continuous A-weighted sound pressure level, the LAeq. The $LAeq_{(1hour)}$ is the constant sound level which, if it persisted over 1 hour, would have the same energy as the varying sound. LAeq is measured over varying lengths of time depending on the frequency of noise under evaluation – infrequent loud noises may not be apparent if a long period is used, whereas highly variable but fairly continuous noise may require a fairly long measurement period.

Further consideration of noise measurement terms including percentile readings and their application in noise monitoring are considered in the Glossary and in Chapter 7.

5.16.2 Noise propagation

Noise will generally radiate in all directions from a source, and will bend around and over walls and buildings. It will also reflect off solid surfaces. Some noise sources generate more noise in one direction than another, while some physical structures can result in an amplification of noise at certain receptors due to focused reflection.

The impact of noise on people both in terms of hearing damage and nuisance depends not only on the loudness but also on the characteristics (acoustic properties) of the noise and other factors such as its frequency and duration. A further complication is that in any given location, some individuals will be more sensitive to noise disturbance than others.

Table 5.10 Examples of noises at different levels of perceived loudness

Sound level dB(A)	Source	Perceived loudness
150	Jet plane at take off	Painful
130		Threshold of pain
110	Sheet piling at 10m distance	Uncomfortably loud
95	Pneumatic drill at 7m distance	Very loud
80	Busy street	Loud
50	Light traffic at 30m distance	Moderately loud
47	Normal conversation	
40	Living room	Quiet
20	Broadcasting studio	Very quiet
0	Threshold of hearing	

Source: Adapted from HSE guidance.

5.16.3 Noise limits/nuisance levels

Noise is the most common cause of public complaint against industrial and construction activities. As well as recourse to civil action, complainants can seek redress through their Local Authorities as noise may also constitute a statutory nuisance. Where the Local Authority is satisfied that a statutory nuisance exists or could exist, it must serve an abatement notice on the appropriate person or persons. Failure to comply with the terms of such a notice constitutes an offence and may lead to criminal prosecution.

In addition, under Sections 60 and 61 of the Control of Pollution Act 1974, Local Authorities have special powers for controlling noise and vibration arising from construction and demolition works. Normally this is achieved through specification of plant and machinery used, working hours and noise limits.

A fuller explanation of the legislation relating to noise control is included in Chapter 3. In all cases, however, the powers or controls relate to limits or operational constraints.

There are two basic approaches to setting noise limit levels, either:

▶ at the site boundary, or
▶ at the affected noise-sensitive property.

Both have advantages and disadvantages. Boundaries are easier for access and monitoring, but do not guarantee control of noise at properties; while monitoring can be difficult at the property because of problems of access and low noise levels or mixed sources i.e. it is sometimes difficult to determine whether all the noise recorded at a receptor is coming from the source in question or from other sources e.g. traffic.

5.16.3.1 Reconciling limits at the boundary with limits at the dwelling

Current practice in setting limits is rather variable and should always be highly context-specific. However, as an example, typical daytime limits might be 57dB(A) at the façade of dwellings, or 65dB(A) at the site boundary.

5.16.3.2 Noise limits at the site boundary

The control of noise nuisance via planning conditions or Section 61 consents issued by the Local Authority is often based on noise limits set at the site boundary. Ideally, site boundary noise limits would be set in order to ensure that the site operator works within the noise levels, which have been confirmed as acceptable for the local environment.

Many authorities base their selection of an appropriate noise limit on British Standard BS 4142 'Method of rating industrial noise affecting mixed residential and industrial areas'. In simple terms, BS 4142 rates the likelihood of complaints in terms of how far the intruding noise is above or below the background noise. The process is explained in Chapter 7.

5.16.3.3 Noise limits at noise-sensitive properties

Since the primary objective of any environmental noise limit is to avoid nuisance, it is not surprising to find that where complaints or nuisance have previously occurred, the most common approach adopted by planning authorities is to define a noise limit at the façade of such properties.

Night-time limits for industrial noise applied at the boundary of receptor properties are based on the World Health Organisation (WHO) guidelines. The 2000 guidelines recommend a level of 30 dB LAeq for undisturbed sleep, and recommend that to prevent people from becoming moderately annoyed during the daytime, outdoor sound levels should not exceed 50 dB LAeq. Because noise should be taken into account when determining planning applications, it has been assumed that the minimum amelioration measure available to an occupant at night will be to close bedroom windows. Single glazed windows provide insulation of about 25 dB(A). Therefore, in order to achieve 30 dB(A) inside a bedroom, the façade level should not exceed 55 dB(A). All of these numbers are, however, only relevant where background noise levels are very low and need to be significantly adapted in living environments more affected by traffic noise, street noise, wind and water noise, etc.

5.16.4 Noise monitoring programmes

Noise monitoring may be a planning or operational consent condition or may be undertaken voluntarily in an attempt to avoid nuisance creation. Frequency of monitoring may vary, depending on the degree of assurance that limits are being met or nuisance avoided, e.g. if daily measurement over a period has shown noise levels to be well within specified limits, then frequency may be reduced to weekly intervals (though the frequency should be increased again should significant change take place).

5.16.4.1 How to measure noise

A British Standard (BSEN 61672) classifies sound level meters according to the precision (or tolerance) to which they measure. The highest precision meters are classified as Type 0, whereas the lowest grade are Type 3. It is often the quality of the microphone (the sound transducer) that dictates the grade of performance of the meter. Planning and operational consents usually specify the use of a type 1 or 2 meter.

To measure LAeq values, an integrating-averaging sound level meter must be used. Again, BSEN 61672 provides a precision grading range.

Whatever equipment is used, it must be calibrated before and after a measurement is taken. A foam windshield should be used to protect the microphone and where extended measurements in poor weather conditions are planned, an all-weather protection enclosure should be considered. Finally, though it is possible to hold the meter in the hand for spot checks, it is preferable to mount the meter on a suitable tripod to avoid detecting noise generated by the monitoring personnel.

5.16.4.2 Where to measure noise

The location of measurements may be imposed in regulatory consents, e.g. '1m from the exposed window of the nearest dwelling', or 'at the site boundary' or '1m from the site boundary'.

A decision is also required as to whether the measurements should be made in a 'free field condition' or near a reflecting surface. As a guide, for free field measurements the microphone should be at least 3.5m away from reflecting surfaces. However, if receptors are located near a reflecting surface, then a free field reading may register as lower than the levels being received.

For long-term, and especially unmanned, monitoring, extra care should be taken in locating the measuring equipment in order to minimise the risk of damage and the influence of other external noise sources.

Where flexibility of monitoring location exists, the following points should be considered. In the first place, experience from monitoring fieldwork suggests that measurements at affected properties only give a representative evaluation of a particular noise source if the prevailing noise from other sources is exceptionally low. Monitoring at the site boundary is generally more reliable, provided the noise control points are correctly selected. In particular, the noise control points:

▶ should be related to the general location of noise-sensitive areas;
▶ should not be unduly affected by unrelated noise such as traffic on public roads;
▶ should not be in the acoustic 'shadow' of a baffle mound (they should be on top of any such mound);
▶ should not be disproportionately close to any one noise source unless this is taken into account when setting the noise control level.

5.16.5 Guidance on noise monitoring

5.16.5.1 BS 7445 Description and measurement of environmental noise

Comprehensive guidance on the general principles relating to noise monitoring is provided in the three-part BS 7445. This standard is used by noise consultants and regulators alike to provide consistent methods of measurement of a range of noise sources in a variety of different environments.

5.16.5.2 Environmental Permitting Horizontal guidance note H3 Noise

This guidance note details principles and methodologies for noise monitoring and assessments associated with pollution prevention and control permits.

5.16.5.3 Interpretation of monitoring results

Where noise monitoring is conducted as part of a planning consent or Environmental Permit, then results are often assessed in relation to limits set at locations specified within those legal documents. However, where no limits have been set and especially if the monitoring is being conducted to try to predict whether noise complaints are likely, a risk assessment approach has to be adopted. A British Standard (BS 4142) has been developed to ensure that a consistent methodology is followed in such cases. It is explored in Chapter 7.

5.17 Monitoring nuisance: odour

Odour is a problem in several industrial sectors, notably waste management, waste water treatment, tanneries, slaughter houses, etc. Operational consents may set limits based on chemical analyses or on 'smell tests'. Whether monitoring is conducted to determine whether a nuisance claim is valid or as part of a legal compliance requirement, one of the techniques shown in Table 5.11 is likely to be used. As with noise sampling, monitoring locations may be at site boundaries or at receptors.

5.17.1 Guidance relevant to odour monitoring

The best guidance relevant to odour monitoring is H4 Environmental Permitting horizontal guidance note – odour assessment and control.

Table 5.11 Odour monitoring techniques

Analysis technique	Sampling technique	Advantages	Disadvantages
Dynamic olfactometry – a panel of people assess odour strength under controlled conditions in a laboratory	Grab sampling into inert bags	Produces a numerical value for overall odour strength as perceived by a person (individual subjectivity is reduced through use of a panel of experienced testers)	No information is generated as to the chemical constituents responsible for the odour
Chemical analysis – full analysis by appropriate analytical techniques, e.g. gas chromatography in conjunction with mass spectrometry	Sorbent tube collection or grab sampling	Gives a quantitative chemical breakdown of components – useful for assessing health impacts and abatement options	May underestimate the cumulative nuisance potential of a chemical mixture as people perceive it
Analysis of a single substance associated with odour nuisance, e.g. hydrogen sulphide at a sewage works	As above or using a continuous monitoring instrument	Quick and cheap and may be used as a continuous monitoring parameter for a fixed source	Only suitable if a single known pollutant is the source of the odour
Field odour assessments – 'sniff tests' by a single person in the field	Person stands in specified location and sees what or whether odour is detectable	Human nose is the best instrument for assessing ambient odour. Daily 'detectable odour' checks often specified in permits.	Subjective – so assessors must work to protocols

Source: Adapted from Brady *et al.* (2011).

Table 5.12 Dust monitoring techniques

	Sampling technique	Analysis technique	Advantages	Limitations
Deposit gauge	Gravitational settling into a collection container	Usually gravimetric, i.e. weighing dust collected	Gives an overall measure of dust fall averaged over the sampling period. Well-documented methods and useful for long-term records	Variable collection efficiency and averaging dust fall over long period may conceal nuisance potential of variable deposition rates
Dust slides	Gravitational settling onto a glass microscope slide	Measure reduction in reflectance compared with clean slide	Gives a measure of nuisance at receptor, e.g. cars, window sills, etc. Low cost allows large surveys	Cannot be compared with deposit gauge results and may be misleading as to the dust source unless chemical fingerprint exists
Sticky pads	Gravitational settling onto white, self-adhesive pad	Measure reduction in reflectance compared with clean pad	As above	As above

Source: Adapted from Brady et al. (2011).

5.18 Monitoring nuisance: deposited dust

Dust is a potential nuisance associated with a variety of industrial activities including construction sites, quarries, clay pits, cement works, foundries and steel works, power stations, etc. Table 5.12 gives a summary of the three main techniques used in monitoring dust nuisance. Deposit gauges are particularly appropriate for monitoring accumulations at site boundaries, whereas slides and sticky pads work well at sensitive receptors.

5.18.1 MCERTS guidance relevant to dust monitoring

There is currently no MCERTS guidance relevant to monitoring of deposited dust, however, particulates monitoring is included in the MCERTS guidance document on monitoring ambient air quality.

VII METHODS FOR THE INTERPRETATION OF DATA AND INFORMATION

There are a wide variety of means to assess data and information and to thereby identify problems, solutions and trends in environmental aspects. They range from the technical to the extremely simple and are covered in some detail in Chapter 7.

VIII METHODS FOR THE COMMUNICATION OF DATA AND INFORMATION

Often the most important element of data management is making sure that the right information is communicated at the right time to the right people in a format that is readily accessible. This issue is covered in the communication elements of ISO 14001 (see Chapter 6) as well as in the more general section on communication principles in Chapter 9. There are many ways to communicate data relating to environmental performance, ranging from corporate reports through to colour coding of bins or labelling of products. The key to effective communication in any case is:

▶ clear identification of target audience;
▶ clear identification of communication goals;
▶ tailoring of communication content and method to suit the above.

For further details and examples refer to Chapters 6 and 9.

IX FURTHER RESOURCES

Table 5.13 presents some further resources.

Table 5.13 Further resources

Topic area	Further information sources	Web links (if relevant)
Choosing and presenting data	ISO 14031:2013 Environmental management – Environmental performance evaluation – guidelines AA1000AS (2008)	www.iso.org www.accountability.org
Assurance and verification of data	AA1000 Assurance Standard (2008) The Environment Agency Operator Monitoring Assessment scheme DEFRA, 2013 Environmental reporting guidelines – including mandatory greenhouse gas reporting guidance. DEFRA	www.accountability.org www.environment-agency.gov.uk www.gov.uk/defra
Data collection and analysis	Brady *et al.* (2011) Environmental Management in Organisation – the IEMA Handbook, 2nd edn, Earthscan Reeve, R. (2002) Introduction to Environmental Analysis, John Wiley & Sons Ltd BS 7445 – Description and measurement of environmental noise	www.iema.net

CHAPTER 6

Environmental management and assessment tools

Chapter summary

This chapter covers a wide variety of environmental management tools and techniques. Some such as environmental auditing and environmental management systems are already in common use in many organisations. Others such as the WRI methodology for corporate ecosystems review are still in their infancy but are expected to be increasingly used in coming years. The full list of areas covered may be seen in the contents list but includes the following headline techniques:

- Aspects identification and environmental risk assessment
- Environmental auditing
- Environmental management systems
- Resource efficiency surveys
- Environmental footprinting
- Environmental impact assessment
- Strategic environmental assessment
- Life cycle assessment and eco-design
- Sustainable procurement
- Carbon management
- Corporate ecosystems review

Each technique or approach is described in terms of principles and purpose while the advantages and disadvantages of each are reviewed in Section XII.

I ASPECTS IDENTIFICATION AND ENVIRONMENTAL RISK ASSESSMENT

6.1 Environmental assessment methodologies: generic issues

It is useful to differentiate between the two principal types of environmental assessment:

An Environmental Impact Assessment (EIA), which is generally defined as 'a comprehensive assessment of the impact of a *planned* activity on the environment'. The EIA has a statutory context and established guidance standards.

The environmental audit/review, which may be defined as 'an assessment of the environmental impacts of an *existing* activity that is normally limited in scope or depth of evaluation'.

Although both share some assessment principles, there are some clear differences in terms of purpose, application and, not least, time and resource requirements. In each case, however, some form of assessment of environmental impacts requiring consideration of the following elements:

identification of actual or potential environmental aspects;

consideration of normal, abnormal and emergency conditions;

evaluation of source–pathway–receptor;

assessment of significance.

These headings might be considered as complementary tools, used to ensure that all relevant aspects are identified and systematically prioritised. With experience, we eventually cease to distinguish between them and we simply look at a process and compile a list of aspects. It can be useful initially, however, to consider them as stages, which will each add to the full list of aspects that should be considered within an operational area or organisation. Remember that the approach can be used at whatever level of detail seems appropriate, depending on the scale and complexity of the area/activity/development being assessed. The goal is to get sufficient detail to enable clarity as to where controls and improvement programmes should be focused, without getting into so much detail that we end up being swamped and not being able to see the wood for the trees.

6.2 Aspects identification: the input-output approach

The identification of potential to cause environmental impact may be conducted in relation to both planned and existing activities. In an environmental audit or review, the identification of aspects is easier and more simplistic than in an EIA, not least because the process already exists so we can see what is happening. The normal approach is to consider site plans, process flow diagrams and input–output information as well as material storage and transport requirements against the aspect checklist introduced in Chapter 1, the REALHENS checklist. This mnemonic can help memorise this really useful checklist:

- ▶ **R**aw material usage
- ▶ **E**ffluent discharges
- ▶ **A**ir emissions
- ▶ **L**and usage
- ▶ **H**azardous material usage
- ▶ **E**nergy usage
- ▶ **N**uisance
- ▶ **S**olid waste.

A simplistic approach to aspect evaluation is provided as an example below. Essentially the activities of the organisation are listed (in a logical flow if possible) and then, for each activity, the inputs and outputs are identified (Table 6.1). These represent the detailed aspect sub-categories for the activity in question, e.g. if the production process generates solvent emissions (an output), these are a specific example of air emissions from the generic REALHENS list above. The REALHENS list may be used as a cross-check to ensure all inputs and outputs have been considered for a particular activity.

Table 6.1 The input-output approach

Inputs (aspects)	Activities	Outputs (aspects)
Typically:	For example:	Typically:
• Raw materials	• Production processes	• Products
• Components	• Materials storage and handling	• Packaging
• Water		• Wastes
• Energy	• Product design	• Effluent
• Solvents	• Transport/ distribution	• Emissions
• Packaging		• Waste heat
	• Offices	• Noise
		• Odour

6.3 Consideration of different operating conditions

Having generated an initial list of environmental aspects using the input-output approach (backed up by the REALHENS checklist) described above, there is usually a need to ensure that we have considered all kinds of operating conditions in a particular area of activity. It is easy to focus only on the 'normal' day-to-day operations while consideration of both intermittent planned and even unplanned events may give rise to new environmental aspects. The following operating

187

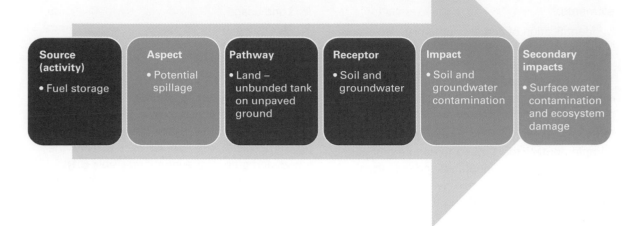

Figure 6.1 The source–pathway–receptor model

conditions are routinely considered for any activity to ensure that all actual and (in the case of emergency scenarios) potential environmental aspects are identified:

▶ *Normal*: routine operations that occur continuously or frequently (giving rise to aspects of the kind described in Table 6.1).

▶ *Abnormal*: planned operations occurring infrequently, e.g. maintenance activities, which may give rise to very different environmental aspects to those occurring under normal operations.

▶ *Emergency*: unplanned occurrences, e.g. spillages which are 'potential aspects' that may be very important if inadequate precautions are in place or insignificant if the likelihood of occurrence is extremely low.

Consideration of these different operating conditions may give rise to some additional actual or potential aspects to include in the list generated by the input-output/REALHENS approach.

6.4 Evaluation of source, pathway and receptor

This stage may again generate new aspects through consideration of how nearby receptors may be influenced under normal, abnormal or emergency conditions. For pollution-based impacts, this stage requires identification of potential pathways and receptors for each aspect (source) (Figure 6.1). In addition to aspect identification, this part of the evaluation begins the process of considering the significance of the environmental aspect in question, as information relating to the pathway and receptor may be crucial in the significance criteria described below. It should be noted that this type of approach is not directly applicable to resource consumption type impacts.

In an environmental audit this can be quite a simple step – merely considering the proximity of receptors such as streams, nearby housing, habitat areas, etc. Recognising the receptors can lead us back to aspects that might affect them. As in the previous step, this may further expand our list of environmental aspects for the area/activity under assessment.

6.5 Environmental risk assessment

The three previous steps may be thought of as overlapping stages in the generation of a complete aspects list for a particular activity or operation. They may be conducted at variable levels of detail depending on the time and goals of the operation or company. In all cases, however, the most important part of the assessment process is the prioritisation of the aspects and impacts identified in order to determine which are 'significant'. The exact definition of significant will vary between activities and locations, e.g. a noise impact which is significant in a village may be totally insignificant in the context of a city or industrial area (and yet in absolute terms the noise is the same).

This stage is essentially an environmental risk assessment process. The basic language and principles of environmental risk management were introduced in Chapter 4 where a three-stage process was presented:

Identify and prioritise risks.
Minimise and mitigate risks.
Monitor and provide assurance that hazards are adequately controlled.

Here we are primarily concerned with the first stage of this process. We have considered the typical approach for identification of environmental aspects (or *hazards* in risk assessment terminology). Now we must decide which of the list of aspects we have identified should be considered most important and be given priority in terms of control and improvement programmes. In ISO 14001 language – 'which are the significant environmental aspects?'

In general terms there are two ways of assessing risk:

▶ qualitative assessments – based on opinion and experience;
▶ quantitative assessments – based on data generated from observation or predictive modelling.

In most circumstances risk assessments comprise a little of each but the balance will vary depending on the goals of the assessment, the levels of harm/impact being considered, and perhaps most importantly, the technical competency of those conducting the assessments. We will consider two contrasting contexts, which share the same basic philosophy but undertake the risk assessment in very different ways.

6.5.1 **Risk assessment in EIA**

Environmental impact assessment is a specialist discipline conducted in relation to proposed developments. The process is discussed in some detail in Section VI below. Here, however, we will focus solely on the risk/impact assessment part of the process. Essentially EIA is conducted by technical specialists, who use quantitative data as far as possible to make 'expert assessments' of the relative significance of impacts on the variety of receptors that might be affected by a particular development. Criteria for significance may include the magnitude and likelihood of an impact and its spatial and temporal extent, the likely degree of recovery and the perceived value of the affected environment, the level of public concern and political repercussions. Key 'sensitive' receptors that may need to be considered include:

▶ Water courses
▶ Residential areas
▶ Areas of designated conservation interest
▶ Aquifers and groundwater extraction points
▶ Cultural heritage sites
▶ Anywhere where protected species of flora or fauna are identified.

The conclusions drawn by the EIA hydrologist, archaeologist, ecologist, etc. are very often qualitative in the final assessment but conclusions are typically backed up with reference data generated by survey work or modelling and by comparison to other developments or receptor locations. These conclusions are then subject to 'peer scrutiny' by the technical experts working for the regulators considering the EIA report as part of a planning application.

The goal of risk assessment in the context of EIA is to highlight actual or potential impacts associated with the proposed development across all stages of the project life cycle:

▶ to enable refinement/modification of construction techniques, design or operational plans to reduce the overall impact of the development on the key receptors identified;
▶ to enable planners and other stakeholders to consider whether or not to support/permit the development;
▶ to enable planners to specify consent conditions associated with the development that aim to ensure that impacts or risks are kept within acceptable limits.

189

6.5.2 Risk assessment in environmental management systems

Generally, in organisations setting up environmental management systems there is not the time, finances or expertise to conduct risk assessment in the manner described for EIA. In addition, for existing operations in such circumstances, our goals for the risk assessment process are slightly different, namely:

▶ to identify those aspects (hazards) that should be the focus of our attention in the EMS, both in terms of controls and improvement initiatives;

▶ to enable a consistent approach to prioritisation over time and in different parts of the organisation;

▶ to make transparent our approach to prioritisation so that stakeholders (including the certification bodies) can see the logic and criteria we have used.

Often key significance criteria are identified and used as part of a risk assessment or significance assessment methodology or procedure. This frequently involves a much simpler approach than in EIA and typically incorporates many qualitative assumptions. If, for example, a particular hazardous substance is involved, then the associated aspect is deemed significant without a complete evaluation of potential environmental impacts in the event of a spillage. It is sufficient from the point of view of the risk assessment to conclude that activities linked to that material *may* have 'significant' environmental consequences and *should* therefore be controlled or modified. The focus thus shifts from the *receptors* that might be affected to the *source* of the actual or potential impacts.

There is huge variety in the significance assessment methodologies used in environmental management systems and environmental auditing. They vary in criteria used, complexity, the use of absolute ranking or banding of issues, etc. Such wide variation is due to a number of factors including the following:

▶ variable audit objectives;

▶ variable company priorities;

▶ variable stakeholder interests;

▶ variable receptor sensitivity.

Sometimes the focus is purely on risk management or pollution control, while in other cases the remit includes resource consumption and product-related impacts, i.e. the widest view of company environmental impact (a specific requirement in ISO 14001).

In all cases, during the development of a formal methodology for prioritising issues, whether simple or complex, there is consideration of what criteria are important to the decision-makers/users of the prioritised list. In EMS risk assessment there are a number of standard principles that should be taken into consideration when deciding on a risk assessment methodology. They are summarised in Figure 6.2.

As well as content, formats for the risk assessment methodologies vary widely, but some common examples are discussed in the following sections.

6.5.3 Risk assessment matrices

Matrices are a very common format used in many environmental management systems, not least because of their familiarity to anyone who has had involvement in health and safety risk assessment. They tend to be quick and easy to apply and range in complexity from the totally subjective and overly simplistic to well guided and rigorous in categorisation. Figure 6.3 presents a very simple risk assessment matrix and a slightly more robust version might look something like Figure 6.4.

Each row has some guidance but is still fairly open to interpretation (what constitutes significant cost, for example). However, this type of matrix can work as long as consensus-based training and peer review of results are included in the way the methodology is applied.

A high reliability version would need supporting guidance on the classification of each box within the matrix, preferably with examples specific to the organisation in which it will be used. And even then there would need to be consensus-based training and peer review as the risk of this type of methodology is that users revert to subjective selection of high/medium/

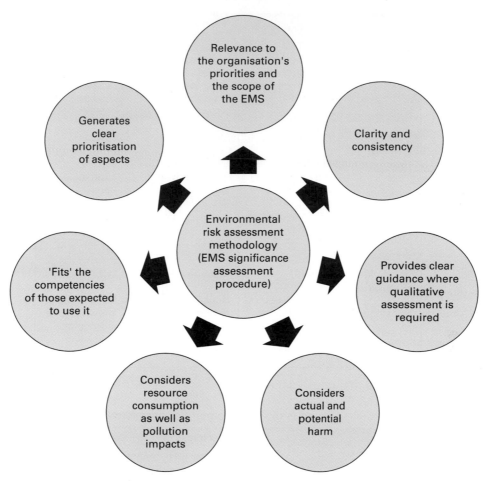

Figure 6.2 Environmental risk assessment

	Likelihood ➡ Consequence ⬇	Low	Medium	High
The simplest matrix – quick and easy to use but totally subjective and hence unreliable in terms of consistency of results	Low	Low significance	Low significance	Medium significance
	Medium	Low significance	Medium significance	High significance
	High	Medium significance	High significance	High significance

Figure 6.3 A very simple risk assessment matrix

191

Likelihood	High (3)	Medium (2)	Low (1)
	Poor or no controls in place or an intended activity	Some controls in place but some doubt as to whether they are adequate	High standard of controls in place giving minimal risk of harm
Consequence – legal and otherwise	High (3)	Medium (2)	Low (1)
	Legal breach is or would occur and/or off site remediation/ compensation is or would be required	Legally acceptable impact but with significant costs to company	Legally acceptable impact
Stakeholder concerns	High (3)	Medium (2)	Low(1)
	Issue has been subject to previous stakeholder discussion	Issue is of interest to one or more stakeholders	Issue is not known to be a stakeholder concern

Figure 6.4 A more detailed risk assessment matrix

Table 6.2 Advantages and disadvantages of matrix format risk assessments

Advantages of matrix formats	Disadvantages of matrix formats
• Familiar format to many users	• They are often overly subjective
• Quick and easy to apply	• Users have a tendency to 'default' to subjective classification
• With simple but clear phrases can be reasonably objective	• With odd number matrices users often over-select the middle choice
	• May demand significant knowledge requirements especially where legal compliance is used as a consequence standard

low categories. Having used these sorts of methodologies with student groups over many years, we can testify to the frequency that groups begin by using a methodology rigorously and then after sometimes a very short period of time drift into subjective classification. There are both advantages and disadvantages to matrix formats (Table 6.2).

6.5.4 Risk assessment flow charts

Flow charts (or questionnaires which are essentially variations on this theme) may encourage more specific assessment criteria than matrices, and allow for multiple stages of classification, providing a short cut assessment

for those aspects or risks that fall into specific categories. A simple example is shown below. In all cases – a YES answer takes you right and a No answer takes you further down the flowchart (Figure 6.5). Again, flow chart risk assessments have their advantages and disadvantages (Table 6.3).

6.5.5 Risk assessment registers

Whatever format and specific criteria we use for conducting our risk assessment/ significance evaluation, we normally need some way to present the results in an easily understood manner. Tabular presentation in the form of a 'register' is by far the most common

Is the aspect subject to a permit condition or in breach of legislation?	➡	High significance
⬇		
Is the aspect leading or imminently likely to lead to off-site pollution or nuisance?	➡	High significance
⬇		
Does the aspect involve resource consumption of non-renewable resources in quantities costing in excess of £10,000 per annum?	➡	High significance
⬇		
Does the aspect contribute directly to climate change, ozone depletion or nuisance impacts?	➡	Medium significance
⬇		
Is the aspect subject to robust control and assurance mechanisms?	➡	Low significance
⬇		
Has any internal or external stakeholder ever expressed a concern or interest in the aspect?	➡	Medium significance
⬇		
Not significant		

Figure 6.5 A risk assessment flow chart

Table 6.3 Advantages and disadvantages of flow chart format risk assessments

Advantages of flow chart formats	Disadvantages of flow chart formats
• Easy to follow if well phrased • Low subjectivity if well phrased • May provide a short cut assessment for those aspects of high significance	• May be difficult to see how the final conclusion has been reached when results are recorded in a register • It is difficult to phrase questions to cover all eventualities • It is difficult to afford multiple assessment criteria that combine to form an overall significance rating as they are based on yes-no answers

method used. An example is provided in Figure 6.6.

II ENVIRONMENTAL AUDITING

6.6 Aims of an environmental audit

An audit may have one or all of the aims shown in Figure 6.7.

Given the range of aims, there is significant variation in the scope and focus of environmental audits, covering a variety of topics and levels of technical and organisational detail. Examples are included in Table 6.4.

The range of audits listed in Table 6.4 indicates that a variety of issues may be addressed during the audit process, however, it is possible to identify a fairly consistent approach which, if applied to any of the categories listed, may be

Activity	Normal, abnormal or emergency conditions	Aspect	Impact/ potential impact	Likelihood score	Consequence score	Stakeholder score	Combined priority score
Manufacturing operations – glass etch process	Normal	Acid gas (predominantly hydrofluoric acid fume) emissions	Human health impacts in local area	3 Planned activity (although high control standards described)	2 Operation of control systems is deemed significant cost	3 As a permitted process there is regular regulator liaison	18
Rain water run-off	Emergency– involving spilled oil entering storm water system	Discharges to estuary of contaminated rainwater	Estuary pollution and possible damage to mussel farm	2 Interceptor system would not catch non hydrocarbons	3 A discharge of contaminated water would be a legal offence	2 Of interest to EA	12
Use of cleaning solvents	Normal	VOC emissions	Local air quality impacts	3 Planned usage and extraction – poor source management practices in place	2 Probably legally acceptable although quantities used may require permit	2 Solvent usage is an issue monitored and may be regulated under permit	12
Rain water run-off	Normal	Discharges to estuary of uncontaminated rainwater	Volume increase to estuary flow	3 Planned activity	1 Legally compliant discharge	2 Although no permit required – of interest to EA	6

Figure 6.6 An example risk assessment register

Box 6.1 Definitions

The term *environmental audit* is used in a number of contexts to describe a variety of investigations into the environmental performance of a site or organisation. The scope of investigation and level of detail vary widely.

In generic terms, environmental audit is the systematic examination of how an organisation affects the environment. It may include all emissions to air, land and water, legal constraints, the effect on the neighbouring community, landscape and ecology and considers products as well as processes. However, it may also concentrate on only one or a few of this range of environmental issues depending on the purpose of the audit and the priorities of the organisation.

expected to facilitate the planning and carrying out of the audit.

6.7 Common audit elements

All audits, to a certain extent, share common elements. Although something of a simplification, these may be summarised as shown in Figure 6.8.

It should be noted that in practice the middle three steps tend to be conducted in parallel, often with loops of inquiry incorporating reference to data, interview or investigation to corroborate data, with provisional and eventually confirmation of findings.

Typically the last of these steps (report compilation) is followed by a number of 'management responsibilities' related to the identification, allocation and completion review of corrective actions. These elements are discussed further in Section III. The rest of this section will describe the principal audit types in common usage with reference to this standard audit model.

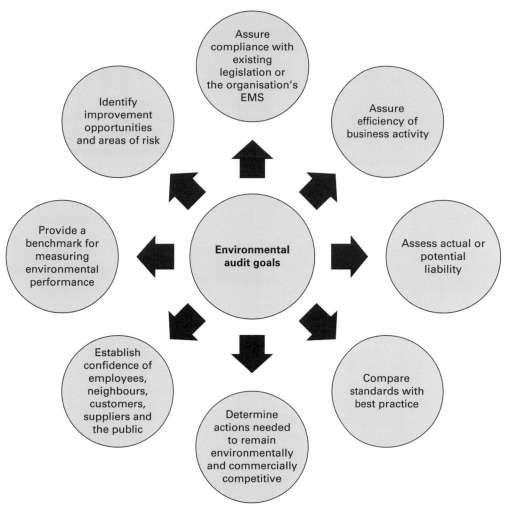

Figure 6.7 Environmental audit goals

Table 6.4 Examples of environmental audits

EMS internal audits	Preliminary environmental review
EMS certification audits	Waste reviews
Duty of care audits	Energy audits
Supply chain audits	Contaminated land (Stage 1) assessments
	Due diligence/pre-acquisition audits

6.8 EMS auditing (internal and certification)

6.8.1 ISO 14001 requirements

The role of the environmental audit within a management system designed to deliver continual environmental improvement is undoubtedly important; as the eyes and ears of the EMS, it is the process which, combined with the management review, helps to give direction to the environmental management process.

ISO 14001 requires audits to determine that the EMS conforms to planned arrangements and has been properly implemented and maintained. It specifies that audit frequency should consider the significance of environmental aspects and the

195

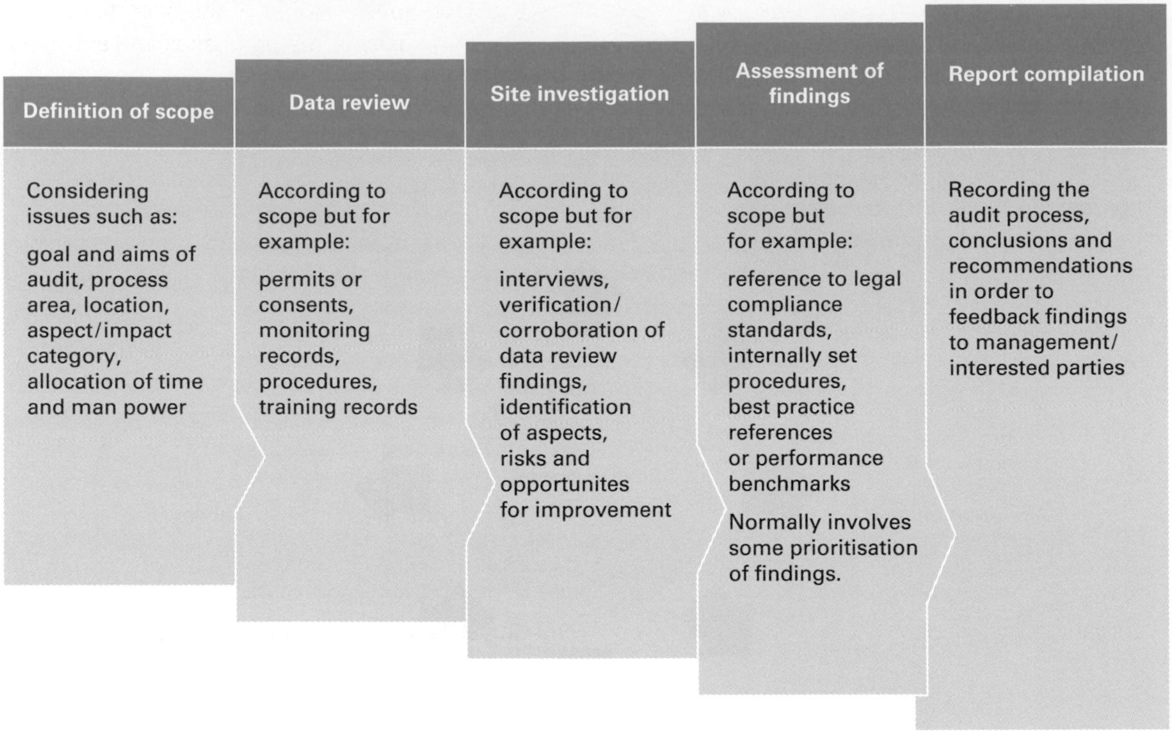

Definition of scope	Data review	Site investigation	Assessment of findings	Report compilation
Considering issues such as: goal and aims of audit, process area, location, aspect/impact category, allocation of time and man power	According to scope but for example: permits or consents, monitoring records, procedures, training records	According to scope but for example: interviews, verification/corroboration of data review findings, identification of aspects, risks and opportunites for improvement	According to scope but for example: reference to legal compliance standards, internally set procedures, best practice references or performance benchmarks. Normally involves some prioritisation of findings.	Recording the audit process, conclusions and recommendations in order to feedback findings to management/interested parties

Figure 6.8 Common audit elements

findings of previous audits, i.e. those areas where the scope for environmental impact is greatest should be audited more frequently.

Additional guidance on audit principles, techniques and auditor competence is provided by ISO 19011 (Guidelines for quality and/or environmental management systems auditing) which defines EMS auditing as: 'the systematic, independent and documented process for obtaining audit evidence and evaluating it objectively to determine the extent to which the audit criteria are fulfilled'. Both internal and certification EMS audits are considered below against the standard audit steps introduced in Section 6.7.

6.8.2 **EMS internal audit**

6.8.2.1 Scope

Depends on the company audit protocol but is typically all or part of the EMS evaluated across all or part of the organisation. An alternative approach used by some companies is to audit the EMS with reference to a particular environmental issue, e.g. waste management.

6.8.2.2 **Data review**

Typically the data reviewed as part of an internal audit includes:

▶ procedures, manuals and policies;
▶ records associated with operational controls, monitoring, training, etc.;
▶ non-conformance reports or complaints;
▶ previous audit reports;
▶ environmental programmes or action plans.

6.8.2.3 **Site investigation**

The site investigation is generally aimed at some or all of the following objectives:

▶ observation of aspects and environmental risks to ensure that the aspects register and significance evaluations remain up to date;
▶ interviews aimed at assessing awareness of issues and current control mechanisms, with particular emphasis on adherence to documented procedures;

▶ confirmation of overall compliance with the organisation's environmental policy through a combination of the above;

▶ the identification of improvement opportunities based on the audit team's experience of best practice both within and outside the organisation.

6.8.2.4 Assessment of findings

Within the EMS internal audit the assessment of findings is typically aimed at:

▶ Identification of non-conformance;

▶ identification of appropriate corrective action;

▶ identification of good practice that should be communicated throughout the organisation;

▶ recommendations for actions to achieve performance improvement/risk reduction;

▶ Emphasis may vary depending on the organisation's audit procedure.

6.8.2.5 Reporting

Generally the internal audit report is a simple summary aimed at generating and recording any necessary action. The precise format is generally defined within audit procedures with, where appropriate, links to the generation of documentation associated with non-conformance or corrective action reporting. It is often important to note those areas assessed which demonstrated acceptable standards of performance, as well as highlighting those areas where improvement is required. Without such information there will be no record of those elements of the EMS that are working well. Such information is equally worthy of a feedback loop.

6.8.3 EMS certification audits

6.8.3.1 Scope

The full coverage of an organisation's environmental management system to assess compliance with the requirements of ISO 14001 or EMAS. Time constraints normally dictate that a 'sampling' approach is employed with the level of audit time committed largely being linked to:

▶ the geographical extent of the operation;

▶ the complexity of the processes and activities involved;

▶ the severity of the environmental hazards associated with the operation.

The increasing magnitude in any of these factors points towards an increase in audit sampling to provide a satisfactory level of confidence that the findings are representative of practice across the whole of the organisation.

6.8.3.2 Data review

The data coverage of an EMS certification audit may include any sources of information related to the organisation's environmental aspects (see the preliminary environmental review below for examples) plus all EMS documentation. Particular emphasis is normally given during the initial certification audit to information relating to the identification and assessment of environmental aspects and regulatory compliance.

6.8.3.3 Site investigation

Essentially the role of the site investigation within an EMS certification audit is to collate objective evidence that confirms whether or not compliance exists both in relation to:

▶ the organisation's own stated standards;

▶ the requirements of ISO 14001/EMAS.

The investigation may include direct observation, interviews and informal discussions across the whole of the organisation under assessment.

6.8.3.4 Assessment of findings

There is a single benchmark within certification audits against which all findings must be assessed, namely conformance with the requirements of the standard against which the audit is being conducted (typically ISO 14001 or EMAS). The assessment outcomes are split into the following categories:

▶ *Major non-conformances* – a non-conformity, which raises significant doubts as to whether the environmental performance of the organisation will meet its stated targets. Persistent or frequent, related

197

minor non-conformities may collectively be considered a major non-conformity.

▶ *Minor non-conformances* – practices or processes (or the omission of the same) within the organisation that do not meet the requirements of the standard.

▶ *Observations* – fall into two categories: practices or trends which, if left unattended, are likely to become non-conformances; or; positive comments linked to procedures, behaviour or processes observed during the audit.

6.8.3.5 Reporting

The focus tends to be on a concise presentation of:

▶ areas of the EMS and organisation assessed;
▶ non-conformances and/or observations made;
▶ supporting evidence for the stated findings;
▶ an overall compliance assessment (certification standards demand that there be no major non-conformances and that all minor non-conformances have been dealt with appropriately).

The report itself is normally formal confirmation of the verbal feedback provided during a close-out meeting.

6.8.4 Duty of Care auditing

Duty of Care audits are essentially legal compliance checks undertaken by a waste producer to ensure that the waste carrier and disposal site are licensed to receive their waste and are operated in a acceptable manner. Although clearly not as complex as an EMS audit, the common audit elements apply and can be illustrated as follows:

6.8.4.1 Scope

For example, to assess the legal compliance of an organisation with specific reference to the Duty of Care requirements set out in the Environmental Protection Act 1990 Part II.

Audit to include evaluation of company waste production and storage areas, waste contractor's operation at collection point, in transit and at disposal site.

6.8.4.2 Data review

Audit to include a review of waste management procedures, operating licences, waste transfer notes and consignment notes, etc.

6.8.4.3 Site investigation

Audit to include a visit to the disposal site and vehicle depot. Site investigation checklist to include: nuisance issues (litter and odour), site security, general housekeeping and waste containment (and, if necessary, labelling), etc.

6.8.4.4 Assessment of findings

Audit findings assessment focuses on the fulfilment of the waste producer's responsibilities – legally compliant or actual/potential non-compliance.

6.8.4.5 Reporting

A summary assessment record will be compiled recording the compliance status and, if necessary, requesting/requiring improvement actions by the organisation and/or contractor(s).

6.8.5 Preliminary environmental review (PER)

The preliminary environmental review (PER) is sometimes considered to be the first EMS audit conducted by an organisation. It should be stressed, however, that the PER is not an audit in the sense of checking and verifying compliance with a system. Instead it defines the organisation's starting point and measures basic environmental performance upon which the EMS and environmental action plan should be based. It is important, therefore, that the review is conducted properly, comprehensively and systematically. Like the scoping study in an EIA, the PER may be conducted internally but frequently some external expertise is brought into the organisation.

If conducted properly, the benefits of the preliminary environmental review are wide-ranging and include:

▶ the identification of relevant environmental legislation and an indication of compliance status;

- the identification of potential cost savings through improvements in areas such as waste and energy management;
- the identification and prioritisation of areas of significant environmental risk – both in terms of legal non-compliance and business losses/reputation;
- the provision of management information relating to significant strategic problems and/or opportunities;
- the identification of current standards and practices that provide a baseline for improvement programmes and define the requirements of an EMS development programme.

6.8.5.1 Scope

It is worth noting that, whereas in an EMS audit the basic starting point is the comparison with the documented EMS, in a PER there may be little or no documented description of the organisation's environmental issues or activities. Many consultants, therefore, begin the PER by comparison of the site or organisation's activities with a standard checklist of issues that is more akin to the scoping phase of an EIA than an environmental audit. The aim is to highlight those activities that may have actual or potential environmental aspects and thus allow further consideration of legal liability and management standards against the key issues.

A PER is typically conducted for a fairly sizable management entity and the detailed consideration of activities and aspects is not normally feasible. Particularly on large sites, further detailed evaluations (e.g. at department level) are often necessary as part of the EMS development. The scope of a PER is defined in relation to:

- the operational/geographical extent;
- the assessment criteria;
- the level of detail.

A typical scope might, therefore, be as follows:

- confined primarily to site-based activities including in-situ contractor operations but excluding supply and distribution issues; assessment criteria to include:
  - legal requirements
  - identification of environmental aspects

- compliance with the requirements of ISO 14001
- identification of environmental risks
- identification of existing environmental management practices
- systems or activities that can enable or impede environmental performance
- assessment to take place at site level only and be confined to the level of detail feasible within a time constraint of 6 man-days.

The detailed coverage of a PER varies in accordance with the required outcomes and the resources available to complete the review. ISO 14004 (the guidance note that accompanies ISO 14001) recommends the following coverage for a PER:

- identification of legal requirements;
- identification of environmental aspects;
- evaluation of performance (against set criteria);
- existing environmental practices and procedures;
- procurement and contracting procedures;
- review of previous incidents;
- opportunities for competitive advantage;
- views of interested parties;
- obstacles and assistance to EMS development.

6.8.5.2 Data review

A very wide variety of data may be considered as part of a PER as illustrated by the examples in Table 6.5.

6.8.5.3 Site investigation

Generally the site investigation phase of a PER focuses on:

- observations related to aspects identification and risk evaluation;
- interviews aimed at assessing awareness of issues and current control mechanisms, whether formal or informal;
- confirmation of compliance with set criteria through a combination of the above;
- the identification of improvement opportunities based on the audit team's experience of best practice.

Table 6.5 Data/documents reviewed during a PER

Aspect/Issue	Documents for data review
Material inputs	Substance Safety Data Sheets, COSHH records, purchasing records
Energy inputs	Supplier invoices, meter reading records
Processes	Process flow diagrams, maintenance records, environmental permits, consents, relevant legislation and guidance notes, solvent inventory data, relevant operational instructions, COSHH risk assessments
Materials handling and storage activities	Site plan showing facilities, emergency procedures and plans, site drainage plans, incident reports
Atmospheric emissions	Monitoring records
Effluent	Discharge consent to sewer, environmental permit to discharge to controlled waters, site drainage plans
Waste	Duty of care transfer notes, hazardous waste consignment notes, environmental permits and/or carrier registrations of site or contractors, certificates of technical competence, contracts with waste carrier/disposer, waste stream tonnage data records, cost records
Nuisance	Complaint records, abatement notices
Product	Output data records, any life cycle studies
General management Issues	Environmental policy statement, organisation chart, site plan, map of surrounding area, training records, correspondence with regulators and other stakeholders, interviews and questionnaires

6.8.5.4 Assessment of findings

In a PER the assessment of findings aims to do the following:

▶ prioritise the environmental aspects identified using a standardised significance assessment methodology;
▶ assess compliance with legislation, EMS specification (often some kind of gap analysis is made here in relation to the requirements of ISO 14001 or EMAS), organisation policy, etc.;
▶ review of current management practices, including those relating to specific operations that have, or could have, a direct environmental impact (waste management, raw materials management and storage, transportation, product and process design and planning, pollution control and incident response) and more general issues (e.g. contractor control, purchasing, training and employee involvement, third party relations). Any existing formal management systems should be evaluated to assess their value in relation to EMS development.
▶ identify recommendations for action linked to the PER objectives – may include specific changes to process or procedure or be an implementation plan aimed at achieving ISO 14001 certification.

6.8.5.5 Reporting

The PER report is normally required to fulfil two functions:

▶ act as a policy guide to senior members of the organisation by clearly presenting identified risks, recommendations and priority in order to facilitate informed decision-making;
▶ act as a reference text/implementation plan for those tasked with acting on the recommendations adopted.

The two functions require rather different reporting styles and for that reason the inclusion of a clear but concise management summary is normally an essential part of PER reporting.

6.8.6 Legal compliance audit

A legal compliance audit is typically undertaken to assess legal exposure or as part of an environmental management system. In some ways a legal compliance audit is an extremely simple assessment – best practice assessment is not normally required and findings are ideally limited to three potential conclusions against relevant legislation: compliant, non-compliant or risk of non-compliance. For a large organisation, however, a thorough compliance assessment is a significant task. The challenge, in such cases, is determining a level of detail and sampling that provides the required level of assurance that all legal requirements have been identified and potential breaches assessed.

6.8.6.1 Scope

Legal compliance assessment of all or part of an organisation's activities. Boundaries may be set by organisation (e.g. department A only), location (e.g. site A only) or legislation (e.g. discharge consents only).

6.8.6.2 Data review

Typically includes a review of process descriptions, operating licences, waste transfer notes and consignment notes, monitoring data relevant to licences (e.g. emissions/discharges monitoring), etc. Also a review of legal coverage and forthcoming legislation/policy relevant to the organisation. Key aim at this stage is to identify compliance requirements.

6.8.6.3 Site investigation

Essentially 'proving' compliance status in relation to the requirements identified during the data review. Also looking for poor control of environmental hazards that could give rise to legal breaches under emergency conditions.

6.8.6.4 Assessment of findings

Audit findings assessment focuses classification under the three key headings of: compliant, non-compliant or risk of non-compliance against key legal requirements.

6.8.6.5 Reporting

A summary assessment record will be compiled recording the compliance status and, if necessary, requesting/requiring improvement actions.

In 2005, the IEMA produced a guide in their practitioner series entitled *Managing Compliance with Environmental Law: A Good Practice Guide*.

6.8.7 Supply chain audits

Supply chain audits vary significantly in the level of detail applied, from a strictly legal compliance

evaluation to a full environmental audit encompassing legal compliance, EMS and best practice assessments. In most cases, the latter is only applied to key suppliers who have consented to enter into an environmental management agreement with the client company.

6.8.7.1 Scope

Varies hugely but may encompass:

- ▶ contractor/supplier policies and action plans relating to environmental performance improvement;
- ▶ contractor environmental performance on site;
- ▶ contractor/supplier performance associated with supply of goods and services dedicated to client (related to activities both on and off the client site).

Product characteristics and raw material sources – this may involve not just the direct supplier but the whole chain of organisations back to the source of primary materials

6.8.7.2 Data review

Dependent on scope and, therefore, ranging from a simple review of key policies and action plans, through to the variety of information sources that might be considered during a PER.

6.8.7.3 Site investigation

A site investigation may, or may not, take place depending on scope but, where it does, it may range from a simple inspection aimed at getting a feel for company standards and practices, to a more formal investigation akin to the PER inspection and interview process.

6.8.7.4 Assessment of findings

Whatever the scope, the assessment is geared to identifying risk to the client organisation. Risk, in this case, may mean a variety of things including:

- ▶ poor standards associated with activities undertaken directly on behalf of the client company;
- ▶ poor standards associated with other activities which, in the event of an incident, may tarnish

the reputation of the client company by association;

▶ significant environmental impacts associated with the raw material supply, processing or distribution of products supplied to the client company.

6.8.7.5 Reporting

Required to fulfil two primary purposes:

▶ reference summary for client company personnel to enable a reputation risk or performance comparison to be made;
▶ to provide the basis for any actions on the part of the supplier/contractor audited as agreed with the client organisation.

A waste management Duty of Care audit is representative of the simplest type of supplier (contractor) audit. Clients wishing to conduct an audit with wider scope than this should ensure that definition and agreement of scope with the supplier/contractor are part of the audit process.

Although a relatively new area of environmental auditing, supply chain audits frequently focus on key stakeholder issues and may be administered under an industry sector scheme. A good example of this would be the Forestry Stewardship Scheme applied by the building materials retailers to suppliers of timber products in an attempt to selectively support companies practising sustainable forestry.

6.8.8 **Waste reviews**

6.8.8.1 Scope

Generally a waste review is the starting point in a waste minimisation programme and as such the scope is, normally, to consider all sources of waste within a defined area and to attempt to identify mechanisms whereby the quantity of waste could be reduced, both in absolute terms and in relation to the proportion sent to landfill.

At whatever scale has been defined, there are generally three stages:

▶ process stream analysis – identifying input and output points;

▶ quantification of input and outputs – mass balance evaluations;
▶ analysis of losses and identification of waste reduction, re-use and recycling opportunities.

6.8.8.2 Data review

The types of data typically referred to during a waste review include:

▶ management reports (especially related to product or process specifications);
▶ production statistics (especially yield rates);
▶ material use reports;
▶ bills of materials and other purchasing records;
▶ by-product and waste disposal records
▶ effluent analysis;
▶ product specifications (outgoing and incoming);
▶ packaging specifications (outgoing and incoming).

6.8.8.3 Site investigation

Varies in complexity from walk-through inspections to in-process trials aimed at quantifying specific losses at particular locations.

6.8.8.4 Assessment of findings

Focus on identifying clear priorities for action ranked on the basis of:

▶ potential savings;
▶ relative ease of saving with reference to time, cost and any associated implications;
▶ nature of material/waste saved (hazardous or non-hazardous, etc.).

6.8.8.5 Reporting

Required to fulfil two functions:

▶ detailed reference to enable clarity of action by those involved in implementing any actions arising;
▶ a clear presentation of opportunities with financial as well as environmental costs and benefits outlined. This is essentially a sales function aimed at securing support for changes required as part of the waste minimisation process.

6.8.9 Energy audits

6.8.9.1 Scope

Generally undertaken as part of an energy-efficiency programme with a typical scope being to obtain an accurate picture of energy consumption including breakdown of costs, usage and efficiencies for the area under assessment.

6.8.9.2 Data review

Typically includes some or all of the following:

▶ utilities bills, including subdivisions and consumption distribution;
▶ site energy records/sub-metering;
▶ equipment design specifications;
▶ equipment operating and shutdown procedures.

6.8.9.3 Site investigation

Depending on the level of detail specified, this may be a simple walk-through to carry out observations of things such as lighting usage, heating control and losses, compressed air leaks etc. Alternatively it may involve temporary sub-metering, equipment surveys, etc.

6.8.9.4 Assessment of findings

The assessment is focused on identifying inefficiencies and hence, opportunities for improvement. Comparison with the DECC best practice guides may provide the basis for assessment in relation to each energy-use group. As with waste, priorities should be identified with reference to:

▶ potential savings;
▶ relative ease of saving with reference to time, cost and any associated implications.

6.8.9.5 Reporting

As for waste auditing the report is required to serve two functions:

▶ detailed reference to enable clarity of action by those involved in implementing any actions arising;
▶ a clear presentation of opportunities with financial as well as environmental costs

and benefits outlined. This is essentially a sales function aimed at securing support for changes required as part of the energy efficiency programme.

6.8.10 Contaminated land (Stage 1) assessments

6.8.10.1 Scope

Generally the aim is to evaluate the likelihood of contamination at a defined site. Assessments are sometimes classified as Stage 1 and Stage 2 investigations. Stage 1 is a desk and visual inspection study aimed at identifying the potential for, and likelihood location of, contamination. Stage 2 is a physical investigation to prove the findings of Stage 1 and determine the nature and extent of any contamination. From an audit perspective the focus is clearly the Stage 1 assessment.

6.8.10.2 Data review

Typically some or all of the following sources of information will be consulted:

▶ current and historic topographic and land use maps;
▶ current and historic site plans;
▶ regulatory registers;
▶ geological and hydrogeological surveys;
▶ any available borehole quality data associated with aquifers potentially connected to the site;
▶ current registers of hazardous material usage and any associated procedures relating to delivery, storage, use and disposal.

6.8.10.3 Site investigation

As a minimum, an observation and discussion-based visit looking for sources of current pollution (chemical/fuel stores, underground tanks, poor site drainage controls, evidence of leaks or spills, etc.) and anecdotal evidence of past sources of contaminants. (Long-term employees can be a valuable source of information where standards have improved considerably over the years of operation.)

6.8.10.4 Assessment of findings

Essentially a risk assessment based on the identification of contaminant sources, likelihood of release and pollution potential through consideration of pathways and receptors. The outcome at this stage can only be an educated 'guess' based on current standards and available historical evidence. Some indication of the level of confidence of assessment may be given based largely on the amount of risk associated with historic contamination.

6.8.10.5 Reporting

By their nature, these assessments require detailed reports that clearly lay out findings and indicate the basis of any conclusions, as well as highlighting assumptions made, and any gaps in the historic record. Recommendations may also be included that relate either to Stage 2 assessment requirements or to remediation works or to actions required to prevent contamination from current or future activities.

6.8.11 Due diligence/ pre-acquisition audits

6.8.11.1 Scope

Normally commissioned prior to the purchase of companies, properties or industrial land or prior to the merger with another company where property or land holdings are involved. The aim is typically to identify environmental liabilities associated with the purchase or merger.

The scale of assessment varies widely but may include:

▶ Actual or potential legal non-compliance in relation to site activities – clearly this will entail some assessment of management controls and current practice.
▶ Actual or potential land contamination – normally a current practice and site history check as a minimum.
▶ Actual or potential polluting activities or structures, e.g. storage of materials, underground storage tanks, bunding, process emissions to air, water and land, etc.

▶ Specific legal obligations and relevant trends, e.g. considerations of Local Authority development plans, consideration of material usage where key process raw materials have been identified as priority substances for use reduction or phase-out, e.g. HCFCs, halons, etc.
▶ Potential litigation or other stakeholder issues – likelihood of nuisance, local community relations, history of relations with regulators, NGOs, etc.
▶ The potential costs associated with any of the above.

6.8.11.2 Data review

May be simply a due diligence inquiry questionnaire completed by the vendor and structured so as to identify potential liabilities and associated risk.

Alternatively, for an existing organisation, the sorts of data reviewed may be similar to the PER to identify the full environmental implications associated with the company, property or site.

6.8.11.3 Site investigation

Again, may range from a simple visual assessment to a full-blown PER style investigation and, in some instances, even an intrusive survey to confirm the presence of contaminated land.

6.8.11.4 Assessment of findings

A risk assessment linked to the actual or potential liabilities identified. Attempts should normally be made to identify the associated costs and foreseeable developments in relation to policy or legislation that may have implications for future liabilities. ISO 14015 has been produced as a guide to conducting due diligence audits and may be used as a reference standard.

6.8.11.5 Reporting

As with the contaminated land assessment reports, these are likely to significantly influence major commercial decisions. They must, therefore, be comprehensive and transparent as well as clear and, if possible, concise.

Conclusions, justifications, and any recommendations made must be supported by evidence and legal references. Any limitations or outstanding risks must be clearly identified.

III ENVIRONMENTAL MANAGEMENT SYSTEMS

6.9 Advantages and disadvantages of a systems approach

Environmental Management Systems were developed in the 1990s in response to the pressures on business to have effective systems in place to reduce the environmental impact from their business activity, and to be able to demonstrate this commitment to third parties.

6.9.1 Advantages of the EMS approach

Business benefits from effective environmental management include:

6.9.1.1 Licences to operate

▶ Assist in obtaining authorisations, consents and permits.
▶ Avoid enforcement action.
▶ Improve business and the regulatory relationship.
▶ Maintain and enhance community relations.

6.9.1.2 Market size and share

▶ Assure customers of commitment to responsible environmental practices.
▶ Identify customer concerns and opportunities for developing and sharing environmental solutions.
▶ Meet or exceed vendor certification criteria.
▶ Develop products and services which will improve the environmental performance.
▶ Retain and/or improve market share.
▶ Create new products and new market opportunities.

6.9.1.3 Cost control

▶ Improve process and operational efficiency.
▶ On-going savings in materials, energy and waste costs.
▶ Avoid the liability costs through criminal proceedings and/or civil actions.

6.9.1.4 Access to financial services

▶ Satisfy investor criteria.
▶ Obtain insurance at reasonable costs.

6.9.1.5 Public image

▶ Enhance public image.
▶ Demonstrate responsible care.

Effective environmental management should be about maximising environmental and business benefits, and, therefore, about turning threats associated with environmental developments, into business opportunities. An effective, well-implemented and accredited environmental management system will assist in this process by ensuring a consistent, comprehensive and planned approach to control and improvements, as well as providing the on-going focus and credibility of third party assessment. It will also help (though not necessarily guarantee) that standards are maintained when key company personnel change.

6.9.2 Disadvantages of the EMS approach

Like every system, there are also potential drawbacks and these should be considered and balanced with the potential advantages before an organisation commits to the implementation of an Environmental Management System.

In the case of an EMS, the drawbacks focus on additional operational costs which may occur in relation to:

▶ preparation and implementation of the EMS system itself, for example, the proprietary review project co-ordination costs;
▶ on-going management administration of the EMS system, e.g. Document Control and auditing;
▶ investment in the continuous improvement process, e.g. training and cleaner technology.

It should be noted that, in most well-implemented systems, these costs are significantly exceeded by the business benefits in both direct and indirect commercial terms. It should also be noted that EMS development is for most companies a one-way street – business standards improvements, once developed, must be kept up. This involves an on-going commitment and cost, as well as building an expectation of performance improvement. While not necessarily a disadvantage, this is an issue that should be recognised prior to embarking on an EMS certification programme.

6.10 EMS standards and schemes

Figure 6.9 presents an EMS model based on ISO 14001.

The two most important Environmental Management Schemes are:

▶ ISO 14001 Environmental Management System
▶ EMAS Eco Management and Audit Scheme.

Both of these standards are described in more detail below.

6.10.1 ISO 14000 Series

ISO, the International Organisation for Standardisation, which is based in Geneva, is responsible for the International Standardisation Process. ISO has been working on a series of standards on Environmental Management Tools & Systems through a number of Technical Committees (TCs).

The ISO 14000 Series of standards relevant to Environmental Management are constantly being added to but include those shown in Table 6.6.

ISO 19011 is the international standard providing combined guidance on quality and environmental management systems auditing. Of all these standards, companies may only be assessed and certified against the requirements of ISO 14001. ISO 14001 was revised and reissued in 2004. No significant changes were made but there was a tightening up in a few areas, notably:

▶ definition of EMS scope;
▶ coverage of competency requirements for those with EMS functions;
▶ periodic legal compliance assessment.

A second revision is currently underway with publication due in 2015.

6.10.2 EC Eco Management and Audit Scheme

This scheme was introduced by EC Regulation in 1993 and has been revised and reissued twice

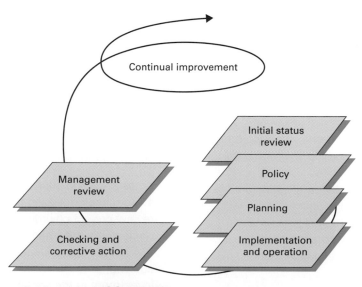

Figure 6.9 An EMS model (based on ISO 14001)
Source: ©Permission to reproduce extracts from ISO 14001 granted by BSI.

Table 6.6 The ISO 14000 series

Number	Title
ISO 14001	EMS: a specification with guidance (against which certification takes place)
ISO 14004	EMS: general guidelines on principles, systems and supporting techniques
ISO 14010–13	Environmental Auditing – now superseded by ISO 19011 (joint EMS/QMS auditing guidelines)
ISO 14015	Environmental assessment of sites and organisations – EASO (due diligence or initial reviews)
ISO 14020–25	Environmental labels and declarations
ISO 14031	Environmental Performance Evaluation – guidelines for environmental management
ISO 14040–43	Life cycle analysis
ISO 14050	Environmental Management: Vocabulary
ISO 14062	Integrating environmental aspects into product design and development
ISO 14063	Environmental communication guidelines and examples
ISO 14064	Standards for greenhouse gas accounting and verification

in 2001 and 2009. Participation in the scheme is entirely voluntary and is not a legal requirement.

ISO 14001 specifies requirements for development, implementation and maintenance of Environmental Management Systems. It does not include certification provisions, but provides a benchmark against which certification can be performed.

EMAS encompasses a total scheme, which in addition to specifying requirements for environmental management systems, requires independent verification of compliance and the production of a public environmental statement. The *EMAS Statement* must include the following information:

▶ a description of the company's activities;
▶ an assessment of all the relevant significant environmental issues;
▶ a summary of figures on emissions; waste arising; consumption of raw materials, energy and water; noise and other significant aspects;
▶ other factors regarding environmental performance and policy;
▶ the deadline for submission of the next statement;
▶ the name of the accredited environmental verifier;
▶ any significant changes since the previous statement.

Such statements are to be prepared at least every three years, with simplified statements usually required annually.

Although originally significant differences existed between the EMS requirements of the two standards, the revisions of EMAS have brought them closely in line. The EMS element of EMAS is now aligned to ISO 14001. The key difference remains the environmental statement and the verification standards associated with its production. Globally, ISO 14001 has proved by far the most popular standard, largely due to a general reticence to publish environmental performance data on an annual basis. Pressure from governments and other stakeholders for organisations to report on environmental performance may, however, mean a strengthening of interest in EMAS in the future.

6.11 Main elements of environmental management standards

Both principal Environmental Management Systems are based on the same model. The content is consistent with the 'Plan – Act – Check – Review' cycle common to all formal management systems. With EMAS, as indicated above, there is the additional stage of public reporting.

An outline of the implementation requirements typically associated with each stage is provided in Figure 6.9 and Table 6.7.

The development of an EMS in an organisation normally follows the format of the ISO 14001 model with an assessment phase,

Table 6.7 The main elements of an EMS

Principal stage	Main components
Commitment and policy	Top management commitment
	Completion of an initial environmental review
	Development of an outline environmental policy
Planning	Identification of environmental aspects and impacts and their relative significance
	Identification of legal and other requirements
	Confirmation of environmental policy
	Identification of internal performance criteria and associated objectives and targets
	Development of a management programme aimed at achieving the objectives and targets (including EMS implementation)
Implementation and operation	Definition of EMS structure and responsibilities including the identification and provision of physical, human and financial resources
	Ensure that employee awareness, motivation and competence is commensurate with their environmental management role and/or potential impact – this normally includes completion of a site environmental awareness programme
	Establish effective communication and reporting procedures
	Develop appropriate documentation and document control processes
	Develop operational control and emergency response procedures that focus on the significant environmental aspects
Checking and corrective action	Develop monitoring and performance evaluation procedures linked to key performance indicators
	Establish non-conformance, corrective and preventative action methods
	Identify and ensure the effective maintenance and storage of EMS records
	Establish an internal audit programme
Review and improvement	Establish a mechanism for review and improvement of the EMS that has as its central goal the continual improvement of the organisation's overall environmental performance

an implementation phase (which normally includes an extensive training and awareness programme) and a monitoring phase linked to a management review and subsequent external certification. A common approach is to appoint 'environmental champions' in logical operational areas to lead the EMS implementation process. Apart from the need to spread responsibilities across the organisation, this approach can help to promote a two-way flow of information, ideas and recommendations regarding environmental management and improvement opportunities.

6.12 Consideration of the requirements of the standard

It may be useful to consider the requirements of each element of the standard in the context of an EMS implementation plan. Although this is presented as a single simplified plan, in reality,

in many organisations, implementing an EMS involves the co-ordination of numerous small projects, each of which is a component of the larger whole. The co-ordination role tends to fall to the Environment Manager who guides and co-ordinates the inputs made by an EMS implementation team made up of representatives from the different operational areas of the organisation.

6.12.1 An implementation plan

The development of an EMS may be broken down into key steps that relate to the corresponding ISO 14001 elements. The steps relate to the development of the whole EMS over a period of time and, therefore, describe a plan to arrive at the on-going cycle of actions covered by the standard. The reason for this description is that some confusion may arise from directly adopting the ISO 14001 model as an implementation plan. For example, it is better to write the company

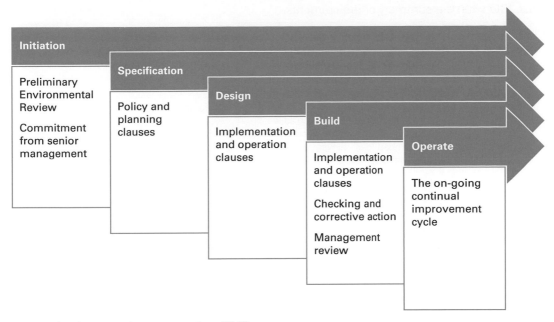

Figure 6.10 Implementation stages of an EMS

environmental policy having first completed the aspects evaluation and legislative review. It is useful, therefore, to think in terms of the steps shown in Figure 6.10.

For simplicity of presentation, however, the individual clauses of the standard are considered in Table 6.8 in the order presented in ISO 14001: 2004. It should be noted that the requirements of the standard have been paraphrased and summarised in these notes. Readers should refer to the full text of the ISO 14001 standard for implementation and certification reference.

6.13 Accredited certification and verification

The UK national accreditation body responsible for monitoring certification standards to both ISO 14001 and EMAS is UKAS – the UK Accreditation Service. UKAS sets out both accreditation (i.e. how organisations become certification bodies) and certification criteria (i.e. how certification bodies certify organisations to ISO 14001).

The requirement for assessment placed upon certification bodies such as the British Standards Institute, SGS Yarsley, etc. by UKAS is that

certification should follow a multi-stage approach. Normally this will require at least two stages: initial assessment and main assessment.

6.13.1 Initial assessment

The objectives of the initial assessment are to 'provide a focus for planning the main assessment by gaining an understanding of the EMS and the client's state of preparedness' by, in particular:

▶ ensuring that the structure of the EMS is based on an evaluation of the environmental effects, development and management of objectives and targets, audit, review and policy, and that the structure is auditable;
▶ assessing the reliance that can be placed on document review and determining whether this can be best performed on or off site;
▶ assessing the reliance that can be placed on the internal audit;
▶ planning and allocating resources for the document review and main assessment stages;
▶ providing an opportunity for immediate feedback of information to the client which may assist in the remainder of the assessment process.

Table 6.8 ISO 14001: a summary of requirements

Element/clause	ISO 14001 Requirements	Implementation tips
4.1 General requirements	An organisation must identify its purpose for EMS development, consider its 'sphere of influence' and then document the scope of its Environmental Management System.	Scope of EMS relates to activities, services and products covered by the EMS – we cannot choose to simply exclude 'difficult' areas.
4.2 Environmental policy	A policy statement is required that meets a set of required principles such as legal compliance and pollution prevention. It must be communicated to all relevant personnel and be publicly available.	Keep it short and as specific as possible. Communication to employees does not have to be verbatim – deliver it in the most appropriate way. It should be signed and dated by the MD, CEO, etc. to demonstrate commitment.
4.3.1 Environmental aspects	A procedure to identify environmental aspects and assess those that do or could cause a significant environmental impact. Organisations must consider all aspects that fall within the EMS scope. The procedure should also include a mechanism for keeping this information up to date.	Make the significance assessment criteria appropriate to the business. In practice most people keep a 'register' of aspects (even though this is not expressly required).
4.3.2 Legal and other requirements	A procedure to identify relevant legislative requirements and other commitments that the organisation commits to complying with	Make the register a useful reference for training and awareness purposes rather than just a list. Consider your mechanism for keeping up to date.
4.3.3 Objectives, targets and programmes	Documented objectives and targets that focus on significant environmental aspects and, if possible, quantifiable key performance indicators (KPIs). Have a programme for achieving the objectives and targets including designation of action, responsibility and time scales.	Make objectives and targets 'SMART' (specific, measurable, achievable, responsibilities allocated and time-bound). In addition to the criteria set by the standard, give priority to targets where you suspect significant improvements can be easily achieved – 'pick some cherries' to get the process off to a popular start! Include a mechanism for tracking progress against the plan and ensure that large tasks are broken down into sub-tasks so that progress slippage can be detected early.
4.4.1 Resources, roles, responsibility and authority	Roles, responsibilities and authorities must be defined, documented and communicated. All those with responsibilities must have sufficient time and financial resources to fulfil them. A management representative must be appointed to ensure compliance with the standard and report on performance to top management.	As well as specific roles, ensure that general roles are set for managers, supervisors and all employees – the concept should be that everyone has a part to play.
4.4.2 Competence, training and awareness	Training needs should be identified across all employee groups in order to demonstrate the competence of all those working on the organisation's behalf. Particular attention should be given to those whose activities may cause a significant environmental impact or have a specific role in the EMS. A basic level of awareness should be ensured in all employees.	Make training appropriate to the audience rather than applying a blanket approach. Don't forget on-site contractors who may be 'acting employees' in terms of the EMS. Demonstration of competence allows for non-training proof.
4.4.3 Communication	Establish a procedure for communications within the organisation and with relevant third parties.	Consider whether information relating to key performance indicators will be communicated externally and document decision.

Table 6.8 (Continued)

Element/clause	ISO 14001 Requirements	Implementation tips
		Consider a network approach to internal communications – don't underestimate how important this is to long-term success.
		Ensure communications procedures take account of two-way traffic and out-of-hours arrangements (particularly in relation to complaints).
4.4.4 Documentation	The EMS documentation must include – policy, objectives and targets, a manual of some kind, plus procedures and records deemed to be essential to the running of the system.	Tie in to existing procedures if they exist. Keep all documents short and clear to make them useful as a reference. Carefully consider distribution to ensure that all remain 'live' documents and not simply 'dust-gathering tomes'.
4.4.5 Control of documents	A procedure to ensure that documents are adequate for use, kept up to date, clearly legible, etc. (in line with wording in ISO 9000:2008).	Link into existing document control regimes wherever possible.
4.4.6 Operational control	Development of procedures/work instructions where their absence could lead to breaches of environmental policy and/or impede the achievement of the objectives and targets. It may be appropriate to develop procedures relating to significant goods and services used by the organisation and communicate requirements to suppliers and contractors.	Consider normal, abnormal and emergency conditions. Link any procedures to training and awareness programmes on a continuous basis. Don't overdo written procedures – simple and to the point is best. Involve as many people as possible in preparing the procedures especially including those who will have to follow them.
4.4.7 Emergency preparedness and response	Establish a procedure to identify potential emergency situations and potential accidents, set out response arrangements and review processes that will be used following an incident. The procedure should be periodically tested where practicable.	Link to Health and Safety plans where possible, but note that expansion of existing procedures is almost certainly required and includes the identification of emergency scenarios.
4.5.1 Monitoring and measurement	Establish a documented procedure to monitor key performance criteria. Ensure calibration and maintenance of any monitoring equipment used.	Make key performance indicators meaningful and, where possible, local enough to communicate regularly and widely as part of an on-going awareness programme.
4.5.2 Evaluation of compliance	Establish, implement and maintain a procedure(s) for periodically evaluating compliance with applicable legal requirements and other requirements to which the organisation subscribes. These compliance audits must be documented.	May be conducted as part of the internal audit process or as a separate, stand-alone evaluation.
4.5.3 Non-conformity and corrective and preventive action	Establish procedures for defining, handling and investigating actual and potential non-conformities, including actions to mitigate impacts and initiate corrective and preventive action	Two suggestions depending on organisation 'culture': • Link to Quality management system procedures • Link to H&S improvement and prohibition style notices In both cases avoid excessive paperwork but always keep to internal standards set.
4.5.4 Control of records	Establish procedures for the identification, maintenance and disposition of environmental	Link to operational control procedures wherever possible.

Table 6.8 (Continued)

Element/clause	ISO 14001 Requirements	Implementation tips
	records including training records, audits and reviews, as well as KPI data.	Keep a central register of key records and their location.
4.5.5 Internal audit	The minimum requirement is an audit procedure which determines whether the EMS conforms to stated planned arrangements. It should be an impartial process and cover all the EMS but prioritise high impact activities.	Chose and train your internal auditors carefully – they are the EMS ambassadors. Try to develop a positive audit culture where there is as much emphasis on identifying opportunities for improvement as evidence of non-conformance.
4.6 Management review	Top management is required to conduct a periodic review of the EMS to ensure its continuing suitability, adequacy and effectiveness. Reviews should cover a defined list of topics including audit results, stakeholder communications, KPIs and progress against objectives and targets, conformance status and changing circumstances that may influence the EMS.	Ensure that reviews are well structured and consider EMS improvements as well as performance against the objectives and targets. Consider incorporating a positive feedback mechanism, e.g. an award scheme for 'best environmental action' for discussion as part of the review.

6.13.2 Main assessment

The objectives of the main assessment will be to determine whether the EMS is designed to achieve, and is capable of achieving, performance improvement and ensuring regulatory compliance. It must also ensure that the client complies with his own policies and procedures. To do this, the assessment will focus on:

▶ the client's identification and evaluation of environmental effects;
▶ the client's consequent objectives and targets;
▶ the client's performance monitoring, measuring, reporting and review against the objectives and targets;
▶ the client's auditing to ensure compliance.

Additionally it will be necessary for the certification body to satisfy itself that:

▶ All staff have received awareness training regarding the company's environmental effects, objectives and the system.
▶ All key staff (those involved in managing significant effects) must have had training linked to a needs analysis.

If the certification body is satisfied that all elements of the EMS standard have been met, the organisation will be awarded the relevant certification and be able to promote themselves as a certified organisation.

6.13.3 Surveillance and re-certification

Once certification has been received, the surveillance audit programme commences. This typically involves once or twice yearly visits by the auditors, usually of shorter duration than the main assessment. The actual schedule will vary depending on the scale and complexity of the organisation and is agreed between the certification body and the client. Every three years, a full re-certification audit is required and this follows the same sort of format and depth as the main assessment.

6.14 Integrated management systems

In recent years we have seen increasing interest in integration of the various management systems relating to environmental management, occupational health and safety, quality, energy, business continuity, etc. There are now many international standards on which an organisation can model its own systems and then seek certification, namely:

ISO 14001 Environmental Management Systems

ISO 9000 Quality Management Systems

OHSAS 18001 Occupational Health and Safety Management Systems

BS 25999 Business Continuity Management

ISO 50001 Energy Management

ISO 27001 Information Security

SA 8000 Social Accountability

AA1000 Stakeholder Engagement

BS 8900 Sustainability Management Systems

. . . and the list continues to grow!

In practice, it is the environmental, health & safety and quality systems where most effort has been made to link systems (and certification processes) together. The main driver is the view that integration is beneficial in terms of reduced duplication and maintenance costs as well as clarity of business standards. A number of organisations have developed their own integrated HS&E management systems, but these systems, while sharing some common features, are generally not transferable from one organisation to another.

The Chemicals Industries Association has published its own guide entitled 'Responsible care management systems for health, safety and the environment'. This guide supports the approach of integrating both systems. The Health and Safety Executive, the Environment Agency and the Institute of Chemical Engineers support this guide.

It is believed that in time, an internationally accredited, integrated standard will be developed. In the meantime, the British Standards Institute has produced a Publicly Available Specification (PAS 99) as guidance. In addition, though no single certification is available, many certification bodies have procedures to combine/ streamline certification processes for integrated management systems.

6.15 A phased approach to EMS implementation

6.15.1 BS 8555/ISO 14005

▶ BS 8555: 2003 Environmental Management Systems – Guide to the Phased Implementation of an Environmental

Management System including the use of Environmental Performance Evaluation.

▶ 1SO 14005: 2010 Guidelines for the phased implementation of an environmental management system, including the use of environmental performance evaluation.

These two very similar standards were developed primarily but not exclusively for smaller organisations that may find full EMS implementation a daunting task for any, or all, of the following reasons:

▶ limited resources;
▶ difficulty in effectively implementing the requirements of ISO 14001;
▶ costs involved in implementation.

The standards outline an implementation process that can be undertaken in up to six separate phases, and allows for phased acknowledgement of progress towards full EMS implementation. This approach, in addition to helping deal with human and financial resource restrictions, allows organisations to do the following:

▶ to choose to engage in environmental management to a level commensurate with the environmental and business risks that they face;
▶ to identify and focus on those areas providing greatest dual potential in terms of environmental management and financial reward;
▶ to demonstrate to interested parties that progress is being made to the target level of environmental management.

The standard may also provide a structure for larger organisations with full EMS to improve the environmental performance of their supply chain through target setting and implementation support.

The six phases and associated activities (or stages) are illustrated in Table 6.9. Note that Table 6.9 is a summary only and readers should refer to ISO 14005 for a complete view of certification requirements. An external audit may be performed following each phase to determine whether sufficient objective evidence exists that the phase has been completed. Where performed by competent and accredited third

Table 6.9 The stages of BS 8555/ISO 14005

Phase	Associated stages/activities
Phase 1 – securing commitment and establishing the baseline performance	Gaining and maintaining management commitment
	Conducting a baseline environmental assessment
	Developing a draft environmental policy
	Developing environmental indicators
	Developing an initial EMS implementation plan
	Completing a programme of awareness raising and environmental training as a first step in initiating culture change
	Initiation of a continual improvement programme
Phase 2 – identifying and ensuring compliance with legal and other commitments to which the organisations subscribes	Identifying relevant legal requirements
	Identifying 'other' requirements, e.g. trade association standards, customer requirements
	Checking and ensuring compliance with such requirements
	Developing compliance assessment methods and indicators which may be used as objective evidence to demonstrate compliance
Phase 3 – developing objectives, targets and programmes	Evaluation of environmental aspects using a defined significance assessment methodology
	Finalising the environmental policy
	Developing objectives and targets
	Establishing indicators for environmental performance evaluation
	Developing an environmental management programme to ensure delivery of the objectives and targets set
	Developing operational control procedures
	Communicating policy, objectives, targets and environmental indicators to all relevant parties
Phase 4 – implementation and operation of the EMS	Finalise management structure and responsibilities
	Implement training, awareness and competence plans and records
	Establish and maintain formal communication methods
	Develop documentation and record-keeping procedures
	Review and test emergency preparedness and response
	Identify and monitor indicators relating to the effectiveness of the management system
Phase 5 – checking, audit and review	Establish audit programmes
	Develop arrangements for dealing with corrective actions and EMS non-conformances and for initiating preventative action
	Develop a management review process
	Work towards improving environmental performance and the EMS
Phase 6 – EMS acknowledgement	Prepare for third party assessment of the EMS against ISO 14001/EMAS or supply chain standards
	Develop reportable information that can provide objective evidence of EMS operation
	Audit the EMS and the organisation's overall environmental performance

parties, these audits may be used as interim 'certifications'.

Guidance is provided on how to achieve each phase and sets criteria against which an organisation should assess itself to determine whether it has met the requirements. At each phase the activities described are linked to the clauses of ISO 14001 so that organisations can clearly see how the EMS development is progressing.

There are several certification schemes emerging that use BS 8555 as the basis of a phased accreditation programme. Two examples follow.

6.15.2 The IEMA Acorn Scheme

The IEMA Acorn Scheme offers accredited recognition for organisations evaluating and improving their environmental performance through the phased implementation of an environmental management system (EMS). Acorn focuses on environmental improvements that are linked to business competitiveness and is flexible so that all types of organisation, whatever their size, can participate.

IEMA promotes the approach, claiming that the IEMA Acorn Scheme offers participants the following benefits:

- EMS implementation broken down into a series of logical, convenient, manageable phases using the British Standard BS 8555;
- clearly defined route plan to ISO 14001 certification or EMAS registration, with UKAS accredited recognition along the way;
- self-determined time frame for implementation – you decide how fast and how far you want to progress;
- environmental performance improvement which can lead to reduced costs and improved business efficiency;
- opportunities to engage supply chain partners in environmental performance improvement;
- entry on the IEMA Acorn Register.

Participation in the scheme requires an organisation to do the following;

- implement one or more phases of BS 8555;
- have this confirmed by an independently accredited Acorn Inspection Body;
- demonstrate continuous environmental improvement to the Acorn Inspection Body on an annual basis, to maintain registration.

Further information on the Acorn Scheme is available from the IEMA website: www.iema.net.

6.15.3 The Green Dragon Scheme

Developed in Wales by Arena Network and now run by Groundwork Wales, this scheme follows the same pattern as project Acorn with third party accreditation to five stages of EMS implementation, namely:

- commitment to environmental responsibility;
- complying with legislation;
- managing environmental impacts;
- environmental management programme;
- continual environmental improvement.

Further information is available at www.greendragonems.com.

IV RESOURCE EFFICIENCY SURVEYS

Wherever organisations engage in projects that are aimed at reducing wastage of materials or energy, some kind of resource efficiency survey is normally conducted to understand how a resource is used and where there are opportunities for savings. The earlier programmes were materials-focused and a common term used for these surveys was 'mass balance calculations'. The same basic principles apply to energy usage assessments, however, hence we will consider them both under the common heading of 'resource efficiency surveys'.

The basic principle is that for any material or energy stream we should be able to calculate the total input and then track the passage and destination of that resource through our system. Consider the example in Box 6.2.

How far we go in terms of detailed breakdown of usage patterns will depend on the complexity of the organisation and also the likely savings available, i.e. there is little point in sub-dividing and tracing usage and losses of relatively insignificant quantities. Having drawn a picture of usage and flows/losses, we then begin the process of quantification. Input data and some waste data will be relatively easy to obtain while uncaptured (fugitive) emissions may require calculation or estimation based on what's left when we've accounted for all other routes. Essentially we are looking for losses from the system that may represent opportunities for savings. In the solvents usage scenario above, for example, if our calculations seemed to point to a significant percentage of incoming solvents being lost as uncaptured emissions, then it would be worth looking carefully at enhancing our storage and recovery processes.

Box 6.2 Solvent usage

A company uses significant quantities of a solvent within its manufacturing process. The routes through the company's operations may be as shown in Figure 6.11.

A similar picture could be drawn for any material stream coming into an organisation, including different energy sources such as electricity, gas and fuel oil.

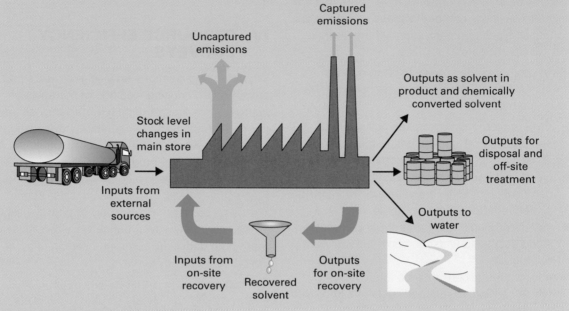

Captured emissions

Uncaptured emissions

Outputs as solvent in product and chemically converted solvent

Stock level changes in main store

Outputs for disposal and off-site treatment

Inputs from external sources

Outputs to water

Inputs from on-site recovery

Recovered solvent

Outputs for on-site recovery

Figure 6.11 An example of the components of a mass balance calculation
Source: Envirowise eco-efficiency guide.

These sorts of resource efficiency surveys have been completed for material streams of many kinds, water usage, energy usage (as a whole or in important sub-divisions such as compressed air usage). Sometimes it is prompted by consumption data and/or cost (e.g. energy usage) and sometimes by waste data or cost (e.g. hazardous waste disposal costs). Wherever in the system we start from, the basic principles are the same:

▶ map out the material or energy usage through the system;

▶ use input data, product/equipment specifications, internal monitoring data, etc. to quantify the key resource movements;

▶ calculate or estimate gaps in the data with as high a degree of accuracy as possible;

▶ look for excessive usage points or pathways and then investigate to see if opportunities for resource efficiency improvements exist.

In Chapter 7 we will consider some of the strategies used to improve resource efficiency that are often the follow-on from the survey work described above.

V ENVIRONMENTAL (ECOLOGICAL) FOOTPRINTING

Environmental footprinting is a technique used to communicate the overall impact of an organisation, country or indeed the whole of humanity in an easily understood and graphic way. Essentially an environmental footprint is the land area calculated as necessary to supply the resources consumed and absorb the wastes generated by an organisation or population. The principle can be used at individual, community, national or even global levels. The calculations are achieved by various methods that include land use requirements for crop production, carbon dioxide emissions, landfill of waste, and so on. In other words the resource demands of the organisation are 'normalised' to an area of productive land.

The Worldwide Fund for Nature has done a lot of work on environmental footprinting at regional and global levels and produces an annual living planet report which summarises its calculations. The UK is currently 15th in the Top 20 consuming nations of the world, and if everyone had an impact – a 'footprint' – as big as the average person's in the UK, we would need three planets to support us.

The 2012 Living Planet report highlights in simple terms the lack of sustainability of current population and resource use trends globally:

> Our demand on the Earth – as represented by our Ecological Footprint – is growing, even as the Earth's capacity for sustainable production, and its ability to absorb CO_2 emissions – its biocapacity – is coming under increasing pressure

At our current rate of consumption, the Earth needs 1.5 years to produce and replenish the natural resources that we consume in a single year. The Living Planet Report 2012 reports an alarming rate of biodiversity loss, with the Living Planet Index showing that, world-wide, biodiversity decreased by around 30 per cent between 1970 and 2008.

The WWF is encouraging the use of footprinting to inform personal awareness and actions in its membership and interested parties. The concept is also promoted to organisations and governments by the WWF under the banner of 'One Planet Living', i.e. where the biocapacity of the Earth is not exceeded. The vision of One Planet Living is a world in which people everywhere can lead happy, healthy lives within their fair share of the Earth's resources (for more information, see www.oneplanetliving.org). One Planet Living is based on a set of 10 guiding principles. These are:

▶ Zero Carbon
▶ Zero Waste
▶ Sustainable Transport
▶ Local and Sustainable Materials
▶ Local and Sustainable Food
▶ Sustainable Water
▶ Land Use and Wildlife

▶ Culture and Community
▶ Equity and Local Economy
▶ Health and Happiness.

Some organisations, e.g. Citigroup, have also used this approach as a way of communicating their overall impact and performance trends to stakeholder groups.

Although as a calculation-based process, the devil is in the detail, and there are bound to be areas where accuracy is less than 100 per cent, as a communication tool, the approach can be very powerful. Perhaps the greatest clarity is achieved when the approach is used for single issues only e.g. carbon/energy footprinting. This area is considered further in the 'carbon footprinting' section below.

You can use an on-line quiz to calculate your individual ecological footprint at www.myfootprint.org. Another good source of information relating to footprinting, at all levels of detail from global to personal, is the Global Footprint Network: www.footprintnetwork.org.

VI ENVIRONMENTAL IMPACT ASSESSMENT (EIA)

This section expands the generic environmental assessment issues discussed in Section I in terms of the specific procedures and coverage associated with EIA as established in UK legislation and government guidance.

In essence, EIA is a planning tool that examines the environmental consequences of development actions in advance. The emphasis, compared with many other mechanisms for environmental protection (including environmental audit), is on prevention. The process involves a number of steps described below. It should be noted that, though outlined in a linear fashion, EIA should be a cyclical activity, with feedback and interaction between the various steps. Indeed, it is often referred to as an 'iterative process', i.e. make a plan, test it, modify it, test again, etc. until we are happy with the results.

The order of the following steps may also vary.

> *Project description and definition of assessment area* – includes the consideration of project alternatives (in terms of design and

location) and, therefore, seeks to ensure that the developer has considered other feasible approaches, including alternative project locations, scales, processes, layouts, operating conditions and the 'no action' option. The description of the project/development action should include clarification of the purpose and rationale of the project, and an understanding of its various characteristics – including stages of development, location and processes.

Impact scoping – seeks to identify at an early stage, from all of a project's possible impacts and from all the alternatives that could be addressed, those that are the key, significant issues.

Baseline environmental descriptions – include the establishment of both the present and future state of the environment, in the absence of the project, taking into account changes resulting from natural events and from other human activities.

Identification of key aspects – bring together the previous steps with the aim of ensuring that all potentially significant environmental impacts (adverse and beneficial) are identified and taken into account in the process.

Impact assessment – aims to identify the magnitude and other dimensions of identified change in the environment with a project/action, by comparison with the situation without that project/action. This stage also aims to assess the relative significance of the predicted impacts to allow a focus on key adverse impacts.

Mitigation proposals – involves the introduction of design changes or mitigation measures to avoid, reduce, remedy or compensate for any significant adverse impacts. This stage may involve public consultation and participation to ensure the views of interested parties are adequately considered in the decision-making process.

Residual impacts – involves definition of the remaining impacts assuming successful implementation of the proposed mitigation measures.

Consideration of cumulative impacts – having established the residual impacts associated with the development, it is important to consider any additional issues that may be created or exacerbated through the interaction of the project impacts with other activities in the vicinity. Cumulative impacts may occur in relation to projects occurring at the same time as the proposal subject (e.g. traffic pressure from the project being assessed may seem to be acceptable but when considered in combination with a road improvement scheme planned to take place at the same time, the impact may be considerably increased). Alternatively, sequential projects may exacerbate nuisance impacts that individually may be insignificant, e.g. increased traffic loads or noise levels from a series of projects/activities.

Preparation of the environmental statement (ES) – this is a vital step in the process – if done badly, much good work in the EIA may be negated. DEFRA guidance exists as to what should be included in an ES. The key requirement is that it contains all relevant information but remains easily accessible to all readers. A common approach, therefore, is a three-section 'summary – main report – technical appendix' format which provides increasing level of detail and allows individuals to review the assessment results in the format most suited to their interest and expertise.

An environmental management system for the project life cycle – the EIA process may be considered the environmental management element of project planning and design. For many developers, the process is now driven, not primarily by legislative requirements, but by a desire to save time and money and avoid delay and conflict with planning authorities and local stakeholders.

Also, there is increasing emphasis on the links between the EIA process and environmental management during the construction and operational phases of a development. Indeed, the *Offshore Petroleum Production and Pipelines (Assessment of Environmental Effects) Regulations 1999 (as amended)* make specific reference to the establishment of a comprehensive, externally verifiable, environmental management system (EMS) for the lifetime of the project. The accompanying guidance notes to these regulations suggest that the details of such an EMS, and the mechanism for its periodic review in the light of experience and technological changes, should be included in the Environmental Statement. This approach highlights the need for transferring knowledge and commitments through the life cycle of a development in a consistent and reliable way. This is particularly important considering that generally there are completely different groups of people involved at different stages, e.g. between planning, construction and into operation.

VII STRATEGIC ENVIRONMENTAL ASSESSMENT (SEA)

The process of environmental assessment has been described in Section VI in relation to individual developments. There has, however, been a recognition for some time that the principles of EIA (and in particular its application as an iterative tool to guide project design) could, and should, be applied to the process of planning and policy-making. This is particularly appropriate at national and local government level but is also being considered by individual organisations interested in finding ways of moving further towards sustainability or at least continual environmental improvement.

The UK government has expressed its commitment to the SEA approach through its Greening Government Initiative which aims to ensure that:

> the environment is at the heart of decision making, throughout Government, at all levels, from the start of the development of policies and right through plans, programmes and projects.

Government guidance has been produced relating to the consideration of environmental impacts in policy formulation. How much has actually been achieved to date is debatable but the radical increase in public consultation is indicative of an approach that is seen as encouraging stakeholder participation and with it, the representation of environmental issues. Similarly, the UK Sustainable Development Strategy stresses

integrated policies which combine environmental, economic and social objectives.

The *EU Strategic Environmental Assessment Directive* (2001/42/EC), discussed in the development control legislation section in Chapter 3, formalises the extension of current UK environmental assessment requirements from projects to national, regional and local plans and programmes. The Directive has been implemented in the UK via the *Environmental Assessment of Plans and Programmes Regulations 2004* and is aimed at creating a 'framework for future development consent' for 11 specified sectors, namely:

▶ Agriculture

▶ Forestry

▶ Fisheries

▶ Energy

▶ Industry

▶ Transport

▶ Waste

▶ Telecommunications

▶ Tourism

▶ Planning

▶ Land use.

Note, with the exception of water companies in England and Wales, all those required by the Directive to conduct SEA are public bodies, not private organisations. Furthermore, requirements for SEA in the sectors mentioned above are only required where the potential for significant effects on the environment exist in relation to the plan or programme.

UK guidance has been produced to help those responsible meet their obligations. The 'Practical Guide to the SEA Directive' states that assessments are to have several components:

▶ *Production of an environmental report.* Those responsible for the development of the plan will be required to produce an environmental report setting out the significant effects of their plan or programme, reasonable alternatives to it, and measures envisaged for mitigating adverse environmental impacts and monitoring implementation. The scope and detail of the environmental report are to be determined after consultation with statutory authorities (in the UK, it is a similar list to that required for a project-based EIA).

▶ *Public consultation.* The environmental report and plan must be made available for public view and comment. Consultation with neighbouring EU member states may also be required where trans-boundary impacts may be involved.

▶ *Taking feedback into account.* The report and outcomes of consultation must be 'taken into account' prior to adoption of the plan or programme. At adoption, a statement must be published explaining how stakeholder views and environmental considerations have been integrated into the plan or programme.

Under the terms of the draft Directive, member states have a duty to do the following:

▶ ensure that environmental reports meet stated minimum standards;
▶ monitor the implementation of environmental protection measures set out in plans or programmes 'with a view to ensuring . . . the effectiveness of environmental impact corrective measures'.

The implementation of the Directive in the UK marks a logical and yet significant development in the application of Environmental Impact Assessment techniques. Examples of where SEA has already been employed come from a variety of 'public' sector sources including:

▶ the Environment Agency river catchment plans;
▶ forestry area management plans;
▶ Local Authority 'Local Plans', e.g. Mendip District Council;
▶ Thames Water's Water Resources Planning Project.

VIII LIFE CYCLE ASSESSMENT (LCA) AND ECO-DESIGN

6.16 Definitions and overview

The International Standards Organisation has produced a series of standards that attempt to provide a common code of practice for conducting a life cycle assessment (LCA) (ISO 14040 series)

and incorporating eco-design into environmental management systems (ISO 14006). The following definitions are taken from ISO 14040 which describes the principles and framework for conducting and reporting LCA studies.

6.16.1 Life cycle

Life cycle is the consecutive and interlinked stages of a product system, from raw material acquisition or generation of natural resources to final product disposal.

6.16.1.1 Life cycle assessment/ analysis (LCA)

LCA involves the compilation and evaluation of the inputs, outputs and the potential environmental impacts of a product system throughout its life cycle. The terms life cycle assessment and life cycle analysis are used interchangeably.

Thus, LCA involves the evaluation of the environmental impacts of a product from 'cradle to grave', i.e. from the extraction of raw materials and fuel from the earth through to the product's use and, ultimately, disposal.

Figure 6.12 provides a pictorial representation of the life cycle of a very simple product – a flip-chart pad. It provides a good representation of the multiple stages and linkages involved in the life cycle of the product.

LCA differs fundamentally from other environmental management tools such as environmental impact assessment and auditing. The latter tend to focus on a single site or location, and hence only part of the life cycle which occurs on that site – typically the manufacturing process in Figure 6.12. LCA, however, assesses the different environmental issues that occur over time in different geographical sites and at the different stages a product will pass through in its lifetime. By taking such a wide view, LCA aims to avoid the problem often encountered in environmental improvement programmes of shifting impacts from one part of the life cycle to another. For example, a change in materials used during manufacture may create less impact at the production site, but increase

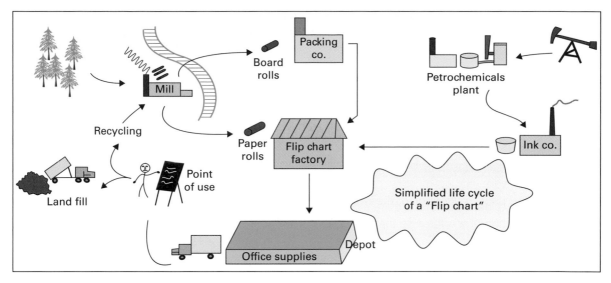

Figure 6.12 A pictorial representation of a simple product life cycle

the impact associated with the extraction of raw materials or with product use or disposal. LCA can help to ensure that any changes made produce a net environmental improvement across the whole lifetime of the product.

It should, however, be noted that LCA does not consider subjective impacts such as nuisance. In addition, there is no consideration of economic or social factors – the standard methodology being a pure environmental rather than sustainability assessment. While acknowledging these limitations, LCA can be a powerful quantitative technique that can be used to inform decision-makers, enabling informed and robust decisions.

6.17 The LCA process

ISO 14040 identifies four stages in the LCA process:

definition of goal and scope
inventory analysis
impact assessment
interpretation of results.

6.17.1 Goal and scope definition

During this phase, decisions need to be made concerning the purpose and coverage of the LCA. Typically this means answering the following sort of questions:

▶ What product is the LCA going to deal with?
▶ Are similar products going to be considered for comparative purposes or will it be a single product assessment?
▶ Are there any design constraints applicable to the product – statutory performance criteria, quality standards, etc.?
▶ What environmental issues are of key concern to the organisation and its stakeholders?

In addition, a key part of the scope definition phase is the establishment of the study's limits, the so-called 'system boundaries'. Theoretically, a full LCA should consider all upstream and downstream processes associated with the product. In practice, this would, in many cases, be unmanageable. Studies are, therefore, scoped (in a similar way to EIA) to identify the key processes of concern or interest. Scoping may lead to a choice to focus the LCA on those processes that seem to be a prime source of a product's environmental impact, or those over which the organisation conducting the LCA may have most influence.

6.17.2 Inventory analysis

Inventory analysis involves data collection and calculation procedures to quantify relevant inputs and outputs of a product system. These inputs and outputs may include the use of resources and releases to air, water and land associated with the system. LCA relates each of these environmental 'burdens' to the same functional unit, e.g. 1 kg of the final product. Examples of inventory data might be:

▶ energy consumed;
▶ emissions to air;
▶ effluent generated;
▶ solid wastes produced.

The process of conducting an inventory analysis is iterative. As data is collected and more is learned about the life cycle, new data requirements or limitations may be identified that require a change in the data collection procedures so that the goals of the study will still be met. Sometimes, issues may be identified that require revisions to the goal or scope of the study.

Data collection can be a resource-intensive process. Practical constraints on data collection should be considered in the scope and documented in the study report. Some significant calculation considerations are outlined in the following examples:

▶ *Allocation procedures* are needed when dealing with systems involving multiple products (e.g. multiple products from petroleum refining). The materials and energy flows, as well as associated environmental releases, must be allocated to the different products according to clearly stated procedures, which must be documented and justified.
▶ *The calculation of energy flow* should take into account the different fuels and electricity sources used, the efficiency of conversion and distribution of energy flow as well as the inputs and outputs associated with the generation and use of that energy flow. Again, conversion to a standard unit is preferred, e.g. carbon dioxide equivalent per kg of finished product.

6.17.3 Life cycle impact assessment

For some LCAs, the completed inventory may be an appropriate point to conclude. It may be possible, from the environmental burdens considered, to identify where a manufacturing process can be improved without recourse to

the impact assessment phase. For other LCAs the inventory phase will have identified a large number of environmental burdens which may be difficult to interpret and compare. The impact assessment phase of LCA is aimed at evaluating the significance of potential environmental impacts identified during the inventory analysis.

In general, this process involves associating inventory data with common environmental impacts and attempting to understand those impacts. The level of detail, choice of impacts evaluated and methodologies used depend on the goal and scope of the study.

The impact assessment phase may include elements such as:

▶ assigning of inventory data to impact categories (classification) such as greenhouse gases, water consumption etc. This is in effect a process of summing up the environmental aspects across or within different life cycle stages.
▶ evaluating the consequences of the inventory data within the assigned impact categories (characterisation) by considering the impact of the environmental aspects either within individual stages or across the whole life cycle. Typically conversion factors have to be used to translate inventory data into changes within environmental receptor groups. For example, we may convert different energy usage types across a life cycle to produce a carbon dioxide equivalent quantity per unit of product (classification). We may then use a standard conversion faction to convert to a figure representing global warming potential (characterisation).

These sorts of calculations may highlight that a particular life cycle stage is the main contributor to an environmental impact, allowing improvements to be targeted in the most critical processes. LCAs undertaken for washing machines, for example, have indicated that the most significant environmental impacts are due to energy and water consumption during use. If one machine uses less water and energy than another, then its environmental impact is the lower of the two.

However, if one machine uses less water and the other uses less energy, it is not immediately obvious which has the lower impact. The purpose of the next phase of the LCA process is to make such comparisons and allow conclusions to be drawn.

6.17.4 Life cycle interpretation

Interpretation is the phase of LCA in which the findings from the inventory analysis and the impact assessment are combined together. The findings of this interpretation may take the form of conclusions and recommendations to decision-makers, consistent with the goal and scope of the study. In a single product assessment, the impact assessment phase will normally provide sufficient information to enable recommendations to be made with regard to improvements in product design or supply chain management in order to produce a net reduction in environmental impact. Associated recommendations may relate to a revised scope of the LCA to provide further information.

In a comparative LCA, the relative performance of products in each impact category can be assessed. Since it is rare that one product performs less well for all impact categories, some kind of prioritisation is necessary. Ranking and even weighting of different impact categories are undertaken on the basis of importance to the environment, company or stakeholders.

6.18 Streamlined LCA

The development of limited coverage LCAs, known as Streamlined LCAs, has focused on the impact scoping incorporated into the definition of goals and scope. The key issue associated with such assessments is how to identify what can be safely omitted from the LCA without overly reducing the reliability of the result. Guidance can sometimes be obtained by reference to completed work for similar products, which identifies the principal life cycle stages or environmental burdens. The streamlined LCA then concentrates solely on the stage(s) or burdens identified as important for the product by the earlier LCA(s). An important source of comparative information is available from the EU Eco-label scheme, which has published LCAs for a number of products in the following sectors:

Bedding	Household appliances	Services
Gardening	Do-it-yourself	Lubricants
Electronic equipment	Cleaning	Textiles
Footwear	Paper	

To illustrate the approach, a light bulb manufacturer could adopt a streamlined LCA having identified from the EU Eco-Label published work that the most significant life cycle stage for a light bulb is the in-use phase. Manufacture and disposal can be reasonably omitted on the basis that 90 per cent or more of the environmental impacts resulting from the manufacture, use and disposal of a light bulb occur during use and are attributable to the electricity consumed. The manufacturer may still want to address other issues concerning manufacture, packaging and distribution through the application of best practice, but these issues can be omitted from the LCA.

6.19 Sources of LCA/comparative data

As the field of LCA has developed, a number of hosts or collation organisations have emerged in an attempt to share information which is of great interest to others carrying out LCA of products that may be similar or have similar components. Examples include:

▶ *Eco-invent* – formally known as the Swiss Centre for Life Cycle Inventories, this organisation hosts a web-based database of over 4,000 LCAs – reports are available freely at www.ecoinvent.org/organisation/ though data sets are available on a subscription basis only.

▶ *Seeds4green* – this is a collaborative website project (www.seeds4green.org) co-sponsored by DEFRA, that aims to gather together LCA style information that is searchable under product categories. Information is presented under two main categories:

   ▶ *life cycle assessments* (essentially individual LCA reports) *environmental product declarations* (EPDs) which are interpretive reports that present findings and information about reducing the impact related to a particular product group. EPDs may draw from several LCA studies and are presented more with the goal of assisting sustainable procurement decisions than facilitating streamlined LCAs.

▶ *WRAP* – the Waste & Resources Action Programme (WRAP) website has a considerable number of LCA reports hosted in the research area of its website.

Globally there are many more organisations collating and making available LCA data – the United Nations Environment Programme has compiled a list of such sources as well as providing a host of guidance and information materials relating to LCA under its Life Cycle Initiative programme.

6.20 LCA applications and benefits

A range of applications and benefits exist in relation to LCA which are summarised in Table 6.10.

6.21 Eco-design

The concept of eco-design is not particularly new but its application in mainstream industry, to date, has been relatively limited until fairly recently. It is closely linked to LCA as much of the information required to make design improvements from a

sustainability perspective require some kind of LCA type of evaluation.

Whether we are considering product development in a manufacturing company or service development in a contracting company, design is what the company does to create and/or express its business strategy. However, the design process is often undervalued. In manufacturing, the design phase absorbs a relatively small resource – typically around 15 per cent of product costs but it commits the other 85 per cent and sometimes for a very long period. It also decides the scale, frequency and nature of associated environmental impacts.

Eco-design has been defined as: 'a method of systematically integrating environmental considerations into the design of products, processes and services throughout their life.' For a product this may be represented as shown in Figure 6.13.

Clearly, the process is closely aligned to life cycle analysis and in many ways it is an extension of the interpretation phase of the LCA process, for eco-design demands an understanding of the environmental impacts through the product cycle in order to develop ways of increasing resource productivity and reducing environmental impact. Eco-design does not, however, require a detailed LCA to be completed as a pre-requisite. A broad understanding of the key product design issues through the life cycle is adequate.

Table 6.10 LCA applications and benefits

Improved efficiency and cost savings	Tracking energy and material inputs as well as waste outputs through production can identify opportunities for efficiency improvements and hence, cost savings.
Product design	As an aid to product design decisions, LCA can help identify the environmental pros and cons of different options. Such information may avoid decisions that could have long-term implications in terms of producer responsibility or simply guide design to maximise benefits of 'green marketing'.
Product marketing	In order to promote the environmental performance of a product with confidence and credibility, claims need to be based on concrete evidence. LCA, particularly where product comparison has been undertaken, is well suited to this function.
Supply chain pressure	For businesses such as retailers, the primary environmental impacts are upstream (with suppliers) or downstream (with consumers). Such organisations are becoming more interested in reducing their environmental impacts, particularly through influencing their supply chain. Suppliers able to demonstrate the environmental performance of their products through LCA may be well placed to win increasing market share with such organisations.
Informs procurement decisions	This is essentially the flip side of the previous item – LCA reports allow interested parties to make informed decisions in order to minimise the impact of their procurement choice.

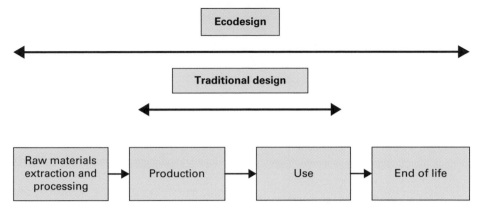

Figure 6.13 The scope of eco-design

As a design philosophy, some commentators advocate consideration of a product as 'a vehicle for the provision of one or more services'. For example, a car should not be considered as a metal box on four wheels powered by a combustion engine but as a unit that delivers customer service in relation to personal transport, status symbol, entertainment, extension of personal space outside the home, comfort zone, etc. By considering the existing design of any product in this way we are more likely to identify new alternatives rather than become stuck in refinement of existing formats.

The company, Shot in the Dark, in their eco-design information pack, describe an approach which uses what they call a design abacus which assists decision-making within the following steps of the design process:

Identify and assess a reference product (our own or somebody else's).

Gather and manage information (relating to the overall impact of the product).

Analyse internal and external drivers and influences including market position (to prioritise the impacts in the eyes of key stakeholders as well as the company itself).

Identify areas for improvement (for an existing market function, i.e. to maintain or expand current share of reference product).

Develop a specification – identify the design issues and set the design goals (by using the abacus as a reference tool for new alternative design modifications).

Systematically generate and brainstorm new ideas, e.g. down weighting, material substitution, etc.

Rate and rank the best ideas with reference to:

▶ price, lifetime benefits, marketing and customer benefits
▶ health and safety – in production, use and end of life stages
▶ costs in production, use and end of life
▶ quality in production, use and end of life
▶ marketing implications down supply chain to end of life
▶ environmental implications throughout the product life cycle
▶ legislative implications
▶ and any other reference criteria that may be appropriate and as identified in step 5.

Select the most promising ideas.

Detail and refine selected concepts, prototype and test.

A template and a completed example of a design abacus are provided in Figures 6.14 and 6.15. It should be noted that in most cases the results of the comparison of two alternative product designs remains subjective to a greater or lesser degree. However, as a planning tool that then enables an organisation to select one or more 'best options' to proceed to prototype/proving stage, the tool may prove extremely useful as it gives an overview of the issues across the whole life cycle of the current product and its potential replacement(s).

Whether this or an alternative decision-making approach is used, the eco-design process is likely

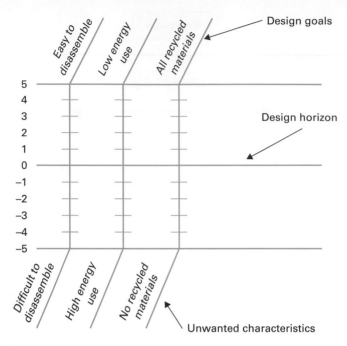

Figure 6.14 A design abacus template
Source: Shot in the Dark / Green Training Works, www.greentrainingworks.co.uk. ©Green Training Works.

to become increasingly important as government pressure increases under the 'sustainable production and consumption' priority area within the UK sustainability strategy.

As part of this process (and with clear overlap to life cycle assessment techniques) various accreditation and consumer communication schemes are appearing in various sectors. Some of the best known are the EU eco-label scheme, the Forestry Stewardship Council and the Fair Trade and Organic schemes in the food sector. Methods of product review and standards/issues on which accreditation is achieved clearly vary widely between the schemes. However, their increasing popularity with consumer groups means that product manufacturers are increasingly turning to eco-design principles to comply or improve their ratings under such schemes.

Another source of guidance on this issue appears in the ISO 14000 series. ISO 14062 was published in 2002 and attempts to integrate life cycle assessment techniques into the traditional steps of the design process. It provides a framework for consideration of environmental aspects into the whole of the product design and development

process including the provision of appropriate information during product launch.

6.21.1 The EC Directive on Eco-design 2009

Eco-design principles are beginning to move into the legislative context with the adoption of the *EU Eco-design Directive* in 2009. This Directive applies to energy-using products primarily but also to energy-related products such as windows, insulation, etc. The Directive requires the European Commission (EC) to conduct studies to produce performance standards for key products. The intention is that this can then be used to drive minimum standards of product manufacture (and import) in all member states.

Essentially the performance standards are set as a result of studies and consultation by the EC which then defines CE marking standards for the products with an associated compliance date. Manufacturers selling to the European market then have to fulfil the design performance standards in order to be able to receive the CE which is a mandatory requirement for sale within the EU.

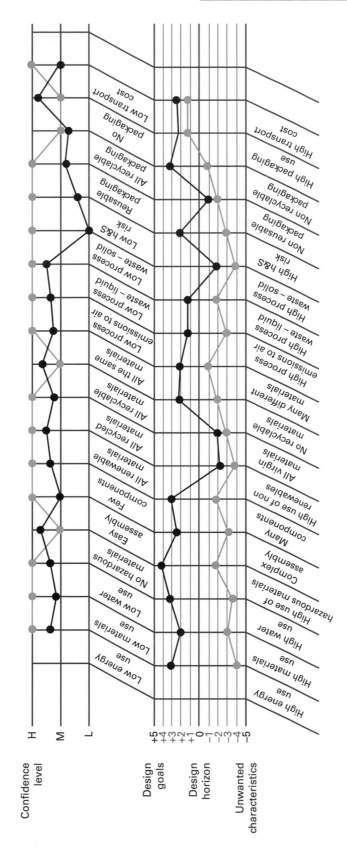

Figure 6.15 An example of a completed design abacus

Source: Shot in the Dark / Green Training Works, www.greentrainingworks.co.uk. ©Green Training Works.

Box 6.3 Example of a completed design abacus for a notional product

In Figure 6.15, the dark scores represent the reference product while the light scores relate to a new product design. The confidence level indicator at the top provides a reliability gauge for each assessment – as with the lower tier it is subjective but based on consensus among the design team.

Table 6.11 Predicted savings from the EU Eco-design Directive

Adopted implementing measures	Estimated savings (annual savings by 2020) in TWh
Standby and off-mode losses of electrical and electronic equipment (household and office)	35
Simple set-top boxes	6
Domestic lighting	39
Tertiary sector lighting	38
External power supplies	9
Televisions	28
Electric motors	135
Circulators	23
Domestic refrigeration	4
Domestic dishwashers	2
Domestic washing machines	1.5
Fans (driven by motors with an electric input power between 125W and 500kW)	34
Air conditioners and comfort fans	11
Total projected savings	Approx. 365TWh
	More than 12% of the 2009 final electricity consumption in the EU

Source: EC eco-design guidance

Significant energy savings are predicted compared with a business as usual scenario as are shown in Table 6.11.

IX SUSTAINABLE PROCUREMENT: POLICY AND GUIDANCE

There has been increasing interest from organisations over the recent years in developing reliable purchasing policies that reflect social and environmental standards. This trend is in line with the expanding view of corporate environmental responsibility associated with sustainability programmes.

6.22 Sources of guidance

Although a notoriously difficult thing to 'standardise', there have been a number of attempts to develop supplier assessment and product selection methodologies that help the organisation reduce the environmental impact associated with the purchase of goods and services. One particular example from DEFRA will be considered below. However, those interested should consider the range of assessment possibilities and decision making references that have been developed by an increasing number of organisations. A few are listed below.

▶ *The Mayor of London's Green Procurement Code.* In 2001, the Mayor of London launched an initiative to promote sustainable procurement via a voluntary 'green procurement code' supported by an awards programme and an information hub linking suppliers of sustainable goods and services to potential customers. See www.londonremade.com for details.

▶ *The US Environmental Protection Agency's green procurement programme.* This includes fact sheets on product selection criteria for a whole range of products and services. See www.epa.gov/epaoswer/non-hw/procure/factshts.htm.

▶ *B&Q's QUEST programme.* One of the longest-running company supplier assessment and development programmes using product and manufacturing sustainability standards to assess and drive supplier performance improvement across the DIY retailer's product range. See www.diy.com for details.

▶ *The Worldwatch Institute's 'good stuff' guide.* This is a freely available and easily accessible guide to the environmental issues that should be considered when selecting goods from a whole range of product groups from cars to dry cleaning and bicycles to electricity. The guide is available for download from: www.worldwatch.org/taxonomy/term/44.

▶ *www.procurementcupboard.org.* This is a primarily public sector information sharing site with a variety of information on procurement policy, tools and case studies. The UK government has made a commitment to be a leader in the field of sustainable procurement and this site plus information via the DEFRA website give something of a sense of the initiatives underway.

▶ *BS 8903 Principles and Framework for procuring sustainably, 2010.* According to Action Sustainability, the technical author, this standard is the first of its kind globally and provides any organisation with guidance on adopting and embedding sustainable procurement practices. Action Sustainability also offers a range of web-based information (www.actionsustainability.com), consultancy and training services in relation to sustainable procurement and the use of BS 8903.

6.22.1 Procurement risk assessment: DEFRA and EA guidance

Supply chain management has been highlighted as a key area for organisations that are moving beyond basic environmental management and into the realm of strategic change and sustainability. For many organisations, initially supply chain management means deciding on which companies or products to target. This may be done generically though a broad prioritisation of industries or products, e.g. production chemicals more important than catering supplies, or as part of a more systematic risk assessment process. A generic approach is described here, while more systematic evaluation methods are discussed in Section IX.

DEFRA, as part of the National Sustainable public procurement programme launched in 2011, fine-tuned and made available a prioritisation methodology that is aimed at assisting organisations to decide where they should focus their efforts in order to reduce the environmental and social impacts caused by their supply chains. The methodology is presented in the form of an active spreadsheet tool with accompanying guidance. Users are led through a series of questions in relation to product groupings which are then ranked in terms of the influence of the purchaser, scope for change and reputation risk to the company with a combined scoring system providing a final list of priority procurement areas.

In addition to this prioritisation tool, the UK Environment Agency has produced a 'sustainable procurement commodities guidance' document which summarises the sustainability issues associated with a range of products/commodities and makes suggestions as to appropriate considerations from a procurement perspective. An example taken from the commodities guide is shown in Figure 6.16.

6.23 Sustainable procurement strategies

6.23.1 Introduction

Most organisations begin their environmental management programme with a focus on direct

229

21. *Environmental issues associated with Mobile Phones*

Overview

Mobile phones require a wide array of raw materials, which have to be sourced from many different suppliers and transported over considerable distances. This document mainly addresses the environmental issues associated with the manufacturing, use and disposal stages as more detail on the impacts of raw materials are covered in separate documents.

Summary of life cycle record

Raw material	Use
Raw materials include plastics, oils, copper, iron, solvents, adhesives, metals, ceramics and glass, etc. Toxic chemicals contained in cell phones, such as arsenic, beryllium, cadmium, copper, lead, nickel and zinc, which can enter the soil and groundwater from landfills. Inks, cables, LCD, flame retardants, components, batteries etc.	Use for contact when no access to landlines. Health and safety issues with prolonged use of phones. Health and safety issues with use of phones whilst driving. Impacts of use of telecommunications services.
Manufacture	**Waste management/disposal**
Various manufacturing processes to produce manufacture electronic components, such as printed circuit boards. Component assembly and packaging.	Hazardous wastes from end-of-life of equipment are a key area. Packaging waste is also a key issue.

Key impacts and priority mitigation measures

The key impacts in relation to telecommunication are:

- Heavy and precious metals are used in the equipment are toxic to humans and animals and persistent in the environment.
- Raw materials that have the potential to impact on the environment during their production or disposal (e.g. plastics, oils, copper, iron, solvents, and adhesives).
- Solvent use releases VOCs into the environment, which can cause ground level ozone, global warming, etc.
- Effluents from production processes could be contaminated with pollutants such as heavy metals, alkalis, acids, spent solvents and heavy metals (production of printed circuit boards).
- Energy use impacts on non-renewable resources and releases CO_2 (contributes towards global warming).
- End of life disposal of equipment is a key issue, because various parts will have impacts, for example components or batteries contain heavy metals, which can contaminate groundwater.
- Various hazardous substances such as lead, flame retardants, arsenic, beryllium, cadmium, copper, lead, nickel and zinc are contained in cell phones. These can enter the soil and groundwater from landfills.
- Coltan and tantaline mining, which can cause negative impacts to wildlife in developing countries.

Control measures – processing/manufacture, use and waste management/disposal:

- Reduce negative environmental impacts of the industry by optimising resource use and the substitution of hazardous materials and incorporating design for environment and establish eco-efficiency
- Design and manufacture equipment with high-energy efficiency, and parts, which can easily be recycled.
- Metal recovery where possible
- Where feasible use recycled components of old equipment in production of new equipment.
- Provision of staff training in environmental, health and safety matters including accident prevention, safe chemical handling practices, and proper control and maintenance of equipment and facilities
- Assess environmental alternatives to lead and flame retardants.

Control measures – procurement action:

- Ensure suppliers used have a high awareness of the potential environmental impacts and are taking appropriate mitigation measures, particularly related to end-of-life waste management.
- Ensure enhancement of design for recycling (the design and production of electronic equipment should take fully into account and facilitate the dismantling, re-use and recovery)
- Encourage implementation of EMS and accreditation to ISO 14001 in supplier companies.
- Specify equipment that incorporates a high proportion of materials which are recycled or that can be recycled.
- Work with suppliers to develop suitable end of life options, and prefer suppliers that provide information and help in collection and waste management of end-of-life equipment.
- Prefer lithium-ion batteries instead of nickel-cadmium with nickel metal-hydride batteries.
- Prefer suppliers that provide information and help with packaging waste, particularly those that take back packaging waste for re-use after delivery.
- Ask suppliers to produce guidelines on safe usage to reduce exposure to electromagnetic radiation for end users.
- Ensure easy removal of batteries for recycling.

Figure 6.16 Extract from the EA Commodity sustainability briefing document (Jan. 2003)
Source: Contains Environment Agency information ©Environment Agency and database right.

environmental aspects. This is partly because these are easiest to understand and influence as they fall directly under the organisation's control, and partly a case of 'getting your own house in order' before looking further afield in terms of environmental improvements. A variety of reasons including customer pressure, reputation management, environment and/or sustainability policies and goals, may expand attention to incorporate indirect environmental issues. The flow chart in Figure 6.17 is intended to provide an idea of the areas that might be considered and also highlights the sphere of influence of the procurement process.

In relation to both raw materials usage and contractor/supplier selection, the first question asked by many organisations is: 'Is there any way we can prioritise certain companies or products/materials to make best use of limited time and resources and yet ensure that we deal with key indirect environmental aspects?' Some common

approaches are considered below. A useful summary of supplier assessment methodologies and principles is an IEMA publication entitled *Environmental Purchasing in Practice: Guidance for Organisations*, which contains case study information from a number of organisations. The publication is available for purchase from IEMA.

6.23.2 Pre-qualification questionnaires (PQQs)

PQQs are widely used but vary enormously between organisations depending on the maturity of their own sustainability programmes and the priority afforded to supply chain management. At their simplest they may simply be a series of questions limited to explorations of legal compliance, the presence of permitted processes and/or possession of environmental policies or environmental management systems. The purpose of such PQQs is often more to do with

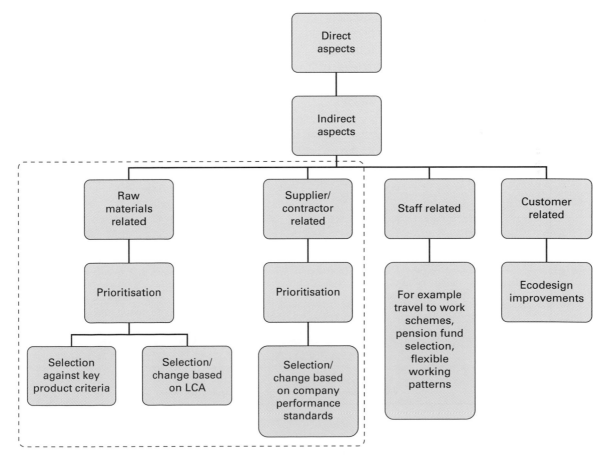

Figure 6.17 The procurement sphere of influence

raising environmental/sustainability awareness among suppliers than as a selection tool.

More advanced PQQs structure questions and call for evidence in such a way as to enable a supplier risk assessment (with results often presented numerically) and hence allow selection or prioritisation of suppliers with whom to do business. They are often focused on the products or services provided by the supplier company to the client organisation.

6.23.3 Prioritisation methods: an overview

As with assessing the significance of direct environmental aspects in an EMS, there is no 'standard' or 'correct' way of prioritising suppliers or raw material procurements. It depends on the perceived priorities of the organisation and its key stakeholders including investors, neighbours and customers.

The DEFRA methodology described in Section 6.22.1 is an example of a fairly simple prioritisation methodology based on procurement groupings. The principle criteria for evaluation were level of spend (and by inference degree of influence), scope for change and risk to the organisation.

The following sections consider these and other potential criteria that should/could be considered in developing a prioritisation methodology or selection procedure. The methodology finally adopted should aim to be objective, repeatable, transparent and as simple as possible to use.

Presentation of findings is also sometimes challenging although the use of some kind of procurement priorities profile/matrix can be useful, see the example in Figure 6.18.

6.23.4 Supplier prioritisation criteria

Supplier selection within sustainable procurement programmes may be divided into prioritisation criteria (the issues) and selection standards (the indicative performance standards linked to the issues). A selection of the issues to be considered i.e. the prioritisation criteria that might be relevant to an organisation, is shown in Table 6.12.

6.23.5 Supplier selection standards

This can either be done in a generic way based on indicators applicable to all organisations or

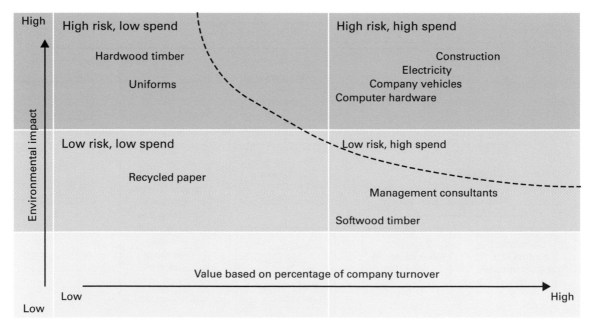

Figure 6.18 A procurement priorities matrix
Source: Adapted from the *IEMA Handbook*.

Table 6.12 Supplier prioritisation criteria

Prioritisation criteria	Reason for inclusion
Degree of influence, e.g. first tier suppliers rather than third or fourth tier	Easier to influence and gauge performance in organisations with whom we deal directly.
Amount of business done (ranking on business spend)	Greater liability links (to us) as well as influence in relation to suppliers for whom our business is a significant proportion of their overall work. Alternatively if a large part of our budget is spent with a small group of suppliers, then it could be argued that in terms of our supply chain impact these represent our greatest responsibility.
Suppliers who operate Permitted processes	By their inclusion under the Environmental Permitting regime we can assume that their activity constitutes a greater environmental impact than a non-regulated process.
Non-European suppliers	Less reliability that the legal standards to which the organisation operates are equivalent to our own.
Non-UK suppliers	Assumption that 'local' represents an environmental preference at least in relation to transport of goods. Also concerns about social and environmental standards in some countries.
Certified environmental management systems	Assumes that companies without a certified EMS represent a bigger risk in terms of pollution and, by association, reputation damage (to us).
Hazardous materials	Assumes suppliers providing us with raw materials or components containing hazardous substances represent a bigger risk in terms of pollution from their activities and, by association, reputation damage (to us).
Packaging	Suppliers that provide us with significant quantities of non-returnable package represent a source of waste that must be disposed of by the organisation.
Suppliers' own supply chain management process	A way of ensuring that raw material or third tier plus supply chain issues are addressed via a responsibility roll-out. In other words, suppliers who are known to have active supplier assessment programme encompassing environmental issues could be considered lower priority than others that do not.
Public environmental reporting	As with EMS, this is a simplified way of deciding whether a supplier is actively involved in environmental management.

it can be done specifically based on company questionnaires or auditing.

Generic standards that could be used as the basis for supplier differentiation and selection include:

▶ ISO 14001 certification;
▶ evidence of environmental policy and/or improvement programme;
▶ environmental law prosecution history, i.e. lack of or limit to ranking in independent schemes such as the BiTC Corporate Responsibility Index or Dow Jones Sustainability Index;
▶ location rating – within 50 miles, UK, Europe, etc.

Specific indicators may require detailed information on each company being assessed. This is commonly obtained using a questionnaire of some kind built around the criteria described in Section 6.23.4. Larger organisations with a

significant degree of supply chain influence may be able to go further and demand performance data and/or undertake an audit or review of supplier operations. The attraction of these kinds of assessment is that they allow weighted scoring of different company attributes and presentation of results in simplified numeric or category ratings. As long as they are transparent and logical, they can also be used to highlight concerns to suppliers and drive performance improvement.

6.23.6 Raw material/product prioritisation criteria

Raw material/product selection within sustainable procurement programmes may also be divided into prioritisation criteria (the issues) and selection standards (the indicative performance standards linked to the issues). A selection of product related procurement criteria are shown in Table 6.13.

233

Table 6.13 Product prioritisation criteria

Prioritisation criteria	Reason for inclusion
Key customer concerns, e.g. wood product manufacturer may be interested in timber from sustainable sources because that is what customers are increasingly demanding	Customer pressure translates into business viability and hence this is a key driver for selection.
Hazardous materials	General principle that we should be looking to substitute less hazardous alternatives but also the assumption that the organisations involved in the supply chain of such materials represent greater pollution risks that those who are not.
Amount of materials purchased/business done (ranking on business spend or quantity of materials)	If a large part of our budget is spent on a small group of raw materials, then it could be argued that in terms of our supply chain impact, these represent our greatest responsibility.
Sustainability of supply	Prioritisation of raw materials where there is no evidence of renewable supply practices or where the product purchased contains wholly or partly non-renewable resources.
Extractive materials	Often associated with habitat/biodiversity damage
Tropical plant materials	Concerns about tropical deforestation and the associated biodiversity impacts
Reputation implications	This is a combination criterion involving consideration of several of the earlier criteria to give an overall 'reputation risk' to the organisation, e.g. raw materials that routinely come from sources in the tropics with low assurance of sustainability of supply and which have previously been the subject of media exposure/public concern would be considered highest priority.

6.23.7 Raw material/product selection standards

The criteria used to prioritise raw materials/products will heavily influence the selection standards used to guide future purchasing decisions. A completely standardised approach is unlikely to be appropriate because different organisations will have different priorities and risks. However, for individual products or raw materials, it may be that similar selection criteria will be adopted by different organisations as the raw material issues will be consistent, e.g. furniture selection criteria are likely to be similar in all organisations in relation to reducing associated environmental impacts.

In an ideal world, all purchasing decisions would be based on complete, comparative life cycle analyses for all product and raw material choices. However, this is unlikely to be the case and organisations will often use simplified selection criteria based on one or a few key environmental issues. Some examples of simplified criteria applicable to different product groups include:

▶ timber and timber products carrying the Forestry Stewardship Council approved logo;
▶ local sourcing, i.e. selecting on the basis of distance from point of purchase to point of use;
▶ selecting on the basis of product weight – assumes reduced material consumption and final disposal when purchasing components or products with lower weight;
▶ expected life span – assumes greater spread of manufacturing impacts over lifetime of product use. This may be judged on the basis of design life or the ability to upgrade or maintain and repair the product.
▶ recycled content – assumes greater resource efficiency the higher the proportion of recycled materials used in the manufacture of a product;
▶ water-based rather than solvent-based inks, adhesives, printed materials, etc.
▶ energy efficiency or water efficiency during operation, e.g. used extensively in white goods sector;
▶ organic food products – assumes lower environmental impact than non-organic equivalents;
▶ eco-labels for product groups assessed under the EU Eco-Label scheme.

X CARBON MANAGEMENT TOOLS

6.24 Introduction

The UK Climate Change Strategy continues to place increasing pressure on organisations to reduce their contributions to this global environmental impact. Key government initiatives are energy focused and this reflects the fact that the majority of greenhouse gases come from emissions generated in the course of power generation, heat production and transport.

Partly in response to these government initiatives, it is increasingly common to see claims or objectives relating to the 'carbon footprint' of an organisation, such as: 'we will strive to be a carbon neutral company'; 'this event/product/service is carbon neutral'; or 'we are seeking to move towards a low carbon economy'. But what do these catchy phrases actually mean? The short answer is – in isolation – not a lot. To date at least, organisations claiming to be 'carbon neutral' generally do not operate without directly or indirectly generating significant quantities of carbon dioxide. However, carbon management and the ideal of a carbon neutral operation do reflect a welcome new approach to the management of energy and emissions within an organisation.

Broadly speaking, there are two distinct approaches to reducing the climate change impact of an organisation:

▶ emissions reduction at source;
▶ offsetting emissions generated.

In both cases an organisation should consider both direct and indirect emissions:

▶ Direct emissions are those created by the organisation itself, e.g. from a site boiler system or production process such as brewing which generates CO_2 as a waste gas.
▶ Indirect emissions are those generated elsewhere typically in the generation of electricity which is supplied to the organisation.

With due consideration of the greenhouse gas hierarchy described in Section 6.24.1, a carbon management programme will typically comprise the elements shown in Figure 6.19.

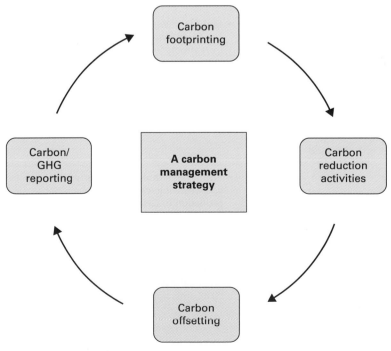

Figure 6.19 A carbon management strategy
Source: Adapted from the *IEMA Handbook*.

Examples of carbon reduction activities are considered in Chapter 7. We will consider the other elements of the carbon management programme here. The starting point for most organisations is an evaluation process commonly known as carbon footprinting which is essentially a specific type of resource usage survey as described in Section IV.

6.24.1 The GHG management hierarchy

The GHG management hierarchy (see Figure 6.20) is a model of carbon management put forward by the IEMA in one of its Practitioner series publications— *A Guidance Framework for GHG Management and Reduction*. Like the waste hierarchy on which it is based, emphasis is put on actions which avoid or reduce emissions rather than actions which compensate for emissions or substitute lower emission practices. While action is needed on all fronts, action lower down the hierarchy should not be taken in isolation or in place of actions higher up the hierarchy.

6.25 Carbon footprinting

6.25.1 An overview

The carbon footprint of any organisation comprises direct and indirect carbon dioxide emissions or, more strictly, greenhouse gas emissions. Examples are shown in Table 6.14 but this is not an exhaustive list.

Typically when an organisation talks about achieving *carbon neutral* status they are referring to the monitoring and management of the *italics* text items only in Table 6.14. But even here neutral does not mean zero emissions. Instead it means that the organisation is taking a two-tier approach to the management of these sources of GHG emissions.

First, it is striving to increase the efficiency of use of energy in all its formats so that the carbon generated per unit of activity is minimised through improvements in energy efficiency and the increased use of renewable energy. Second, the remaining carbon footprint is 'offset' through investment in a credible offset scheme. This means using a 'carbon calculator' to convert the

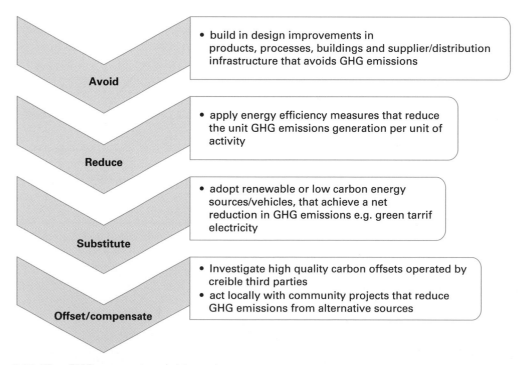

Avoid
- build in design improvements in products, processes, buildings and supplier/distribution infrastructure that avoids GHG emissions

Reduce
- apply energy efficiency measures that reduce the unit GHG emissions generation per unit of activity

Substitute
- adopt renewable or low carbon energy sources/vehicles, that achieve a net reduction in GHG emissions e.g. green tarrif electricity

Offset/compensate
- Investigate high quality carbon offsets operated by creible third parties
- act locally with community projects that reduce GHG emissions from alternative sources

Figure 6.20 The GHG management hierarchy
Source: Adapted from the IEMA e-briefing on green tariff electricity.

Table 6.14 Examples of GHG emissions associated with an organisation

Direct GHG emissions	Indirect GHG emissions
Gas consumption for space heating	*Power station emissions associated with electricity usage*
Transport emissions – company transport	Employee travel to work
	Supplier delivery transport
	Contractor delivery transport
	Supply chain manufacturing and distribution direct and indirect emissions

organisation's GHG emissions into a financial cost which represents one of two things:

▶ the cost of planting and managing an equivalent area of trees (or other biomass) that, over its lifetime, will absorb an equivalent amount of carbon dioxide from the atmosphere;

▶ the cost of investing in a project that would not otherwise have existed and that will result in the reduction or prevention of an equivalent amount of GHG emissions to that generated by the organisation.

6.25.2 Standardising approaches and ensuring good practice

UK government guidance and scientific support are available for the calculation of GHG emissions from a variety of activities into a common denominator of carbon dioxide equivalent. The difficulty is in establishing the 'offset valuation' per unit of carbon dioxide equivalent (i.e. the financial value of each unit of carbon dioxide in terms of offset costs). In response to this issue, a number of bodies have produced guidance and reference standards for those involved in carbon footprint calculations and carbon management programmes. The following are some of the most important:

▶ the DEFRA Guidelines for Company Reporting on Greenhouse Gas Emissions include a number of conversion factors and advice tables which are periodically updated. The guidance is presented for the calculation of emissions as part of a reduction programme and also includes suggestions with regards improvement initiatives, champions and progress review.

▶ The Carbon Trust also provides an easy to read summary and explanation of the stages to footprint calculations in its publication 'Carbon footprinting – an introduction for organisations' – available from their website www.carbontrust.co.uk.

▶ The World Business Council for Sustainable Development (WBCSD), in collaboration with the World Resources Institute, has also produced corporate guidance on the calculation and reporting of GHG emissions. Formats are similar to those presented in the DEFRA guidance but are more detailed in terms of direct emissions from various industry groupings. Details and calculation spreadsheets are available for download from www.ghgprotocol.org.

▶ The International Standards Organisation have produced ISO 14064–69 series which are billed as an international standard and guidance for corporate emissions reporting. See www.iso.org for more information.

▶ Publicly Available Specification (PAS) 2050 – Specification for the assessment of the life cycle greenhouse gas emissions of goods and services was developed by the British Standards Institution in 2008 and updated in 2011.

The approach is still in its infancy and, rather like emissions trading initiatives, if the administration can be made transparent and standardised, then we have a promising new tool at our disposal. It is important to remember, however, that any claims to carbon neutrality are dependent on/tempered by:

▶ the accuracy of offset calculations and the long-term success of such schemes in achieving the carbon 'savings' calculated;

237

▶ the proportion of an organisation's carbon footprint that is considered within the calculation, e.g. if a manufacturer of electronic goods does not include the supply chain carbon dioxide emissions and consumer use emissions of its TV or hi-fi, then a significant proportion of its overall carbon footprint is being ignored and any statement about being 'carbon neutral' becomes, if not meaningless, then severely diminished.

In addition to these best practice and referencing standards, since 2011 we have also had a standard reference from an energy management perspective, namely ISO 50001. This ISO standard provides a template for energy management that is closely aligned with the standard models of control and improvement used in the ISO 14000 and ISO 9000 series.

6.25.3 GHG emission calculations and cost allocations

Both the emissions reduction and emissions offsetting approaches to carbon management rely on accurate and transparent monitoring and calculation of greenhouse emissions by an organisation.

Greenhouse gases are those which contribute to the greenhouse effect when present in the atmosphere. Six greenhouse gases are regulated under the Kyoto Protocol, as they are emitted in significant quantities by human activities and contribute to climate change. The six regulated gases are Carbon dioxide (CO_2), Methane (CH_4), Nitrous oxide (N_2O), Hydrofluorocarbons (HFCs), Perfluorocarbons (PFCs) and Sulphur hexafluoride (SF_6). Emissions of greenhouse gases are commonly converted into carbon dioxide equivalent (CO_2e) based on their 100-year global warming potential. This allows a single figure for the total impact of all emissions sources to be produced in one standard unit. Conversion factors of greenhouse gas to CO_2e are calculated by the International Panel for Climate Change (IPCC) and included in the DEFRA Guidance.

The DEFRA Guidance describes calculations and reporting against three different 'scopes' of GHG emissions as shown in Figure 6.21.

The methods referred to above enable an individual, organisation or even country to calculate its carbon footprint. The DEFRA Guidance recommends that in all cases, there should be transparency in the scope of emissions covered in the calculation and the conversion factors used.

6.25.3.1 Allocating costs/values

For individuals and organisations and even nations, the decision to act to reduce GHG emissions typically involves consideration of the associated cost benefit analysis. The *Stern Review*, published in 2006, provided an 'economist's view of climate change' and considered in detail the mid and long-term financial implications of unchecked climate change. It also considered the financial costs of actions to address the problem and presented a strong economic conclusion in favour of global action to address the problem as soon as possible. Put simply – *it is cheaper to invest now in improvements and solutions than to wait and pay for the problems that will arise from climate change*.

The economic analyses of the Stern Review were conducted with a global perspective. However, the UK government is also adopting a unilateral cost benefit analysis to climate change policy. Guidance that is intended for use in all policy making has been developed and is based on the concept of the '*shadow price of carbon*' (SPC). Essentially, the SPC captures the damage costs of climate change caused by each additional tonne of greenhouse gas emitted, expressed as carbon dioxide equivalent (CO_2e) for ease of comparison. Policy-makers and decision-makers should take account of these costs when considering alternative proposals. The government produces valuation tables based on the emission date of each tonne of CO_2e so that a cost benefit analysis can be conducted over a project lifetime. For more information, see the DECC website.

For private organisations the allocation of costs and values is likely to fall somewhere between the individual (short-term cost focus) and government (longer-term strategic focus) extremes. Decisions on carbon management

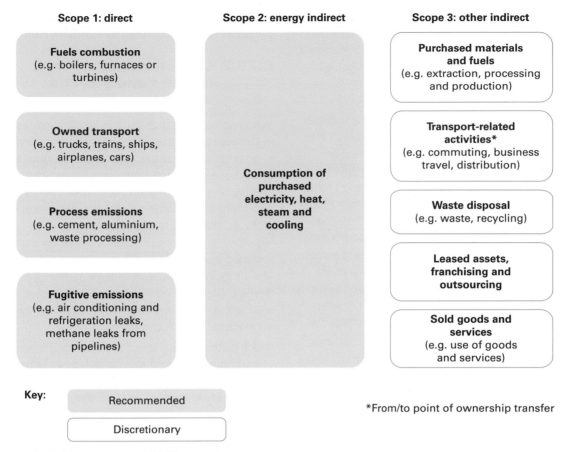

Figure 6.21 Three levels of GHG reporting
Source: Adapted from DEFRA Guidelines for Company Reporting on Greenhouse Gas Emissions.

improvements will reflect corporate commitment and short-term economic viability (as with the individual) but should also consider stakeholder interests and the longer-term financial implications to the company (including trends in energy costs, security of supply, etc.).

6.26 Carbon offsetting

Offsetting is considered the last stage in the GHG management hierarchy but is a critical piece in most carbon neutral programmes. The theory behind emissions offsetting is that an individual, organisation or even national/international community can 'compensate' for the greenhouse gas emissions generated by one of two mechanisms:

▶ investment in a carbon sink (thereby absorbing a quantity of carbon equivalent to that emitted);

▶ investment in initiatives which reduce greenhouse gas emissions from other sources (thereby preventing the future emission of a quantity of carbon equivalent to that emitted).

Although both strategies are the subject of some debate there is increasing interest in the use of carbon offsetting as part of a wider carbon management strategy.

6.26.1 Carbon sinks; management of forests

The classic approach to *carbon sinks* is via investment in the planting of new forestry areas. As climate change is a global phenomenon, it does not matter where in the world such forests are planted as long as they are adequately managed/protected so that they grow to maturity. Tree biomass, when dry, is made up of approximately 50 per cent carbon. Thus if an area

239

of forest is planted and a respective increase in tree biomass is measured, carbon will have been stored, or *sequestered*, in that area. Furthermore, carbon will be stored in the soils and organic litter that surround the forest.

This 'above ground' and 'below ground' increase in carbon storage has been the subject of extensive study over the past 20 years. It has been found that carbon sequestration is measurable and is region- and species-specific. Research has shown that generally tropical forests exhibit the fastest accumulation of 'carbon stocks' during growth of relatively new forests. Boreal and temperate forests show slower accumulation of carbon, but have greater below ground carbon stocks associated with them. Additional social and environmental (including biodiversity) benefits may be achieved through afforestation in areas of the world where deforestation has created a series of negative impacts.

More recent research has also shown that climax or fully mature forests continue to remove carbon from the atmosphere. The mechanisms behind this are not fully understood, but current research may shed some light on this concept and open up future international agreements that consider the protection of standing forests as investment in carbon sinks.

Indeed, there does seem to be growing international interest (under the auspices of the UNFCCC negotiations) in the benefits of initiatives aimed at stopping deforestation. This is a twist on the offsetting theme, in that instead of planting new trees we invest in the protection of existing forests which are under threat of clearance. The *Stern Review* in 2006 concluded that: 'Curbing deforestation is a highly cost-effective way of reducing greenhouse gas emissions.'

One area currently being explored by the international community is the possibility of establishing schemes whereby countries where deforestation is currently at very high levels would be 'paid to protect' those same forests, e.g. Brazil and Indonesia. Clearly such schemes would have to be carefully established and run but the potential gains are great given that CO_2 emissions from deforestation are estimated to represent more than 18 per cent of global emissions (a

share greater than is produced by the global transport sector, according to the *Stern Review*).

6.26.2 Carbon sinks: carbon capture and storage schemes

In addition to forests there is increasing interest in some circles in the use of geological carbon sequestration, i.e. the storage of carbon dioxide in appropriate geological formations. These *carbon capture and sequestration* plants take the CO_2 emissions from a large combustion process and store it rather than releasing to atmosphere. However, the process is itself very energy-intensive and thus the overall benefit remains uncertain. The UK government is investing in research and development that, along with other examples around the world, will attempt to clarify the 'carbon economics' of such strategies.

Much of the debate about the appropriateness of carbon capture and storage schemes is based on whether or not they represent any kind of permanent solution or whether they are simply a short-term measure to reduce the impact of CO_2 emissions while we find alternatives to our current carbon-dependent lifestyle.

6.26.3 Investment in GHG reduction schemes

The alternative emissions offsetting approach to carbon sinks is the investment in low carbon technologies/schemes that will reduce emissions from other sources. Climate Care is one of a number of organisations worldwide who offer the service of calculating the equivalent tonnage of carbon dioxide (CO_2) emissions from information on business or individual space heating, travel and power consumption over a given period. The CO_2 tonnage is then converted into an amount of money relative to the CO_2 emissions reductions achieved through a range of projects run by the organisation in a number of developing countries. Each project is designed to have multiple benefits: first, a reduction in CO_2 emissions and, second, in relation to a social, environmental or conservation benefit. Further information on the Climate Care approach is available from their website at www. climatecare.org.

DEFRA has produced a Code of Practice with a provider accreditation scheme and 'quality mark' that is intended to ensure minimum standards for all accredited schemes. This is intended to help simplify the variation in carbon calculations and offset costs that is currently evident between different offset schemes and to help provide assurance to customers that the money they spend does actually generate additional reductions in carbon emissions.

6.27 Carbon management reporting

This is the final stage in the carbon strategy model introduced in Section 6.24. However, for larger organisations in the UK and in other countries around the world, it may become a key driver of the whole process. This is because since April 2013 all publicly listed companies in the UK are required to report on their annual greenhouse gas emissions. The programme will be reviewed in 2015 with a view to rolling out to all large companies in 2016. The existing DEFRA guidance (*Guidance on How to Measure and Report Your Greenhouse Gas Emissions*, 2009) is the standard that organisations will have to refer to in compiling such reports.

In relation to the scopes referred to in Section 6.25.3, Table 6.15 lists the reporting recommendations made in the DEFRA Guidance.

Some companies have for some time been voluntarily including carbon accounting data, targets and trends in their corporate environmental, sustainability or corporate social responsibility reports. Some go so far as to seek independent verification for the data they use in their calculations. Third party assessment schemes such as the Business in the Community Corporate Responsibility Index have included elements of carbon management in their assessment tools for some time. In addition, the carbon disclosure project described in case study 6.1 focuses purely on carbon reporting.

XI CORPORATE ECOSYSTEMS SERVICES REVIEW (ESR)

6.28 The World Resources Institute methodology

The concept of 'ecosystem services' has grown in use as a way of raising awareness of human dependence on natural systems. As a result, there has been increasing interest in finding ways to guide organisations in carrying out risk assessments to identify key threats and opportunities related to the ecosystem services on which they depend. The World Resources Institute led a project in 2009–10 that culminated in the production of a methodology that helps organisations assess and prioritise impacts and dependencies on ecosystem services. The methodology, which is known as the Ecosystem Services Review (ESR), describes a five-step process as shown in Figure 6.22.

Table 6.15 Recommended reporting coverage and format

Gross emissions data that should be reported	Format of the information
Total annual gross global Scope 1 GHG emissions	In tonnes of CO_2e
Total annual gross global Scope 2 GHG emissions	In tonnes of CO_2e
Discretionary – Total annual gross global Scope 3 GHG emissions	In tonnes of CO_2e
Total annual gross global GHG emissions	In tonnes of CO_2e for all scopes reported
Comparative emissions data from previous reporting year	In tonnes of CO_2e for all scopes reported
Base year data	In tonnes of CO_2e for all scopes reported
An intensity measurement related to the total global gross emissions for scope 1 and scope 2 emissions combined*	Reported separately from total gross global figure and stating intensity measurement used

Note: * Intensity measurements use activity-based denominators to create normalised data that allows emissions comparison over time with variable corporate activity eg tonnes of CO2/tonne of product produced.

Source: DEFRA Guidelines for Company Reporting on Greenhouse Gas Emissions, 2009.

Case study 6.1 The Carbon Disclosure Project (see www.cdproject.net)

The Carbon Disclosure Project (CDP) is an independent not-for-profit organisation that acts as a 'global secretariat for institutional investor collaboration on climate change'. The aim is to provide a mechanism through which corporations are encouraged to measure, report and manage their greenhouse gas emissions and that also provides the investment community with vital information about climate change for incorporation into investment decisions.

Representing more than 700 institutional investors with assets exceeding $80 trillion, the scheme has enough authority to secure information from a large number of the biggest institutions globally (more than 5,000 companies in 2013). Results are scored using a transparent methodology and then presented by company, market sector and financial sector. Since the project began, the percentage of organisations responding and the scores allocated related to carbon management have both shown consistent improvements.

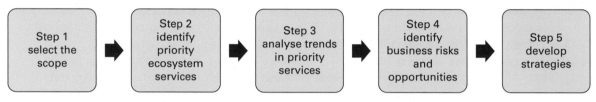

Figure 6.22 The World Resources Institute Corporate Ecosystems Review process
Source: Adapted from World Resource Institute Guidance.

6.29 A five-step approach

6.29.1 Step 1

The scope of the assessment can vary widely, focusing on a single product, the organisation's direct activities, a particular part of the supply chain or a customer group. This stage is very similar to the boundaries-setting stage in LCA which is intended to create a manageable project size that still meets the objectives of the organisation.

6.29.2 Step 2

Within the scope set in step 1, essentially step 2 asks questions corresponding to dependence and impact, in relation to a series of headings organised under the four principle ecosystems services:

▶ Provisioning services, e.g. crops, livestock and capture fisheries
▶ Regulating services, e.g. regulation of water timings and flows
▶ Cultural services, e.g. recreation and tourism
▶ Supporting services, e.g. primary production by organisms at the base of the food chain.

6.29.2.1 Dependence and impact assessment

The categories are listed in a spreadsheet tool provided by the WRI and the dependence questions asked against each category are as shown in Figure 6.23.

6.29.3 Step 3

In step 3 the high dependence and/or high impact ecosystem services are reviewed in relation to the trends in supply and demand affecting the ecosystem service in question, together with the role of the organisation in creating the pressure or demand on the service. This is a research task and, depending on the scope of the ESR, may involve literature searches, consultation and/or survey work by the organisation itself.

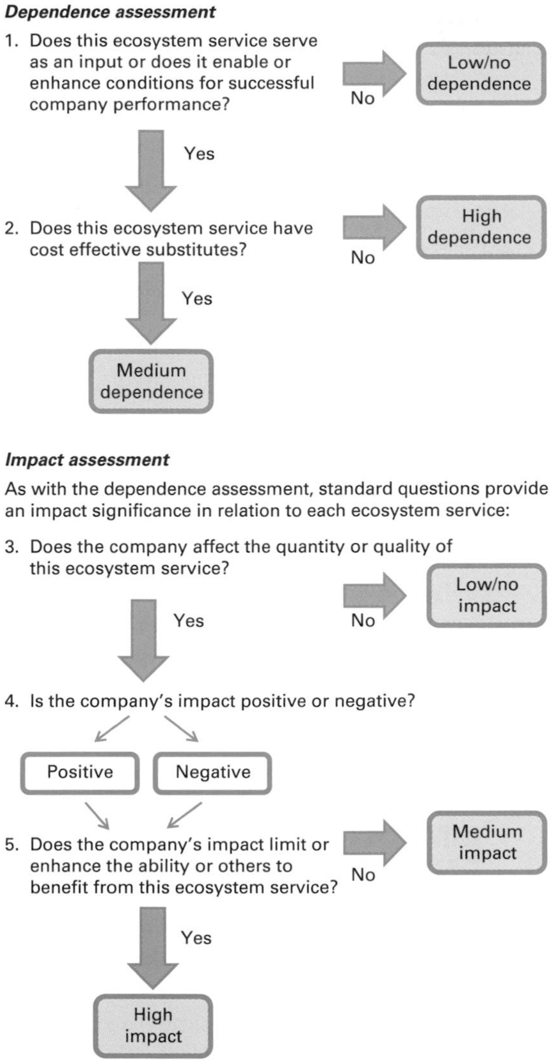

Dependence assessment

1. Does this ecosystem service serve as an input or does it enable or enhance conditions for successful company performance? → **No** → Low/no dependence

   ↓ **Yes**

2. Does this ecosystem service have cost effective substitutes? → **No** → High dependence

   ↓ **Yes**

   Medium dependence

Impact assessment

As with the dependence assessment, standard questions provide an impact significance in relation to each ecosystem service:

3. Does the company affect the quantity or quality of this ecosystem service? → **No** → Low/no impact

   ↓ **Yes**

4. Is the company's impact positive or negative?

   Positive | Negative

5. Does the company's impact limit or enhance the ability or others to benefit from this ecosystem service? → **No** → Medium impact

   ↓ **Yes**

   High impact

Figure 6.23 The WRI Corporate Ecosystems Review Step 2 assessment

6.29.4 Step 4

On the basis of the findings from step 3, business risks and opportunities are considered in relation to the five categories in the left-hand column of Table 6.16. Examples of risks and opportunities are provided under the relevant headings in Table 6.16.

6.29.5 Step 5

Essentially the identification of strategies to help minimise the risks and maximise the opportunities identified in step 4. Three categories of strategies are identified:

▶ Internal changes, i.e. things the organisation can do itself.
▶ Sector or stakeholder engagement, i.e. things that the organisation can only achieve in consultation or cooperation with suppliers, customers, competitors and other stakeholders.
▶ Policy-maker engagement, i.e. things that require wider change in society and demand government engagement in the form of lobbying or consultation.

The methodology recommends a classic management systems approach to these strategies with prioritisation followed by the establishment of management programme initiatives to drive implementation.

For more information on the ESR methodology, case studies, and to download a whole range of associated resources, visit http://www.wri.org. As a follow-up to this approach, in 2011 the World Business Council for Sustainable Development

Table 6.16 Example risks and opportunities arising from an ecosystems services review

Types	Risks	Opportunity
Operational	• Increased scarcity or cost of inputs • Disruption to business operations	• Increased efficiency • Low impact business processes
Legal/regulatory	• Permit limitations • Quotas • Prohibitions on activities or material usage	• New products to meet new regulatory requirements
Reputational	• Damage to brand or image	• Differentiation of brand
Market & product	• Changes in customer preferences	• New product/service opportunities • New revenue streams from company owned or operated ecosystem services
Financing	• Higher cost of capital • More rigorous lending requirements	• Increased investment by progressive investors and ethical investment funds

released the Guide to Corporate Ecosystem Valuation (CEV), which provides information on how to quantitatively, or in some cases monetarily, assess risks and opportunities related to ecosystem services. CEV can therefore be a logical next step after undertaking an ESR.

6.30 Business risk categories

An even more business risk-based approach was undertaken in 2004 by F&C Asset Management who produced a report which considered the impact or risk to business of biodiversity loss either in general terms or when directly attributable to them. The report presented the following seven categories of what they termed 'biodiversity risk'. Some are sector-specific and others are applicable to all organisations:

Access to land – access to new sites may be affected by a company's track record on protecting/restoring biodiversity and water resources.

Access to capital – poor biodiversity track record may led to an organisation being seen by lenders/investors as credit risk.

Security of supply – biodiversity loss may lead to a reduction in the quality or availability of materials essential to the organisation, e.g. fish.

Liabilities – unforeseen impacts of an organisation's activities on biodiversity may generate financial liabilities (in terms of civil claims or clean-up costs) even where no legal breach has occurred.

Relations with regulators – concerns about an organisation's track record on biodiversity management may lead to permit delays, increasing regulator costs (e.g. EP OPRA) or prosecutions, etc.

Reputation (and sales) – media coverage over issues such as genetically modified organisms, dolphin-friendly tuna, etc. can cause major, overnight reduction in consumer confidence in a brand or company, resulting in loss of sales.

Access to markets – an inability to meet specifications from substantial buyers such as government departments and agencies for sustainably sourced raw materials such as timber restricts access to a major market.

The clear ethos of the WRI approach and consideration of the above risks is that, though biodiversity varies considerably between ecosystems, governments and indeed organisations should always endeavour to ensure maximum biodiversity at whatever level of consideration is appropriate.

XII ADVANTAGES AND LIMITATIONS OF ENVIRONMENTAL TOOLS

All of the tools and techniques introduced above have strengths and weaknesses. Most have been mentioned directly or indirectly in the relevant sections. Table 6.17 provides a simple summary of some key pros and cons associated with each. It is not intended to be an exhaustive list:

Table 6.17 Advantages and limitations of environmental tools

Environmental management and assessment tool	Advantages	Disadvantages
Environmental risk assessment/aspects prioritisation	• Can be a simplified approach requiring minimal environmental knowledge • May provide systematic prioritisation of issues • Can allow benchmarking across an organisation and over time	• May be overly subjective and hence unreliable for comparative purposes • May be overly simplistic and fail to give due consideration to actual environmental impacts
Environmental auditing	• May provide an effective assurance mechanism for risk control or management systems	• May be overly bureaucratic and systems focused, providing little concrete benefit in terms of risk control or impact reduction
	• May provide ideas and opportunities for improvement • May highlight risks, liabilities and legal threats	• May be demanding in terms of staff time and disruption to core activities

Table 6.17 (continued)

Environmental management and assessment tool	Advantages	Disadvantages
Environmental management systems	• May provide a systematic and consistent approach to environmental improvement and control • May provide independently verified credibility of control standards • May act as a driver for continual improvement and best practice achievement	• May be overly bureaucratic and systems focused, providing little concrete benefit in terms of risk control or impact reduction • May be demanding in terms of staff time and disruption to core activities • May be misleading – masking mediocre actual environmental performance
Resource efficiency surveys	• May provide a clear picture of usage patterns and hence opportunities for improvement • Can provide a benchmark for reference at the start of an improvement programme • Can help quantify savings and hence justify investment in cost-benefit analysis terms	• May lead to an excessively narrow-viewed focus on one particular issue with improvements in one area leading to increased environmental impacts in other areas, e.g. supply chain, disposal, etc.
Environmental footprinting	• May provide an easily accessible and high impact communication tool	• Contains many built in assumptions and inaccuracies • The global perspective may detract from local priorities
Environmental impact assessment	• May provide early warning of key environmental impacts allowing cost effective design and planning decision making • May be used as a solid basis for stakeholder consultation and thereby ease concerns and reduce the likelihood of objections to development	• May be costly and very time consuming • May be 'diverted' by key stakeholders to achieve research or investigative goals beyond the scope of the development project • May overlook key issues due to the reliance on the scoping process
Strategic environmental assessment	• May provide 'joined up' thinking reducing the likelihood of cumulative impact from multiple smaller developments or management plans • Encourages stakeholder engagement in detailed public planning processes	• May be too high level to accurately predict the impacts of plans and policies • May be costly and very time consuming
Lifecycle assessment and eco-design	• May avoid shifting one impact to another when making design or supply chain improvements • May provide robust evidence for environmental claims • May provide information on supply chain liabilities or risks in the short or longer term	• May be highly intensive in terms of time and human resources • May be limited by lack of access to information • May have a skewed focus based on immediate stakeholder concerns rather than informed environmental evaluation
Supplier assessment methodologies	• May provide a systematic and consistent method for identifying liability and driving performance improvement through the supply chain • May provide assurance to downstream customers of the impact potential associated with upstream activities	• May be overly simplistic and symbol driven e.g. presence of a policy or even an EMS does not guarantee high standards of environmental performance • Can be time consuming for suppliers and assessors alike without clear benefits in terms of differentiation and selection
Carbon footprinting	• As for resource efficiency surveys above, plus; • May provide a clear sense of the cumulative contribution of an organisation to climate change • May meet customer/regulator requirements	• Conversion factors may lead to inaccuracies in calculations • Limited scope definition may provide misleading statements about 'carbon neutrality'

Table 6.17   (continued)

Environmental management and assessment tool	Advantages	Disadvantages
Corporate ecosystem services review	• May provide a systematic approach to raise understanding and awareness of an organisation's dependence on natural systems	• May be costly and very time consuming • May be limited by available information in the public domain
	• May be used to drive direct and indirect impact reduction measures • May help highlight supply chain risk particularly in relation to resources and hence enable effective long term sustainability planning	• May demand environmental expertise beyond the scope of many organisations

XIII   FURTHER RESOURCES

Table 6.18 presents some further resources.

6.18 Further resources

Topic area	Further information sources	Web links (if relevant)
Aspects identification and environmental risk assessment	ISO 14004:2004 – Environmental management systems – general guidelines on principles, systems and support techniques IEMA practitioner guide, 2006. *Risk management for the Environmental Practitioner*. IEMA	www.iso.org www.iema.net/shop
Environmental auditing	ISO 19011:2011 – Guidelines for auditing management systems	www.iso.org
Environmental management systems	ISO 14001:2004 – Environmental management systems – requirements with guidance for use EMAS:2009 – EC Eco-management and audit scheme	www.iso.org www.ec.europa.eu/environment
Resource efficiency surveys	WRAP and carbon trust case studies and assessment tools	www.wrap.org.uk www.carbontrust.com
Environmental footprinting	The living planet report and footprint calculators available from WWF	www.wwf.panda.org/
Life cycle assessment and eco-design	ISO 14040:2006 Environmental management – Life cycle assessment – principles and framework ISO 14006:2011 – Environmental management systems – guidelines for incorporating eco-design	www.iso.org
Sustainable procurement	DEFRA (2006) *Procuring the Future*. DEFRA BS 8903:2010 – Principles and Framework for Procuring Sustainably – guide Environment Agency, 2003 – commodity sustainability briefings document	www.gov.uk www.bsigroup.com www.gov.uk/environment-agency.
Carbon management	Carbon trust guidance and tools DEFRA (2013) *Environmental Reporting Guidelines – Including Mandatory Greenhouse Gas Reporting Guidance*. DEFRA	www.carbontrust.com www.gov.uk/defra
Corporate ecosystems review	Hanson *et al.* (2012) *The Corporate Ecosystems Review: Guidelines for Identifying Business Risks and Opportunities Arising from Ecosystem Change*. WRI publication. WBCSD (2011) *Guide to corporate ecosystem valuation*. WBCSD.	www.wri.org www.wbcsd.org

Analysing problems and opportunities to deliver sustainable solutions

Chapter summary

This chapter covers a wide spectrum of tools and techniques related to the analysis of environmental threats and opportunities and the development of strategies to deal with the issues identified. In many ways this chapter should be considered a follow-on from Chapter 6 but while that chapter focused on assessment techniques and linked to control strategies, this chapter has control strategies as its main focus. That said, in Section II of this chapter, we consider methods of evaluation and prediction used for specific issues that complement the various assessment methods covered in Chapter 6. The areas covered include:

- Environmental modelling

- Statutory contaminated land assessment

- Noise nuisance prediction using BS 4142

- Planning guidance relating to noise nuisance

- Predicting dust and odour nuisance

- Ecotoxicity assessments and the use of indicator species.

Section III then covers control strategies relating to pollution prevention and resource protection. This section too ties in to earlier coverage on management strategies covered in Chapter 6 and elsewhere and links are highlighted in the text. It is organised into the following areas:

- a general overview of pollution control techniques and emergency planning;
- materials and waste management to reduce environmental impact;
- carbon management strategies focusing on building design and transport planning;
- air pollution control techniques for particulates and gaseous pollutants;
- effluent treatment techniques in both public main sewer and industrial pre-treatment plants;
- contaminated land remediation techniques;
- solid waste disposal options;
- control approaches for noise, odour and dust nuisance;
- sustainable procurement strategies;
- biodiversity protection and enhancement strategies.

I   INTRODUCTION

It is impossible for an individual, let alone a company, to exist without causing some kind of environmental change or impact. Therefore, it is important to have a sensible way of deciding on what impacts are significant and which fall within the bounds of acceptability. There are many ways used to make such assessments of the significance of pollution. Some of the key assessment tools have been covered in Chapter 6 including:

▶ environmental risk assessment;
▶ environmental impact assessment;
▶ life cycle analysis;
▶ environmental auditing;
▶ resource efficiency evaluations;
▶ carbon footprinting;
▶ ecosystems services review.

In this section we will add some further 'issue-specific' assessment techniques, explore some examples of the benchmark standards used to determine the significance of an impact or risk and then provide an overview of sustainable solutions and programmes to address environmental problems and opportunities.

II   FURTHER IMPACT/ OPPORTUNITY ASSESSMENT TECHNIQUES

7.1 Statutory limits as the basis for acceptable risk/ impact

In the case of the more common and/or serious pollutants to air, land or water, local, national and even international limits may have been set in relation to the point of discharge or the receiving environment. Expressed as concentrations or total quantities over a defined time period, these limits provide the simplest significance assessment methodology:

Are the pollutants found in the discharge stream or receiving environment above defined statutory limits?

No – pollution level considered acceptable, i.e. not significant.

Yes – pollution level considered unacceptable, i.e. significant.

Limits appear in a wide variety of legal contexts and/or guidance documents. Some examples are provided in Table 7.1. Units used include:

▶ µg – microgram – 1 millionth of a kilogram;
▶ 10 percentile measurement – concentration exceeded by 90 per cent of the tests;

▶ 90 percentile measurement – concentration below which 90 per cent of the tests are measured.

Although attractive as a simple determination of significance, there are clearly some major flaws in the use of legal standards alone to determine significance of pollution. These include the fact that standards are usually negotiated rather than arrived at through scientific evaluation of what is harmful to individual receptors or the wider environment. This applies at a variety of levels, e.g.:

Table 7.1 Examples of statutory limits

Pollutant category	Legislative context	Example statutory limits
Air emissions	Clean Air Act, 1993	Ringlemann shade 2 or 4
	Environmental Permitting regime	Defined by BREF notes for industrial sector, e.g. electric arc furnaces particulate emissions from extraction plant – average concentration 10mg/m³ peak concentration 20mg/m³ mass emission 50g/tonne of product
	Ambient air quality regime – Environment Act 1995 Part IV	Annual mean concentrations: Benzene – 5µg/m³ Lead – 0.5µg/m³ Nitrogen dioxide – 40µg/m³ PM$_{10}$ – 40µg/m³ 1 hour mean concentrations: nitrogen dioxide – 200µg/m³ PM$_{10}$ – 50µg/m³ Sulphur dioxide – 350µg/m³
Effluent discharges	Water quality objectives (for receiving water bodies) set by the Environment Agency under powers granted by the Water Resources Act 1991	River Ecosystem Class 1 (cleanest) – Dissolved oxygen (% saturation) – 80% (10 percentile) BOD – 2.5 mg/l (90 percentile) Unionised ammonia – 0.021 mgN/l (95 percentile)
	Water Resources Act, 1991 effluent discharge consents	Specific to source and receiving water body but common standards used for treated sewage are: BOD – 20mg/l Suspended solids – 30mg/l
Land contaminants	Soil guideline values used as guidance by regulators when making decisions under the Town and Country Planning regime and the Contaminated Land regime	For residential sites with plant uptake: Arsenic – 20 mg/kg dry soil Lead – 450 mg/kg dry soil
Noise	World Health Organisation limits – measured at the receptor	The 2000 guidelines recommend a level of 30dB LAeq for undisturbed sleep and, to prevent people from becoming moderately annoyed during the daytime outdoor sound levels should not exceed 50 dB LAeq

industry–regulator liaison to establish BAT for a Permitted process;

international negotiation to agree concentrations of long-range pollutants in ambient air;

'planning gain' debates that take place in relation to a development proposal.

They take little account of the sensitivity of the local environment or cumulative pollutant loading from multiple sources.

7.2 Environmental modelling techniques

The term environmental modelling is used to describe a range of methods for predicting the physical, ecological and social outcomes of an activity. The approach is used extensively within environmental impact assessment (EIA) to facilitate decision-making with regards the acceptability of impacts associated with a development proposal and/or the need for project modification. There are two basic categories of modelling used: (1) physical modelling; and (2) mathematical modelling.

7.2.1 Physical modelling

Physical modelling, as the term suggests, involves building a representative model of the environment in which the development is planned and then subjecting it to the changes likely to be associated with the proposed activity. The approach can be extremely costly because of the scale of models that may need to be constructed but has been used to a high degree of accuracy in relation to hydrological and geological impacts, e.g. the prediction of changes to flow rates, erosion/sedimentation rates, etc. associated with a development such as a barrage or hydroelectricity scheme in an estuary.

7.2.2 Mathematical modelling

Mathematical modelling is much more commonly employed than physical modelling because of its flexibility and relatively low cost. Computer models are developed using mathematical parameters that can predict the outcomes of

events entered into the programme. Typical applications are emissions plume models (to predict dispersion characteristics from chimneys), acoustic models (to predict noise transfer and attenuation) and traffic flow models (to predict the impact of increased vehicle movements at various times of day).

This approach too has its limitations though, primarily linked to the number of variables that may be practically considered in any single model. However, for single issues with a manageable number of impact variables, the approach is very popular. It is even used at a global scale with highly complex models using advanced high speed computers to predict the impacts of variation in the carbon cycle on climate change. Such models must consider the impacts and inter-relationships between oceanic and atmospheric levels of carbon dioxide.

It is worth noting that the application of environmental modelling to ecological impacts is much less well developed, largely because of the complexity of inter-relationships between living organisms and their physical environment.

7.3 Contaminated land: an assessment methodology

Under the terms of the Environment Protection Act, 1990, Part 2A local authorities are required to carry out assessments to identify land within their jurisdiction that falls under the legal definition of 'contaminated land'. Guidance produced in 2012 provides a simplified set of definitions for contaminated land and aims to reduce the incidence of unnecessary and costly remediation works.

7.3.1 Legal definition of contamination

The general definition of contaminated land is:
Any land which appears to the local authority in whose area it is situated to be in such a condition, by reason of substances in, on or under the land that –
(a) significant harm is being caused or there is a significant possibility of such harm being caused; or

(b) significant pollution of controlled waters is being caused, or there is a significant possibility of such pollution being caused; (The Environmental Protection Act 1990, Part IIA)

The 2012 Statutory Guidance, however, goes on to provide four distinct grounds for designation as 'contaminated land' and then identifies four categories of contamination which are then sub-divided into four sub-categories of risk which are then used in deciding remediation requirements. The grounds for designation are as follows:

Significant harm is being caused to a human, or relevant non-human, receptor.
There is a significant possibility of significant harm being caused to a human, or relevant non-human, receptor.
Significant pollution of controlled waters is being caused.
There is a significant possibility of significant pollution of controlled waters being caused.

7.3.2 Categories of contaminated land

The *categories of contamination* are:

Significant harm to human health.
Significant possibility of significant harm to human health.

Significant harm and significant possibility of such harm (non-human receptors).
Significant pollution of controlled waters and significant possibility of such pollution.

These categories are used to classify the results of a risk assessment process conducted by the relevant Local Authority. Within each category the statutory guidance defines the benchmarks and the interpretation methods that should be used in assigning a risk category to a particular site. For example, in relation to the third category relating to non-human receptors, the Local Authority should use all appropriate information including consultation with the relevant countryside body to determine harm and harm potential to listed receptor groups as shown in Tables 7.2 and 7.3.

This approach to the risk assessment of contaminated land has taken a definitive step away from previous approaches that attempted to identify standard concentrations of contaminants in soil as a key measure in both human health risk assessment and remediation standards.

This risk-based approach is also applied to establishing remediation requirements, whether as part of statutory notices or voluntary agreements, aiming broadly:

to remove identified significant contaminant linkages, or permanently to disrupt them to ensure they are no longer significant and that

Table 7.2 Ecological system effects reference table

Relevant types of receptor	Significant harm	Significant possibility of significant harm
Any ecological system, or living organism forming part of such a system, within a location which is: • A site of special scientific interest • A national nature reserve • A marine nature reserve • An area of special protection for birds • A special area of conservation (a European site) • A candidate special area of conservation, potential special protection area or listed Ramsar site • Any nature reserve established under the National Parks and Access to the Countryside Act 1949	Significant harm includes: • Harm which results in an irreversible adverse change, or in some other substantial adverse change; or • Harm which significantly affects any species of special interest within that location and which endangers the long-term maintenance of that species at that location	Conditions would exist for significant possibility of significant harm where: • Significant harm is more likely than not to occur from the pollution linkage in question; or • There is a reasonable possibility of significant harm, and, if that harm were to occur, it would result in such a degree of damage to features of special interest that they would be beyond any practical possibility of restoration.

Source: Adapted from DEFRA Contaminated Land Statutory Guidance (2012).

Table 7.3 Property effects reference table

Relevant types of receptor	Significant harm	Significant possibility of significant harm
Property in the form of: • Crops, including timber • Produce grown for consumption domestically, or on allotments, • Livestock and other owned or domesticated animals • Wild animals which are the subject of shooting or fishing rights	For crops, a substantial reduction in yield or other loss in value resulting from death, disease or other physical damage. For domestic pets, death, serious disease or serious physical damage. For other property in this category, a substantial loss in value resulting from death, disease or other serious damage.	Conditions would exist for considering that a significant possibility of significant harm exists to the relevant types of receptor where harm is more likely than not to result from the contaminant linkage in question (a risk based approach using probabilities based on pollutant linkage and ecotoxicity assessments).
Property in the form of buildings	Structural failure, substantial damage or substantial interference with any right of occupation. Substantial damage or substantial interference may be said to be occurring when any part of the building ceases to be capable of being used for the purpose for which it is or was intended. In the case of a schedule Ancient Monument, substantial damage should also be regarded as occurring when the damage significantly influences those attributes by reason of which the monument was scheduled.	Conditions would exist for considering that a significant possibility of significant harm exists to the relevant types of receptor where significant harm is more likely than not to result from the contaminant linkage in question during the expected economic life of the building (or in the case of a scheduled Ancient Monument, the foreseeable future).

Source: Adapted from DEFRA Contaminated Land Statutory Guidance (2012)

risks are reduced to below an unacceptable level;

to take reasonable measures to remedy harm or pollution that has been caused by a significant contaminant linkage.

This reflects the philosophy of 'avoiding harm' rather than meeting any particular concentration of contaminants in soil or groundwater. The 2012 statutory guidance still interfaces to a degree with earlier guidance and methodologies produced by the Environment Agency and DEFRA, notably the CLEA methodology. The latter is a risk assessment process based on a consideration of the health risks posed to the most sensitive human receptor likely to be present on the land under consideration given the land use classification. The critical receptor for residential land/allotments is a female child aged up to 6 years, whereas the critical receptor for commercial/industrial land use is a female adult over a working lifetime of 42 years (ages 17–59).

7.4 Predicting noise nuisance: BS 4142

In Chapter 3, we discussed the use of noise limits in defining operational standards in relation to the management of noise from a nuisance perspective. Often the setting of limits is preceded by one of two things:

▶ a complaint that leads to an investigation by the generator of the noise or a regulator;

▶ a standard assessment methodology conducted as part of an EIA, planning or operational consent application aimed at predicting the likelihood of nuisance associated with a particular development.

It is the second scenario that will be considered initially in this section. A British Standard, BS 4142, provides us with a standard methodology for measuring noise levels and predicting the likelihood of related nuisance, based on a comparison with background noise levels.

7.4.1 **An overview of BS 4142**

This section is not intended to be a full explanation or representation of the BS 4142 methodology but merely provides a summary of the key steps undertaken. Readers interested in using the methodology to assess noise nuisance potential should refer to BS 4142. In layman's terms there are three essential stages to the BS 4142 methodology as shown in Figure 7.1.

As noise nuisance potential is affected by the context in which it occurs and the characteristics of the noise, as well as the absolute intensity (loudness), there are adjustment factors introduced in the methodology that prioritise noise sources that might be considered particularly annoying or that occur in areas where they stand out as being much louder than other noise sources locally.

There is a lot of jargon used in BS 4142 and it is important to ensure that the correct measures are used at the different stages in the calculation. For delegates not familiar with noise terminology, the glossary of terms in Section 7.4.6 will prove useful. It is useful to remember that the methodology is not really an acoustic appraisal, so much as a risk assessment using noise monitoring data. As such, to a lay person, it is often confusing to try to understand why a particular noise monitoring term or measure is used. Our advice is – don't try – simply follow the methodology closely and recognise that this is a planning and design tool to facilitate decision-making, rather than a procedure to calculate particular noise levels.

The methodology is used by regulators and operators alike in reviewing the likely nuisance implications of a new noise source. An overview of the calculation and then the stages of the methodology are provided below:

Rating noise level – background noise level = X

Where:

X = a number that can be compared with the table in BS 4142 (and reproduced in Table 7.5) to determine the likelihood of complaints were the noise source to be introduced at the location as proposed.

Rating noise level = the noise level of the new source (known as *the specific noise level*) adjusted for average background context and annoyance characteristics.

Background noise level = *not* the average background noise but the background levels exceeded for 90 per cent of the time, i.e. when most other sources of noise are 'off'. This could be thought of as the 'real' background noise level.

The example in Box 7.1 provides an illustration of how the terms apply in practice.
Try to keep sight of this overview of the methodology as we describe the stages in a little more detail.

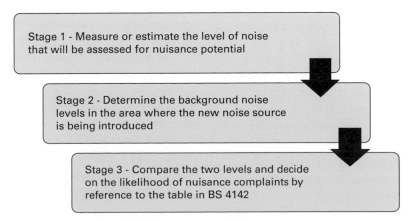

Stage 1 - Measure or estimate the level of noise that will be assessed for nuisance potential

Stage 2 - Determine the background noise levels in the area where the new noise source is being introduced

Stage 3 - Compare the two levels and decide on the likelihood of nuisance complaints by reference to the table in BS 4142

Figure 7.1 The three basic stages of BS 4142

Box 7.1 Water pumping station noise level

Imagine a proposal to build a water pumping station in a rural area near a main road and next to a residential dwelling.

▶ The specific noise level would be the noise measured (in LAeq terms, see Section 7.4.6) with the pumping station operating and would include any noise generated by the road too. It is this number that would be 'adjusted' to provide the rating noise level.

▶ The background average noise level would be the noise measured (in LAeq terms)

with the pumping station absent but with road noise included (as it is likely to be a significant source of noise in this rural location).

▶ The background noise level exceeded 90 per cent of the time (in noise monitoring terms this is the LA_{90} reading on the noise meter) might be expected to be much lower than the average level as it would perhaps represent those times during the day when road noise dies down to a minimum.

7.4.2 Stage 1 Measure or estimate the level of noise that will be assessed for nuisance potential

In BS 4142 terminology this is the determination of the *specific noise level* (the sound levels with the potential nuisance noise present) and conversion into a *rating noise level* (which involves a series of adjustments or weightings).

With due regard to the monitoring equipment and location guidance referred to earlier, noise levels from the source under investigation should be determined. Specific reference time intervals for monitoring are required to be used by BS 4142 and are as follows:

▶ 1 hour during the day
▶ 5 minutes during the night.

These intervals may not, however, be appropriate measurement periods in some cases because of variation in levels or characteristics in the noise source being assessed. For this reason BS 4142 provides a method for converting LAeq measurements made over different monitoring periods into equivalent readings for the defined reference periods. The methodology may be summarised as follows:

Measure (or estimate) specific noise over a representative period (Tm) that reflects variations in the specific noise source, e.g. if the noise being assessed varies over a 2-hour cycle then it should be measured over the whole of

that 2-hour period to give an LAeq(2 hours). If Tm is the same as the reference period provided in BS 4142 (Tr = 1 hour in the day and 5 mins at night) then no time correction is required.

If the specific noise is continuous, then Tm may be less or more than the reference period without a time correction being required.

If, as in our example, Tm differs from Tr, then BS 4142 provides calculation methods to convert LAeq(Tm) into LAeq(Tr).

Following conversion we are left with an LAeq (day time) and an LAeq (night time).

It should be noted that as BS 4142 is often used as a development planning tool, measurement of the proposed noise source at the place where it is proposed to be is frequently not possible. Often manufacturer's data or noise monitoring data from similar sites that are already in operation are used as a 'guesstimate' of the specific noise source. Clearly this introduces scope for error as it assumes that the measurements are taken with comparable background noise levels in place. Care must be taken if using monitoring data from comparable sites that the influence from other noise sources is as low as possible. Using our pumping station example, we would want to use measurements taken at a similar existing site with nearby road noise (or other significant sources) at levels that most closely correlate to the location of the new development.

Now that we have our specific noise level (normalised against the reference time periods) we must make two adjustments to convert it into the rating noise level which is what we need for the Stage 3 comparison.

7.4.2.1 Adjustment 1: considering average background noise levels

The *average background noise* (what we hear) comprises *residual noise* (e.g. traffic, wind, electricity pylons, nearby school, etc.) and the *specific noise* (e.g. the pumping station) when present. When taking measurements to determine the specific noise level, it is important to distinguish between the specific noise and the residual noise. Care should be taken to minimise the influence of other noise sources, by measuring at times when other sources (the residual noise) are at low levels. BS 4142 provides us with a compensation table to take account of the influence of residual noise.

To use the example provided earlier in Box 7.1 relating to the pumping station – we would subtract the average noise level measured with the pumping station off from the noise level measured with the pumping station on. The difference between the two is then compared to Table 7.4 and, where appropriate, a correction subtracted from the specific noise level.

We can assume here that the noise is likely to be fairly continuous and therefore we probably will not need a time correction to match the reference time periods demanded by BS 4142, i.e. monitoring for the reference time periods is likely to be representative of the noise generated by the pumping station.

The reason for this first adjustment is so that noise sources that are significantly higher than the average background levels are weighted in terms of nuisance potential, or more strictly speaking, noise sources closer to the average background levels are 'downgraded' in terms of nuisance potential.

To add numbers to our example:

> Specific noise level (with pumping station operating) – 65dBA
>
> Residual noise level (with pumping station off) – 58dBA

Difference = 7dBA so by reference to Table 7.4 we need to subtract 1dBA from the specific noise level hence:

> Following adjustment 1 the specific noise level is considered to be – 64dBA

7.4.2.2 Adjustment 2: considering annoying noise characteristics

In addition, we may also need to adjust or weight the corrected specific noise level to take into account the 'complaint potential' of certain noise characteristics. Certain acoustic features increase the likelihood of complaint beyond that which could be expected from a simple comparison between the specific noise level and the background noise level. A +5dB correction factor is added to the specific noise level where one or more of the following features is present in the specific noise:

▶ The noise contains a distinguishable, discrete, continuous note (whine, hiss screech, hum, etc.).

Table 7.4 Specific noise level adjustment 1

Corrections to noise level readings	
Difference between noise level readings with specific noise present and absent (dB)	Correction to be *subtracted* from noise level reading with specific noise present (dB)
>9	0
6 to 9	1
4 to 5	2
3	3
<3	Detailed measurements required.

Note: When measuring the residual noise level, all other conditions should be similar to those existing when the measurements were taken with the specific noise present.

Source: Adapted from BS 4142.

- ▶ The noise contains distinct impulses (bangs, clicks, clatters or thumps).
- ▶ The noise is irregular enough to attract attention.

Again, remember that the purpose of both these adjustments is to weight those noise sources most likely to generate nuisance complaints.

The resulting number after both of these adjustments have been applied is referred to as the *rating noise level*.

Again to continue with our example calculation:

> Following adjustment 1, the specific noise level is considered to be – 64dBA

Let's assume there is a continuous hum associated with the pumping station operation, so we add 5dBA:

> Following adjustment 2, the specific noise level is considered to be – 69dBA

It is now known as the *rating noise level* in BS 4142 terminology.

7.4.3 Stage 2 Determination of background noise level

Background noise levels should be measured at the same location as the specific noise measurements if possible or at a location where the background levels may be presumed to be equivalent. The measurement time interval should be sufficient to obtain a representative sample of background noise, which may vary over time (e.g. peak time traffic) or be significantly influenced by weather conditions (e.g. wind direction).

The measurement time should also correspond to the days of the week and times of day that the specific source will be operating.

Background noise levels are not measured as LA_{eq} but as LA_{90}, i.e. the A-weighted noise level exceeded for 90 per cent of the specified measurement period. This is the difference between the residual noise level (in LA_{eq}) mentioned in stage 1 and the background noise level (in LA_{90}) referred to here. Averaging sound level meters have both of these as optional settings. While the LA_{90} reading is not the absolute lowest level measured in any short-term samples, it gives a clear indication of the underlying noise level, or the level that is almost always there in between intermittent noisy events.

7.4.4 Stage 3 Assessment of complaint potential

Once sound levels been obtained for both background (expressed as the LA_{90} reading) and the adjusted specific source (known as the rating noise level), a comparison of the two is undertaken. The likelihood of complaints is obtained by subtracting the measured background noise level from the rating level. Logically the greater this difference, the greater the likelihood of complaints. The BS 4142 likelihood scale is shown in Table 7.5.

Table 7.5 BS 4142 likelihood of complaints reference table

Difference between rating level and background level	Likelihood of complaints
+10 dB or more	Complaints are likely
Around +5 dB	Marginal significance
More than 10dB below the measured background noise level	Complaints are unlikely

Source: Adapted from BS 4142.

7.4.5 Usefulness and limitations of BS 4142

Local Authorities frequently use BS 4142 to help set noise limits or consent conditions in planning conditions, abatement notices and Section 61 consents (under the Control of Pollution Act) issued in relation to construction sites. A developer or site operator could similarly use the methodology to help design noise control measures or determine where noisy equipment or operations should be located.

Remember, however, that the methodology cannot be used as a defence against nuisance complaints. That is partly because it is a 'risk assessment methodology' and also because there are no universally acceptable levels at which nuisance occurs, each case being considered in relation to the nuisance factors described in the civil law section of Chapter 3.

7.4.6 Glossary of noise terms

Table 7.6 presents a glossary of noise terms.

7.5 Planning and noise nuisance

BS 4142 is used in relation to the introduction of a new industrial noise source to an area. In contrast, the 'National Planning Practice Guidance – Noise' is used by planning authorities to guide planning decisions that may create noise nuisance. This may occur either by the introduction of new noise-generating developments which could affect existing receptors or by allowing residential development in areas where existing noise levels are likely to lead to disturbance.

Essentially the guidance works by classifying the effect of noise levels and applying a hierarchy of response from the perspective of the planning department. A summary of the classifications and the hierarchy of response is shown in Table 7.7.

7.6 Predicting dust and odour nuisance

Unlike noise, there are no British Standard methodologies for assessing the nuisance

Table 7.6 Glossary of terms of noise level measurement

Frequency of sound	Sound is the quickly varying pressure wave travelling through a medium. When sound travels through air, the atmospheric pressure varies periodically. The number of pressure variations per second is called the frequency of sound, and is measured in Hertz (Hz) which is defined as cycles per second. The higher the frequency, the more high-pitched a sound is perceived. The sounds produced by drums have much lower frequencies than those produced by a whistle.
Hertz (Hz):	Unit of frequency, equal to one cycle per second. Frequency is related to the pitch of a sound.
Loudness and the decibel scale	Another property of sound or noise is its loudness. A loud noise usually has a larger pressure variation and a weak one has smaller pressure variation. Pressure and pressure variations are expressed in Pascal, abbreviated as Pa, which is defined as N/m^2 (Newton per square metre). The human ear can perceive a very wide range of sound pressure. To express sound or noise in terms of Pa is thus quite inconvenient because we have to deal with numbers from as small as 20 to as big as 2,000,000,000. A simpler way is to use a logarithmic scale for the loudness of sound or noise, using 10 as the base.
dB(A):	Decibels are measured on a sound level meter incorporating a frequency weighting (A weighting) which differentiates between sounds of different frequency (pitch) in a similar way to the human ear. Measurements in dB(A) broadly agree with people's assessment of loudness. A change of 3 dB(A) corresponds roughly to halving or doubling the loudness of a sound. The background noise level in a living room may be about 30 dB(A); normal conversation about 60 dB(A) at 1 metre; heavy road traffic about 80 dB(A) at 10 metres; the level near a pneumatic drill about 100 dB(A).
Equivalent continuous sound pressure level	The steady state dB(A) level which would produce the same A-weighted sound energy over a stated period of time as a specified time-varying sound. Equivalent continuous A-weighted sound pressure level is widely used around the world as an index for noise. It is defined as 'the A-weighted sound pressure level of a noise fluctuating over a period of time T, expressed as the amount of average energy'.

Table 7.6 (continued)

Percentile sound levels	Percentile levels are used greatly when measuring environmental noise, ie that which may pollute the environment.
	Ln, where n may be anything from 1 to 99, is that noise level exceeded for n% of the measurement time. By definition of percentiles, L1 must be greater than or equal to L2, which must be greater than or equal to L3, etc. It is often the case that only a few Ln values are ever used.
	An example of how Ln values look in a graphical format:
	L10–L90 is often used to give a quantitative measure as to the spread or 'how choppy' the sound was.
L_{A10}, T	The A weighted level of noise exceeded for 10 per cent of the specified measurement period (T). It gives an indication of the upper limit of fluctuating noise such as that from road traffic.
L_{A90}, T	The A weighted noise level exceeded for 90 per cent of the specified measurement period (T). In BS 4142: 1990 it is used to define background noise level.
L_{Aeq}, T	The equivalent continuous sound level of a notionally steady sound having the same energy as a fluctuating sound over a specified measurement period (T).
L_{Amax}	The highest A weighted noise level recorded during a noise event.
Rating level	The noise level of an industrial noise source which includes an adjustment for the character of the noise. Used in BS 4142: 1990.

potential of dust or odour. Essentially a risk assessment approach is demanded by regulators in order to determine the scale of nuisance potential. Factors that would need to be considered include:

▶ the nature of the dust or odour (chemical constituents, health and property harm potential, etc.);

▶ any relevant ambient air limits, e.g. PM_{10} set under the UK air quality strategy;

▶ the sensitivity of receptors/neighbourhood characteristics, e.g. residential location or heavy industry area;

▶ dilution and dispersion rates and conversely deposition rates and concentration points;

▶ the presence of other sources of dust and odour in the same vicinity;

▶ the frequency and scale of the dust/odour generation.

For both dust and odour, guidance and/or indicator tests are available to support the risk assessment approach. These are referred to in the sections below.

7.6.1 Odour

Guidance in relation to the assessment of odour nuisance appears in both the H4 Environmental Permitting horizontal guidance note and in the DEFRA Odour Guidance for Local Authorities. Odour nuisance is extremely subjective and, therefore, inherently difficult to predict. For development proposals, the approach to odour prediction is essentially similar to the mathematical modelling approach used for other air emissions. The questions asked are:

▶ Given the probable weather conditions and background air quality, what will be the dispersal pattern of the odorous gases?

▶ Will any human receptors be within detectable range of the odour source?

For both development proposals and existing odour complaints, the DEFRA guidance then

Table 7.7 Planning guidance and noise nuisance

Perception	Examples of outcomes	Increasing effect level	Action
Not noticeable	No Effect	No Observed Effect	No specific measures required
Noticeable and not intrusive	Noise can be heard, but does not cause any change in behaviour or attitude. The background noise levels of the area may be affected but not so much as to lead to a perceived change in the quality of life.	No Observed Adverse Effect	No specific measures required
		Lowest Observed Adverse Effect Level	
Noticeable and intrusive	Noise can be heard and causes small changes in behaviours and/or attitudes of those affected, e.g. turning up volume of television; speaking more loudly; closing windows for some of the time because of the noise. Potential for sleep disturbance but not to the extent of being kept awake. Affects the acoustic character of the area such that there is a perceived change in the quality of life.	Observed Adverse Effect	Mitigate and reduce to a minimum
		Significant Observed Adverse Effect Level	
Noticeable and disruptive	The noise causes a material change in behaviour and/or attitude in those affected, e.g. having to keep windows closed most of the time, avoiding certain activities during periods of intrusion. Potential for sleep disturbance resulting in difficulty in getting to sleep, premature awakening and difficulty in getting back to sleep. Quality of life diminished due to change in acoustic character of the area.	Significant Observed Adverse Effect	Avoid
Noticeable and very disruptive	Extensive and regular changes in behaviour of those affected and/or an inability to mitigate effect of noise leading to psychological stress or physiological effects, e.g. regular sleep deprivation/awakening; loss of appetite, stress symptoms, etc.	Unacceptable Adverse Effect	Prevent

Source: Adapted from DEFRA, the National Planning Practice Guidance – Noise.

proposes consideration of what it refers to as the FIDOL factors determining offensiveness, as shown in Table 7.8.

Essentially if modelling or on site 'sniff tests' suggest odours will be (or are) detectable by receptors and if the FIDOL factors point towards the odour being offensive in nature, complaints are likely. In such cases Local Authorities may be expected to support claims of statutory nuisance and/or the imposition of planning or permitting conditions.

7.6.2 Dust; the Beaman and Kingsbury method

The use of sticky pads to assess airborne dust nuisance was first devised and described by Beaman and Kingsbury (1981), copies of which can be obtained from the National Society for Clean Air (NSCA). The methodology they describe is a

cheap and easy way of assessing dust nuisance, but should not be regarded as a substitute for quantitative monitoring for health effects.

The use of sticky pads has certain drawbacks, such as their susceptibility to losing dust in heavy rain, but as this is precisely what happens to cars, window sills, etc. in similar situations, it can be regarded as a fairly good method of assessing perceived dust nuisance. Alternatively, if the sticky pad is exposed for too long, then it can become saturated with dust, since new dust will not stick on top of dust already trapped. Suitable exposure periods are usually between 2 and 7 days.

After exposure, pads should be covered to prevent further pick-up prior to analysis. Analysis is conducted using a sticky pad reader which provides a measure of reflectivity of the exposed pad compared with a clean pad. The Effective Area Coverage (EAC) is calculated by subtracting

Table 7.8 FIDOL factors determining the offensiveness of a particular odour

The FIDOL factors determining offensiveness	Comments
Frequency (how often an individual is exposed to a particular odour)	Even an odour generally perceived to be quite pleasant can be considered a nuisance if exposure is frequent. At low concentrations a rapidly fluctuating odour is more noticeable than a steady background odour.
Intensity (the perceived strength of the odour)	Generally the higher the intensity, the more likely the nuisance.
Duration (the length of a particular odour event or episode, i.e. the duration of exposure to the odour)	Generally the longer the exposure the more likely the nuisance although see the comments about fluctuating odour above.
Offensiveness (based on a mixture of odour character and offensiveness/attractiveness at a given concentration/ intensity – there is a standard 'offensiveness' scale used.)	Some odours are universally considered offensive, such as decaying animal matter. Others may be offensive only to those suffering unwanted exposure in residential proximity, e.g. coffee roasting odour.
Location (the type of land use and nature of human activities in the vicinity of an odour source, i.e. the tolerance and expectation of the receptor)	Relates to the characteristics of the neighbourhood and the sensitivity of the receptors, e.g. complaints about the odour of animal manure are less likely to be justifiable in a rural area than in an urban or predominantly residential area.

Source: After the DEFRA Odour Guidance for Local Authorities.

the sticky pad reader result from 100, and then dividing by the number of days' exposure to give per cent EAC/day. Beaman and Kingsbury conducted research that provided the typical levels shown in Table 7.9.

Their research also provided the complaint thresholds shown in Table 7.10, though these

varied to some extent depending on the colour of the dust.

As with odour, this approach can be used as part of a predictive modelling assessment to determine the significance of a new development in terms of likely dust nuisance, or to monitor existing activities where dust generation potential exists in order to assess when additional controls might be necessary.

Table 7.9 Typical dust level readings in different land use areas

% EAC/day	Situation
0.01	Rural
0.02	Suburban
0.3–0.4	Urban
0.5	Rural summertime
0.8–1.0	Industrial

Source: Based on Beaman and Kingsley (1981).

Table 7.10 Dust level readings linked to probability of nuisance complaints

% EAC/day	Response
0.2	Noticeable
0.5	Possible complaints
0.7	Objectionable
2.0	Probable complaints
5.0	Serious complaints

Source: Based on Beaman and Kingsley (1981).

7.7 Assessment of environmental toxicity

Unlike human toxicity testing which is dealing with a single receptor species, environmental pollution must consider a wide variety of potential receptors organisms. In order to determine the likely impact of a pollutant release in the environment, a number of stages are commonly required.

7.7.1 Choosing indicator species

Biological indicator species (bio-indicators) are selected for a number of reasons to be representative of pollution impacts to their ecosystem. The following criteria might be considered in selecting an appropriate bio-indicator to be tested or observed to predict or monitor the pollution potential in a given ecosystem:

- The species is relatively abundant and typically present in the ecosystem likely to be affected, e.g. freshwater microorganisms such as the caddis fly larvae in UK streams.
- The species displays sensitivity to pollutants of interest, e.g. lichens are particularly sensitive to air pollution.
- The species manifests observable changes in relation to pollutants of interest, e.g. some micro-organisms will produce new proteins, called stress proteins, when exposed to contaminants like cadmium and benzene. These stress proteins can be used as an early warning system to detect high levels of pollution.

Different indicator species may be required, depending on the primary pollution pathway, as well as the specific ecosystem or pollutant characteristics. For the aqueous environment, indicator species might include a freshwater fish, a freshwater invertebrate and freshwater green algae. For sediments and soils, they might be sediment dwelling organisms, earthworms or terrestrial plants. For air, the indicator species would often be an appropriate bird species.

7.7.2 Consider critical levels on target species

This is more difficult than it might at first appear to be. There has been extensive (though hardly exhaustive) testing of environmental pollutants in relation to key target organisms (those likely to be selected as bio-indicators). Tests have traditionally been done with acute concentrations of pollutants used to assess the lethal dose (LD) on the test organisms. Often expressed as LD_{50} – the concentration at which 50 per cent of the target population will die. Such acute exposures have limited value in terms of environmental pollution where exposure levels will typically be much lower but often experienced over extended periods. Lethal dose measurements can be useful 'worst case' markers for major incident planning or monitoring. However, they are less useful with considerations of sub-lethal chronic exposure which may lead to reductions in growth rate, reproduction rates, immune suppression or mutagenic effects,

etc. – all of which will also affect the target species populations in the longer term. This highlights the need for on-going monitoring of bio-indicator species to look at the cumulative impact of chronic exposure to sub-lethal levels of pollutants. The issue is further complicated by the potential for enhancement or mitigation interactions between two or more pollutants, e.g. the health impacts of sulphur dioxide are much greater when PM_{10} particulates are present in smog. Conversely, the toxicity of ammonia in water decreases with a decrease in pH. Even varying climatic conditions such as ambient temperatures in the receiving environment may affect the tolerance of receptors or the toxicity of pollutants.

7.7.3 Consider critical loads on target ecosystems

Due to the limitations of usage of lethal dose data, other methods of environmental toxicity assessments have been and continue to be developed. In Europe the most widely used approach involves the estimation of *Predicted Environmental Concentration (PEC)* and the *Predicted No Effect Concentration (PNEC)*.

Calculation of the PEC of a pollutant or group of pollutants might involve the sort of environmental modelling assessments detailed in an earlier part of the book with due consideration being given to criteria such as:

- discharge concentrations
- dilution and dispersal rates and directions
- biodegradation rates of the pollutants
- temporal variation of pollutants
- physical and biological characteristics of the receiving environment.

The calculation of the PNEC involves consideration of exposure testing data, including lethal dose testing and data from ecological monitoring. It is a 'best guess' of the environmental toxicity threshold for a pollutant or group of pollutants under specific environmental conditions in the presence of specific target organisms. The use of such estimates, while not scientifically reliable, is based on the precautionary principle and the need to determine appropriate levels of action. The approach has been used to map pollution loads and/or limits, taking into account various physical, chemical, ecological, geological and hydrological factors that may affect deposition, dilution, etc. of pollutants and/or the sensitivity of receptors. One example of where this approach has found its way into the legislative regimes is the use of air quality limit values within the National Air Quality Strategy, which are essentially PNECs in relation to human receptors.

As indicated in Section 7.7.2 on critical exposure levels, the PEC/PNEC approach has significant limitations in terms of accuracy because of the uncertainty introduced by pollutant interactions, chronic exposure effects and the likelihood of bio-accumulation, etc. However, as a best estimate approach applying the precautionary principle, it provides a basis for decision-making and may help inform appropriate on-going monitoring of key bio-indicators or environmental condition indicators.

III CONTROL STRATEGIES AND SUSTAINABLE SOLUTIONS

7.8 Pollution prevention techniques: an overview

Technical pollution control and arrestment techniques are covered in Sections 7.2–7.17, however, the basic principles of preventing pollution to air, surface water, groundwater and soil as well as noise nuisance prevention are outlined below. The Environment Agency and SEPA produce a series of Pollution Prevention Guidance (PPG) Notes covering a variety of pollution prevention and response issues. The full list of PPG Notes is available for download from the Environment Agency and SEPA websites.

7.8.1 Air pollution control techniques

The choice of technique used depends on the nature of the air pollutant and the standard of 'clean-up' or emissions control required, however, some commonly used techniques include:

▶ *Filtration*, e.g. bag filters which are useful for dust or particulates collection. They may be either fitted to a single machine or point of emission or used as a whole area collection system.
▶ *Separation systems*, e.g. gravity settlement chamber – used for heavy particles which can be contained within a chamber designed to capture them as the air emissions stream passes through.
▶ *Scrubber systems*, e.g. acid gas scrubber which uses fine water mists to capture acid gases which dissolve in the water vapour and can then be collected and neutralised before disposal as an effluent.
▶ *Water curtains*, used in situations such as paint spray booths, to prevent the escape of the airborne pollutants from an enclosed area.

7.8.2 Water pollution control techniques

These may be broadly divided into pollution prevention and waste water treatment techniques.

7.8.2.1 Pollution prevention techniques

▶ *Surface water treatment*. Rainwater run-off can be contaminated with silt, heavy metals, chemicals and oil, which can be damaging in watercourses and groundwater. Best practice is to control pollution at its source by preventing the contamination of rainwater run-off. However, in some cases, such run-off will require treatment before discharging from a site. A common example is found in areas where there is a high risk of oil pollution, where oil separators (interceptors) are installed to capture hydrocarbons and thereby reduce the risk of pollution leaving the site.
▶ *Wrong connections*. Wrongly connected effluents can cause severe pollution problems which can be difficult to remedy. Sources of dirty water, such as sinks and toilets, should always be connected to the foul sewer (leading to an effluent plant or sewage treatment facility). Manhole covers and gullies should be clearly marked, for example, by

colour coding with red for foul and blue for surface water, and site drainage plans should be readily accessible.

▶ *Cleaning activities*. Wash water from mobile pressure washers should not be discharged to surface water drains, watercourses or soakaways. No detergents are suitable for discharge to surface drains, even if described as biodegradable, so such activities should be carried out in designated, contained areas draining to the foul sewer (subject to the approval of the local sewerage undertaker). Alternatively, closed loop vehicle wash recycling systems are available.

▶ *Deliveries*. Special care should be taken during deliveries, particularly when hazardous materials are involved. Deliveries should be supervised at all times, tanks and containers should be labelled with the nature and volume of their contents, and the content levels should be checked prior to delivery to prevent overfilling. Loading and unloading areas should be clearly marked and isolated from the surface water drainage system either by catch pits or sumps with isolating valves, or by roofing. Cut-off valves in the drainage system and raised kerb surrounds may be needed with drainage to the foul sewer if possible. Delivery pipes should be fitted with automatic cut-off valves to prevent overfilling. Adequate bunding should be installed to prevent the release of contaminants to soil or drainage systems (this is a legal requirement in some cases).

▶ *Security*. Vandalism and theft are frequent causes of pollution. Lockable valves should be fitted on all storage tanks, fences should be secure and doors and gates kept locked. Where possible, materials should be stored under cover and on delivery, potential pollutants should be transferred into safe storage without delay.

▶ *Flood plain developments*. All drainage manhole covers which lie within a flood plain should be of the sealed, screw-down cover design, and sink waste gullies should be built up above flood level. The construction and use of chemical stores within a flood plain generally require prior consultation with the Environment Agency.

7.8.2.2 Waste water treatment techniques

As with air emissions treatment, the choice of technique used depends on the nature of the effluent pollutants and the standard of 'clean-up' required. Some commonly used techniques include:

▶ *Settlement/flotation chambers*. These are similar in principle to the gravity settlement chambers used for air pollution control. With effluent, chemical additives are sometimes used to accelerate the process of settlement or flotation of the pollutants to the bottom or top of the tanks. Settled solids are removed from the tank bottoms periodically as a sludge, while floating materials are often skimmed off the effluent. In both cases the pollutant stream will need to be treated or disposed of as a solid waste.

▶ *pH correction*. A very simple and very common effluent treatment technique, pH correction relies on the addition of acid or alkali substances to the effluent to bring it back to a neutral pH which is safe to discharge. Hydrochloric acid is a commonly used acid and sodium hydroxide a commonly used alkali.

▶ *Cooling towers*. Cooling towers are a simple physical treatment technique used where effluent is generated at a temperature that would cause harm if discharged to a water course. Normally the volume of effluent is high otherwise the receiving water body would quickly dilute any local temperature increase. Cooling towers work by simply pumping water up high and allowing radiant heat loss to occur into the air as the water falls back to the collection chamber.

7.8.3 Land pollution control techniques

Techniques used to control or prevent the entry of pollutants to land are essentially similar to the water pollution prevention techniques covered above. The 'clean-up' techniques used to remove pollutants from land once it has been contaminated vary widely and, as always, the appropriate techniques depends on the characteristics and extent of contamination, as well as the clean-up standard required. Examples include:

▶ *Dig and dump*. Essentially entails removing contaminated soil or hard standing from site and disposing of a solid waste which may be landfilled or treated off site.
▶ *Barriers*. Where the movement of contaminants in soil is limited or the contact with receptors is unlikely, barriers may be installed to contain the contamination in situ. For example, asbestos found underground on an industrial site may best be dealt with by capping and mapping to ensure that neither release nor human contact is permitted.
▶ *Soil washing*. On major contaminated sites, equipment may be installed to enable the literal washing of contaminated soils. The technology is available to process significant tonnages of spoil but, as might be imagined, this is an extremely expensive process, and in addition, will generate quantities of effluent or sludge from the washing process which will require disposal.

7.8.4 **Noise nuisance control techniques**

Noise nuisance control techniques may be divided into management controls and technical abatement techniques.

▶ *Management controls*. These include measures such as the regulation of working hours to avoid night-time or weekends (when local residents are more likely to be affected). Traffic management plans can choose routes that prevent high volumes of traffic passing sensitive receptors – either completely or at particular times of day. Preventative maintenance can head off issues associated with mechanical wear (bearings on extractor fans, for example, or loose or damaged panels on excavators and other construction plant).
▶ *Technical controls*. These are extremely varied but may include things such as cladding of equipment or buildings, the use of dampers on static equipment fixed to concrete floors, or simply the selection of equipment, plant or vehicles that emit lower noise levels than competitors.

7.8.5 **Emergency planning**

Even well-controlled sites can have accidents and, if storing hazardous materials of any kind, a spill, fire or other emergency scenario may represent not only a human health risk but also a major source of pollution. For this reason, UK legislation such as the COMAH Regulations 2006, the Environmental Permitting Regulations 2010, and even planning law, may require that sites storing quantities of hazardous materials have tried and tested incident prevention and emergency response arrangements in place. Even where there is no legal requirement to have such emergency plans in place, it is generally considered to be good practice. ISO 14001 highlights this fact and requires an organisation to establish a procedure to do the following:

▶ identify potential emergency situations and potential accidents that can have an impact(s) on the environment and how it will respond to them;
▶ identify and respond to accidents and emergencies, and prevent or mitigate the associated environmental impacts;
▶ review procedures following an incident and periodically test where practicable.

Taking each of these requirements in turn, the sections below provide some examples of what might be done to meet the requirements of ISO 14001. This gives an indication of what might be considered good practice. Of course, the degree of involvement and assurance level that an organisation should achieve in its emergency planning will vary depending on the scale of hazard it represents and also on the proximity of key receptors, i.e. on the potential impacts associated with an emergency scenario.

7.8.5.1 Identifying emergency scenarios

In ISO 14001 implementation programmes this is normally done as part of the aspects evaluation process. Various scenarios will be considered in relation to both likelihood and consequence and those with greater likelihood or consequence are the ones that should be carried forward for consideration in terms of controls and emergency planning.

Examples might include leaks from pipework or storage vessels, spillages and even fire-fighting run-off water from a site. All may have potential to cause significant damage to controlled waters or the contamination of land. In addition, a fire may result in uncontrolled release of air emissions. More minor unplanned events may also be considered important, not least if they are more likely to occur. Examples might include the failure of emissions or effluent control equipment or the inappropriate mixing of hazardous and non-hazardous waste streams.

In all cases, the first part of the ISO 14001 requirement with regard to emergency planning demands the introduction of safeguards to try to prevent the identified unplanned scenario occurring, i.e. *take preventative action*. Such actions will vary depending on the scenario but may include:

► technical/physical controls such as bunds or automated alarm systems;
► monitoring or assurance controls that involve checks that things are operating as planned;
► where human error may contribute to the likelihood of an incident, training and supervision may form part of the incident prevention plan.

7.8.5.2 Identify emergency response and impact mitigation measures

Incident response plans often involve some generic measures such as area isolation and personnel evacuation but will normally also contain *response actions specific to the incident* in question. On a high hazard site, an emergency plan may therefore be sub-divided into multiple different incident scenarios, e.g. a hazardous material release, a fire, a failure of emissions control equipment, etc. However, the following elements will often be considered in relation to any defined scenario and for low hazard sites may even be written as a generic emergency response plan:

► on discovery of incident what should be done by the first person on the scene;
► lines of communication and responsibility (both during and outside normal operational hours);

► emergency response actions involving things such as spill kits, emergency equipment, shut-down operations and containment measures such as drainage isolation or the deployment of booms in nearby water courses (to prevent the spread of a spill);
► off-site support and information requirements, e.g. spill response contractors, the emergency services, neighbours or receptors that may be affected by the incident, regulator notification;
► follow-up arrangements such as material recovery, clean-up and waste disposal, management of temporary operations, record-keeping, incident investigation and the identification of actions to prevent recurrence.

7.8.5.3 Emergency plan review and test procedures

It is unfortunately all too easy to have a well-written emergency plan that is completely ignored when disaster strikes. The only way to avoid this is to make sure, through training and regular drills, that critical people know what should be done when things go wrong. This must include not just emergency response teams but also groups that are likely to be 'first on scene' or initial coordinators in an incident – often site security personnel and team leaders when dealing with out-of-hours emergencies. If spill kits or other response and containment equipment are to be used, people need to be able to access and use them and regular checks need to be scheduled to ensure that supplies are not depleted. Key information needs to be readily accessible – not just the emergency plans themselves but things like site drainage plans, contact telephone numbers, etc.

Regular exercises should be undertaken to 'test the response plans' and allow the organisation to learn from mistakes made in controlled conditions. Depending on the level of risk associated with a site, the exercises will vary from simple evacuations and shut-down procedures through to a whole schedule of desk-top drills involving key decision-makers, in-company response exercises using staff and equipment on site, and full-scale drills involving off site actions such as boom deployment in water courses, etc. which may even require the involvement of regulators and the emergency services.

The review part of the ISO 14001 requirement in this respect is to ensure that lessons are learned and *corrective actions* adopted to either improve incident prevention measures or the emergency response abilities of the organisation following an actual or a dummy incident.

7.9 Managing materials to reduce environmental impact

The management of raw materials is important in terms of an organisation's overall environmental impact for a number of reasons including:

▶ the quantity of materials consumed (especially where the materials are made from non-renewable resources);

▶ the supply chain impacts associated with the raw material supply;

▶ the pollution potential associated with the use of hazardous raw materials.

Once material flows are understood, environmental impacts associated with materials usage may be reduced by doing three things:

▶ use less materials for the same level of activity;

▶ use equivalent materials from suppliers that for some reason (e.g. location, operational standards etc.) have less environmental impact;

▶ replace existing materials with high pollution potential with new materials with low pollution potential.

Within each of these categories there are a number of steps or tasks that may be appropriate. An overview (although far from exhaustive list) of such tasks is provided in Tables 7.11, 7.12 and 7.13.

7.10 Managing waste to reduce environmental impact

The environmental impact of waste may be reduced in one of three ways, as presented in the waste hierarchy seen in Chapter 3. Waste reduction/elimination has been considered above

Table 7.11 Reducing material consumption

Improvement	Examples/comments
Avoid over-ordering	Materials management systems (including first in, first out systems) that reduce waste arising from out-of-date or excess materials disposal.
Minimising or returning packaging	Sony Manufacturing saved millions of pounds per year at its Bridgend tube plant through the introduction of returnable packaging to replace the high value packaging for glass parts that had previously been disposed of after a single use.
Ensuring that production specifications are accurate/appropriate	Whether in relation to dosage/application rates for coating processes, introducing timed rather than continuous release of materials or rinse water or changing the size, strength or materials used in a particular application – there are many ways that fewer materials can be used to achieve the desired goal in a manufacturing process. Often the improvements are obvious and easy to achieve once the production activities are reviewed in this manner. It is simply that the improvement emphasis in most production environments is on quality and production rates rather than material consumption efficiencies. A dramatic industry example of improvements in this category comes from the Paper and Board sector. In 1900 paper manufacturers would typically use a tonne of water per kilogramme of paper produced. Today the best mills use as little as 1.5kg of water per kilogramme of paper.
Weight-down initiatives	Working with suppliers to reduce the weight of component parts through redesign or process improvements has been used in a number of industries to radically reduce the quantity of materials used to produce the same functional product. Examples include milk bottles, steel cans and cathode ray tubes for TVs.

Table 7.12 Reducing supply chain impacts

Improvement	Examples/comments
Selection of local suppliers	Basic environmental benefit is a reduction in the fuel consumption and air emissions associated with transport of goods. However, hidden environmental (and potentially financial) benefits may also exist through things such as the capability to deliver in smaller consignments (reducing stock holding), return packaging or reject goods, etc.
Selection of high performing suppliers	Companies as diverse as BP, B&Q, Ford and Bodyshop utilise supplier selection to a greater or lesser degree as a way of influencing overall environmental impact. Selection may be on a performance review basis for contract award, for example, or it may be part of an on-going supplier assessment methodology such as that used by B&Q to highlight the issue the company sees as important and encourage suppliers to work towards set standards though a combination of carrot and stick incentives. Organisations with less purchasing clout may still use this approach simply by selecting on the basis of reputation or product accreditation, e.g. white goods energy efficiency labels, forestry stewardship scheme marks, etc.
Optimising supplier deliveries	Improved loading and packaging design, route planning, driver training, efficient vehicle selection and 'milk round' deliveries are all ways that the environmental impacts of road transport may be reduced. Shifts from road to rail (and avoiding air freight completely) for all or part of a journey will also produce a significant impact reduction.

Table 7.13 Reducing pollution potential of materials used

Improvement	Examples/comments
Replacing high hazard materials with lower hazard substances	The printing sector provides some of the best documented examples of this kind with the shift from solvent-based to water-based inks. Although often involving technical challenges, the resulting savings in raw material costs as well as the elimination of air emissions have been proven time and again to be worthwhile.
Optimising storage and transfer arrangements	In contrast to the previous improvement, sometimes an increase in the concentration of a hazardous material may lead to easier handling and storage and, thereby, reduce the risk of spills and leaks. The use of bunding and smart monitoring techniques (e.g. pH alarms and shut-off valves in drainage systems near acid storage tanks) is also relevant here.
Designing out the need for hazardous components	In some instances a hazardous material may be integral to the design and function of a product or service, e.g. lead in a cathode ray tube. Redesign of the product may produce benefits in terms of performance, market differentiation and, at the same time, eliminate the need for a hazardous material content, e.g. plasma or TFT screens.

in the section on material management. Re-use and recovery of waste materials as opposed to disposal are the other options which are considered in Tables 7.14 and 7.15.

7.11 Carbon management strategies

7.11.1 An overview

Carbon footprinting and offsetting were discussed in Chapter 6. Here we will focus on the actions organisations can take to reduce their direct and indirect carbon emissions. There are a number of ways that most organisations can approach this challenge, broadly falling into the following categories:

▶ *Improving consumption efficiency in lighting, heating and power*, i.e. using less overall energy per unit of work completed. This normally involves initial monitoring and analysis to determine where inefficiencies and hence potential savings may be present. Typically actions then fall into one of two camps: technological fixes or operator improvements. Often simple 'switch off' or 'turn down' campaigns can produce very significant reductions in energy usage. Key areas of opportunity and improvement examples applicable in many organisations are shown in Figure 7.2. Particularly with regards lighting and space heating, building design can make efficiencies easy or difficult to

Table 7.14 Re-use of waste

Improvement	Examples/comments
Returnable packaging	Dedicated reusable packaging systems are becoming increasingly common, e.g. pallets, drums and IBCs, slip sheets, plastic boxes, metal crates, stillages. All may be used between an organisation and its suppliers or customers. However, cost and transport considerations must be included in a review to decide whether an environmental benefit will accrue from the introduction of returnable packaging for any individual product stream.
Material reuse – off site	In some instances one organisation's waste may be another's raw material. If the logistics of transfer allow, this may be commercially beneficial to both parties and produce an environmental benefit through both reducing waste and reducing manufacturing impacts associated with the material in question. Numerous examples exist from the reuse of packaging through to the direct use of chemical by-products in unrelated manufacturing activities.
Material reuse – on site	The reuse of things like cleaning solvents or rinse water on production lines in pre-cleaning stages can significantly reduce both overall material consumption and also the volume of waste solvent or effluent generated for disposal.

Table 7.15 Recovery of waste

Improvement	Examples/comments
Recycling of waste materials	Recycling of wastes is something that can be done in almost all work environments, from offices to manufacturing sites, factories to retail outlets. Factors critical to the success of recycling schemes include: effective segregation of materials effective preparation of collected materials to make storage and collection cost effective, e.g. compaction, shredding, containerisation good communication with those involved in waste generation – there are plenty of schemes that have failed through a lack of clarity over disposal routes/the occasional need to landfill segregated materials or listening to the requirements of waste producers for segregation facilities.
Waste to energy	Waste to energy plants, either in the form of incinerators, cement kilns or waste oil/solvent combustion facilities, are a way of obtaining secondary value from waste that cannot be reused or recycled in other ways.
Reprocessing wastes to enable input to other processes	The refurbishment of electrical goods, treatment and reuse of waste water and refilling of ink cartridges are all examples of reprocessing to avoid waste disposal.
Composting of organics	Increasing numbers of landfill sites are now operating commercial scale composting facilities. For organisations generating significant quantities of green waste, composting may be a viable option.

obtain. This aspect of carbon management is considered further in Section 7.11.2.

▶ *Increasing energy efficiency in transport*. By change of fuel type, transport distance or transport type ,e.g. greater use of public transport, rail and sea instead of air freight, etc. Further details and examples in Section 7.11.3.

▶ *Changing energy supply* to reduce associated pollution and increase the efficiency of supply. For example, gas as a primary fuel for electricity and space heating creates less impact than coal or oil, both through lower emissions output and reduced losses during energy or heat transmission. The use of combined heat and power plants may produce more useful energy per unit of fuel burned than using electricity and gas from the national grids for power and space heating. For sites that can adequately utilise the heat generated, they may represent a significant reduction in overall environmental impact.

Clearly, renewable energy sources – wind, solar, tidal, wave, biomass, etc. – are even better in terms of reducing environmental impact. Organisations may invest in their own on-site renewable energy sources such as wind turbines or photovoltaic panels. Alternatively

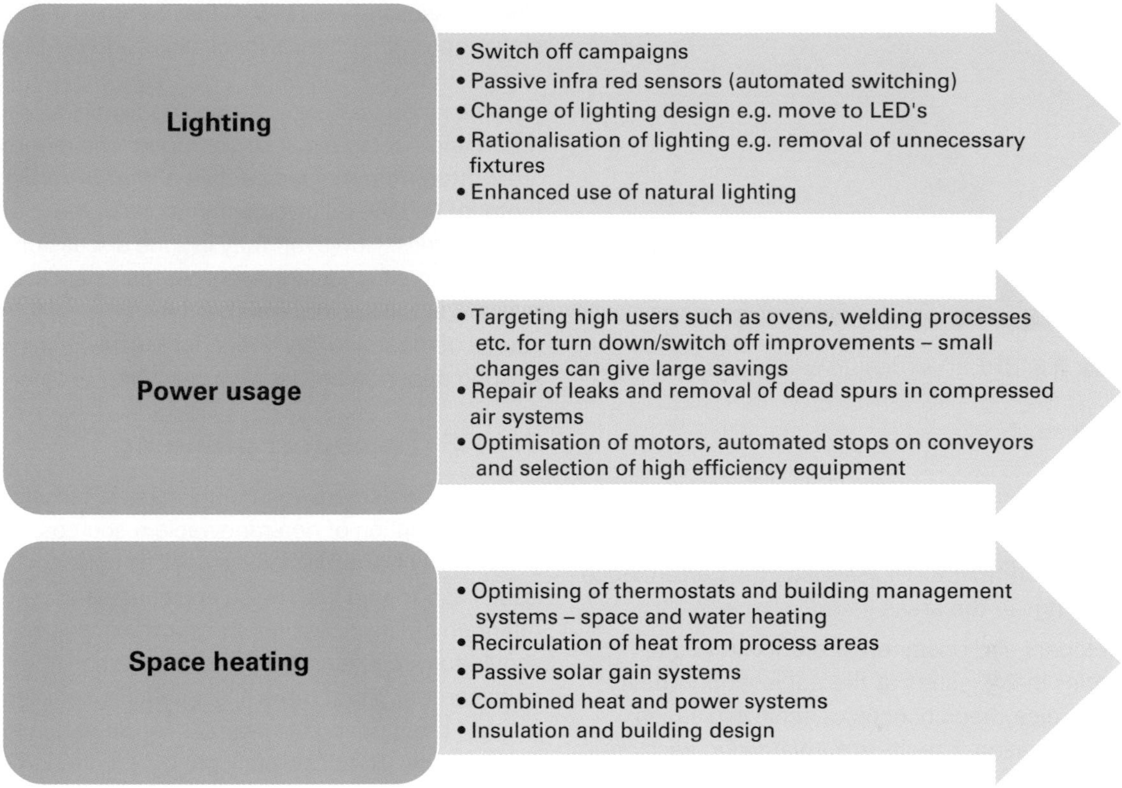

Lighting

- Switch off campaigns
- Passive infra red sensors (automated switching)
- Change of lighting design e.g. move to LED's
- Rationalisation of lighting e.g. removal of unnecessary fixtures
- Enhanced use of natural lighting

Power usage

- Targeting high users such as ovens, welding processes etc. for turn down/switch off improvements – small changes can give large savings
- Repair of leaks and removal of dead spurs in compressed air systems
- Optimisation of motors, automated stops on conveyors and selection of high efficiency equipment

Space heating

- Optimising of thermostats and building management systems – space and water heating
- Recirculation of heat from process areas
- Passive solar gain systems
- Combined heat and power systems
- Insulation and building design

Figure 7.2 Typical energy efficiency actions

electricity supply companies are increasingly offering customers the opportunity to sign up to a 'green tariff' which is linked to their renewable contribution to the national grid. Charges for 'green tariff' electricity are typically higher than for conventional sources but increased costs may be offset against the climate change levy charged on fossil fuel-sourced energy from which the former is exempt.

This is a huge topic area and there is significant guidance, case study and other support material available from the Carbon Trust via their website: www.carbontrust.com.

7.11.2 Building design to reduce environmental impact (including carbon footprint)

For organisations involved in the ownership of buildings, possibilities exist to reduce the environmental footprint as a whole and the carbon footprint in particular through building design at new build or refurbishment stage. Design considerations cut across all elements of an industrial, commercial or retail premises and relate to materials, lighting, heating, water, transport, etc.

The Building Research Establishment (BRE) is a wholly owned subsidiary of the BRE Trust, a charitable company whose objectives are: 'through research and education, to advance knowledge, innovation and communication in all matters concerning the built environment for public benefit'.

The *BRE Environmental Assessment Method (BREEAM)* has become widely recognised as a benchmarking standard for eco-design in construction. Using a standard assessment methodology and certified assessors, buildings and/or project designs are classified into one of four categories as shown in Figure 7.3.

Assessment methodologies are tailored to the nature of project within the following categories:

▶ Offices
▶ EcoHomes

271

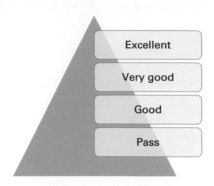

Figure 7.3 BREEAM design categories

▶ Industrial
▶ Retail
▶ Schools
▶ Multi-residential.

In each case, BREEAM assesses the performance of buildings in the areas listed in Table 7.16.

Developers and designers are encouraged to consider these issues at the earliest opportunity to maximise their chances of achieving a high BREEAM rating. Credits are awarded in each area according to performance. A set of environmental weightings then enables the credits to be added together to produce a single overall score. The building is then rated on the Pass to Excellent scale and a certificate awarded that can be used for promotional purposes.

BRE also produces best practice guides in relation to the assessment of development projects both in terms of materials specification and waste minimisation during the construction process. Further details are available at: www.breeam.org.

In addition to the BRE guidance, the Carbon Trust has produced a series of Technology Innovation

Needs Assessments (TINAs) that compare the status quo with potential innovations and the likely impact on carbon footprint. Two TINAs are of particular interest here – the *non-domestic buildings report* and the *industrial sector report*. The reports are written from the perspective of the UK carbon footprint and the relative contributions of the subject areas. However, they include discussion of the 'cutting edge' design and operation standards relevant to the sectors and may thus be a useful source of ideas and innovation for organisations seeking opportunities for improvement.

7.11.3 Transport planning

Transport-related environmental impacts include the consumption of non-renewable resources, the generation of air emissions as well as nuisance issues associated with noise and congestion. Although strictly speaking, transport management should form one element in a wider carbon management programme it is a sufficient challenge to most organisations to warrant consideration in its own right. The basic principles of monitoring and calculating total fuel consumption and emissions generated are as outlined in the carbon footprinting section in Chapter 6. For most organisations, this is the starting point in a transport management plan. There are then three key areas where improvements can be considered, as discussed below.

7.11.3.1 Transport reduction initiatives

Initiatives under this heading may vary widely. At the simplest level the tightening of management approval requirements on business travel may

Table 7.16 BREEAM performance assessment criteria

Management	Overall management policy, commissioning site management and procedural issues
Energy use	Operational energy and carbon dioxide (CO_2) issues
Health and well-being	Indoor and external issues affecting health and well-being
Pollution	Air and water pollution issues
Transport	Transport-related CO_2 and location-related factors
Land use	Greenfield and brownfield sites
Ecology	Ecological value, conservation and enhancement of the site
Materials	Environmental implication of building materials, including life-cycle impacts
Water	Consumption and water efficiency

be used to discourage unnecessary journeys (especially where air travel or non-routine expenditure is required). Additional planning time may be spent on rationalising journeys made by personnel and freight traffic (in relation to both supply and despatch). Some companies have also reported very significant reductions in travel requirements through the use of tele-video- and web-based conferencing. Clearly all of these initiatives have the additional benefit of saving the organisation money, especially when the cost of personnel time is considered. British Telecom, for example, reported an annual saving of £128 million in 2004/5 from tele-conferencing when travel costs, accommodation fees and unproductive employee time was considered.

7.11.3.2 Reduced impact transport

Broadly there are two options here – reducing the impact of road transport and finding lower impact alternatives to road transport. In the case of the latter, the general rule is to use rail as a preferred alternative in most cases. Examples of initiatives to reduce the impact of road transport include:

▶ *Efficient driving.* Driver attitude is critical here and cultivation of an appropriate approach is likely to involve a combination of awareness-raising and monitoring of fuel consumption. Consumption levels can be reduced significantly for the same vehicle mileage using conservative driving techniques such as speed reduction, avoiding rapid acceleration and braking, avoiding high revving of engines and idling after start-up or when in stationary traffic.

▶ *Efficient vehicle selection.* Fleet management choices can be particularly important here, given the variation in fuel efficiency between models of cars and light goods vehicles in particular. Across the board, adoption of higher fuel efficiency vehicles may have considerable impact in terms of total consumption and emissions generation.

▶ *Use of alternative fuels.* There is growing opportunity in this sector with hybrid vehicles offering the opportunity of gas/petrol combinations which have benefits in terms of total emissions generated. Perhaps of more relevance has been the emergence of a new generation of hybrid vehicles that use

battery storage of power during periods of low engine work which can be used to 'top up' power during peak load periods. The result is improved efficiency per litre of fuel burned.

There is also increasing interest in the use of biofuels either as replacements or as blends with conventional petrol or diesel. These can result in significant reductions in CO_2 emissions depending on the biofuel source and production method.

Longer-term, hydrogen fuel cell technology has been suggested as the answer to the emissions and non-renewables issues related to transport. However, commercial manufacture of the technology that uses hydrogen from renewable sources (rather than manufactured during fossil fuel processing) is still some way off.

7.11.3.3 Indirect travel impacts

There are three key areas to be considered here, namely:

▶ *Employee transport.* Travel to work by employees may constitute a significant part of an organisation's total travel-related impact. A travel plan developed to reduce the impact of this category might include measures such as:
  ▶ provision of cycle storage and showers (to encourage cycling);
  ▶ promotion of car sharing schemes;
  ▶ provision of public transport information or even negotiation on routes and stops;
  ▶ provision of pool cars and bicycles (for business journeys);
  ▶ flexible home-working arrangements for staff;
  ▶ loans/subsidy for rail and bus season tickets.

▶ *Supplier transport.* All organisations buy in considerable quantities of goods and services, and in some, if not many cases, little consideration is given to the transport implications of purchasing choice or contractor selection. Options to reduce the carbon intensity of supplier transport include local sourcing, specification of milk-round style deliveries, outsourcing of transport to minimise the amount of 'empty vehicle' miles, etc.

▶ *Customer transport.* Although for most organisations there is little possibility or

273

incentive for influencing customer transport, in some sectors (most notably retail), consideration of alternatives to road transport access is demanded as part of the planning process.

There is a great web resource provided by Business in the Community available for anyone interested in finding out more about business travel planning: http://ways2work.bitc.org.uk.

7.12 Air pollution control techniques

For an overview of key air pollutants, their primary sources and impacts see Chapter 1. Section 7.8.1 above also provides an overview of control techniques which are dealt with in a little more detail here. In general terms, there are three approaches used in air quality improvement strategies:

▶ *end-of-pipe technology* – including production and emissions control technologies;
▶ *substitution* – the substitution of less damaging substances for those with higher pollution potential;
▶ *source controls* – involving the limitation (and even elimination) of polluting activities, e.g. car transport, energy usage.

From the perspective of the regulators, the UK National Air Pollution Strategy comprises elements of these three strategies in two key areas:

▶ the air quality strategies for England, Scotland, Wales and Northern Ireland, which focus on limiting ambient air concentrations of key pollutants;
▶ the UK climate change programme, which aims to reduce the UK total greenhouse gas emissions by 80 per cent against 1990 baselines by 2050.

From an operator's perspective, the spectrum of strategies responding to the regulatory regimes range from output-oriented (abatement or end-of-pipe controls) to input-oriented (process or materials management). The purpose of this section is to look at some illustrative examples from across this spectrum of abatement techniques.

7.12.1 End-of-pipe technology

Air pollutants can exist in the form of particulate matter or as gases or vapours. The techniques and equipment used to remove particulates are different from those for the recovery of gaseous emissions. The two categories will therefore be considered separately.

7.12.1.1 Control of gaseous emissions

Techniques used to control gaseous emissions include:

▶ adsorption
▶ absorption
▶ condensation
▶ combustion
▶ bio-filtration.

The application of a given technique depends on the physical and chemical properties of the pollutant and the exhaust stream, and more than one technique may be capable of controlling emissions from a given source. For example, vapours generated from handling petrol at a large bulk terminal may be controlled by using any of these techniques. Most often, however, one control technique is used more frequently than others for a given application. For example, absorption is commonly used to remove SO_2 from combustion exhaust gases.

7.12.1.2 Adsorption

Adsorption is a process that involves removing a gaseous contaminate by adhering it to the surface of a solid. Adsorption can be classified as physical or chemical.

Adsorption of organic vapours and odours onto activated carbon provides a means of removing these pollutants from gas streams. Activated carbon has a very high surface area onto which the organic molecules become attached. When the activated carbon filter becomes saturated, it is either changed or regenerated. Regeneration liberates the pollutant in high concentration where it may either be recovered or destroyed by incineration or some other means. Activated carbon is expensive and if filters are replaced and not regenerated, the operating costs of the plant will be high.

7.12.1.3 Absorption

Absorption in this context involves a process in which a gas is dissolved in a liquid. A contaminated exhaust stream contacts a liquid and the contaminant defuses from the gas phase into the liquid phase. The absorption rate is enhanced by high solubility of the contaminant, and good mixing between the liquid and gaseous phases.

The liquid most often used for absorption is water since it is inexpensive, readily available and can dissolve a wide number of contaminates. Chemical reagents can be added to the absorbing water to increase the removal efficiency.

The devices used for gas absorption are often the same as for particulates removal including packed columns, spray towers and venturi scrubbers. Although the devices may be identical, a particulate scrubber is not operated in the same way as a gas absorber. Therefore, optimising both gas and particulate pollution removal in one device is difficult. Gaseous emissions absorption relies upon holding the contaminated gas stream for as long as possible to allow maximum absorption.

A classic application for absorption techniques are acid gas scrubbers which vary in scale from desk top units to multi-storey scale chambers (Figure 7.4).

7.12.1.4 Condensation

Condensation is a process in which volatile gases are removed from a contaminated stream by conversion into their liquid phase. Condensation can be achieved either by reducing the temperature or increasing the pressure of the gas. In pollution control equipment, condensation is usually achieved by reducing the temperature of the vapour mix to the point at which the target gas will condense. Careful temperature and pressure control can be used to separate mixed vapour streams.

7.12.1.5 Combustion

Many industries rely on combustion to break down pollutants that are toxic or have foul odours. Such pollutants are often volatile organic compounds (VOCs) used in a wide variety of chemical processes such as painting and

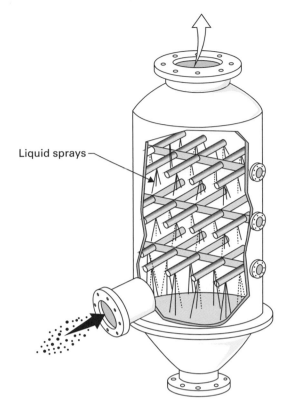

Figure 7.4 Acid gas scrubber
Source: Image from the United States Environmental Protection Agency.

varnishing operations, solvent-based cleaning activities, textile manufacturing, etc.

Such VOC emissions can sometimes be burned directly, or may require the addition of a fuel, usually gas or oil, to effect combustion. The fact that most organic compounds burn to form carbon dioxide and water vapour means that combustion is a very efficient method of emissions removal. Carbon dioxide and water vapour, however, are themselves greenhouse gases and the question may need to be asked as to whether they are regarded as more desirable than the pre-combustion emissions.

There are two principal categories of combustion treatment:

▶ *Thermal oxidation*. Thermal oxidation, or incineration, is a high temperature operation. The process can be self-propagating. The controlling factors in incineration design are temperature, residence time and turbulence (the 3 Ts). A typical specification would be to run a commercial unit at approximately 350 °C with a gas residence time of 0.75 seconds for a 98 per cent conversion efficiency. Turbulence is the third parameter affecting the operating efficiency. The better the mixing of the gases, the higher the conversion efficiency.

▶ *Catalytic converters*. A catalyst is a substance that accelerates a chemical reaction without being changed by that reaction itself. Catalytic conversion uses a catalytic surface to convert reactants at much lower temperatures than is generally required in the thermal oxidation processes described above. They also tend to be safer as there is no flame involved in the process.

▶ Catalysts used in VOC oxidation are principally palladium and platinum (both are used in vehicle exhaust systems). Other catalysts including rhodium, nickel, gold and other noble metals, are used principally for waste streams containing chlorinated hydro-carbons. In all cases the waste gas streams must be pre-heated to temperatures required to initiate the conversion process carried out *on* the catalytic bed. Catalytic burners usually operate in the 400°–500°C temperature range.

Selective catalytic reduction (SCR) is a means of converting nitrogen oxides with the aid of a catalyst into nitrogen and water (see Figure 7.5). Commercial selective catalytic reduction systems are typically found on large utility boilers, industrial boilers and municipal solid waste boilers and have the capacity to reduce NO_x emissions by 70–95 per cent. More recent applications include

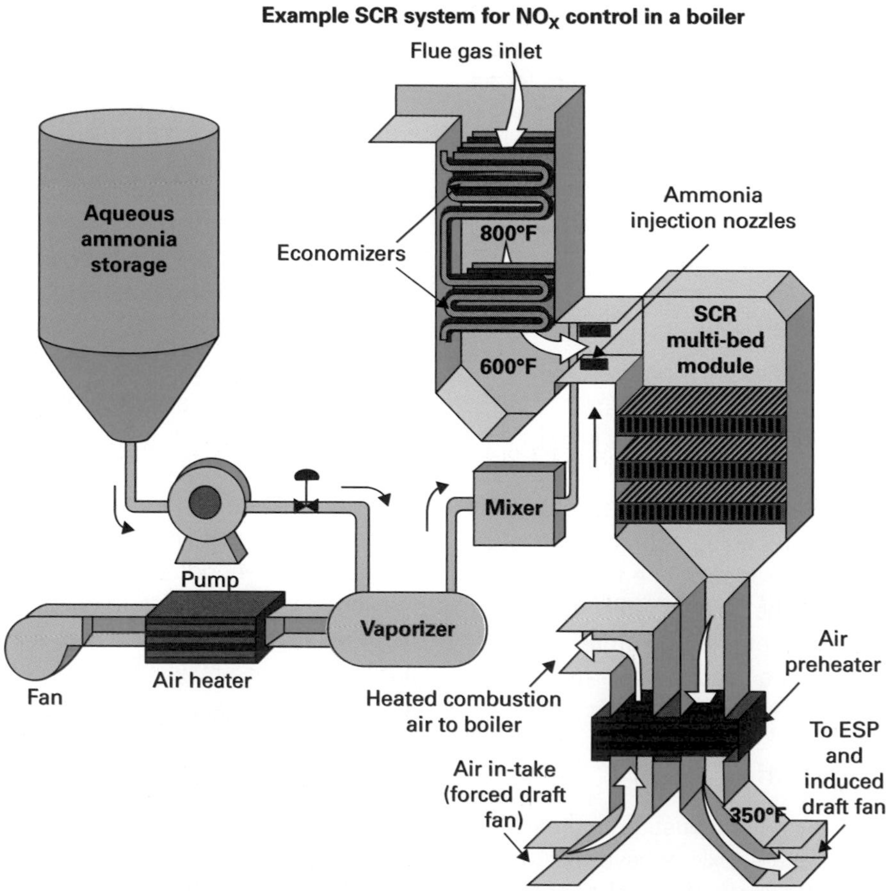

Figure 7.5 Selective catalytic reduction
Source: Image from the United States Department of Energy.

the use of SCR on large diesel engines, such as those found on large ships, diesel locomotives, combustion turbines and even automobiles.

7.12.1.6 Bio-filters

Odorous gases and other volatile organic compounds may be oxidised by microbial activity to less objectionable products. Details of large-scale odour control systems are provided in the odour control section 7.16. However, an increasingly common application of this technology is in the design of in-line 'filters' used for the capture of VOC emissions from industrial processes.

The microbes exist in an aqueous medium, which is brought into contact with the contaminated airstream; the pollutant gases dissolve in the water and are oxidised by microbial digestion with by-products (typically carbon dioxide) passing back out of solution into the exhaust air. When first installed, the microbial population is low unless inoculated with a suitable culture. In general, a natural population of self-selected bacteria will build up in the equipment over a period of weeks reaching a level of colonisation in equilibrium with the supply of nutrient (the organic pollutant). Inoculation will reduce the time required to build up the population of microbes but is not necessary for the ultimate success of the equipment.

7.12.1.7 Control of particulate emissions

A number of factors must be determined before a proper choice of collection equipment can be made. Among the most important data required are the following:

▶ the physical and chemical properties of the particulates;
▶ the range of the volumetric flow rate of the emission stream;
▶ the range of expected particulate concentrations (dust loadings);
▶ the temperature and pressure of the emission stream;
▶ the humidity and the nature of the gas phase (e.g. whether corrosive or not); and
▶ the required condition of the final emission.

The last piece of information may be the most important, for it indicates the collection efficiency that must be met.

Collection equipment falls into five main types:

▶ gravity settling chambers
▶ cyclone separators
▶ wet collectors
▶ electrostatic precipitators
▶ fabric filters.

Each method is applicable to a different range of conditions and will operate with a different efficiency according to the nature of the conditions. Each is subject to its own operational criteria and challenges and will have its own maintenance requirements. Following is a brief description of each category together with their typical efficiencies.

7.12.1.8 Gravity settling chamber

The settling chamber is the simplest form of gas cleaning device and depends on the fact that solids or liquid particles suspended in a gas settle out under the action of gravity at differing rates depending on their size, shape and density. A practical form of settling chamber (incorporating a particulates classifier) is shown in Figure 7.6.

The gas is uniformly expanded to fill the full height of the gas chamber. Below the settling chamber are the hoppers into which the dust falls when it leaves the gas stream. There should be no flow in these chambers and therefore once the dust falls below the level of the hoppers, it is effectively retained in the settling chamber. Since the rate of fall of smaller and/or lighter particles under gravity is low, settling chambers are only practical for the removal of coarser or denser particulates.

An example application is the recovery of grit in usable and non-usable size ranges from a commercial grit blast process.

7.12.1.9 Cyclone separators

Cyclone separators are gas cleaning devices which employ a centrifugal force generated by a spinning gas stream to separate the particulate matter (solid or liquid) from the carrier gas (see Figure 7.7 for a wet version). The dust-laden gas enters the tangential inlet and whirls through

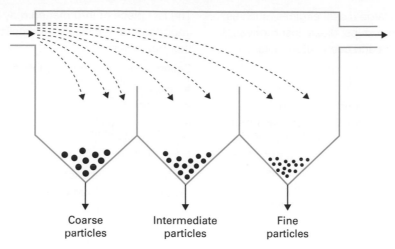

Figure 7.6 A simple gravity settlement chamber

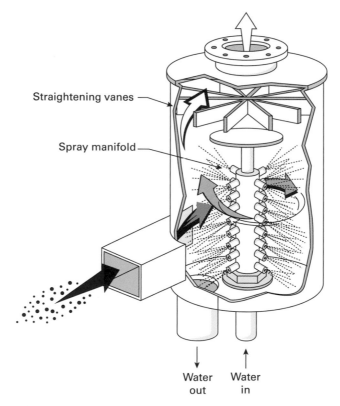

Figure 7.7 A cyclonic scrubber used for particulates removal
Source: Adapted from US Environment Protection Agency.

several revolutions in the body and cone, dropping its dust load. The clean gas is emitted through the axial cylindrical air outlet. The dust particles, which were uniformly dispersed in the entering gas stream, tend to concentrate in the layer of air next to the cyclone wall under the influence

of centrifugal force. The helical motion of the main airstream down and the small quantity of air through the cyclone dust outlet project the separated dust solids into the receiving bin.

Cyclones have long been regarded as one of the simplest and most economical mechanical collectors. Chief advantages include high collection efficiency on large particle sizes, adaptability, low maintenance and relatively low cost.

Their main disadvantages, however, are their relatively low efficiency on the smaller particle sizes and erosion in certain areas of the design due to the high tangential velocity and abrasive nature of the particulate. Collection efficiency of single units ranges from 65 per cent upwards whereas multiple high-efficiency or irrigated cyclones have efficiencies better than 90 per cent.

Example applications include wood dust recovery in mills, workshops and factories generating significant quantities of sawdust.

7.12.1.10 Wet collectors

Water is used for dust collection in wet collectors, of which there are two main types. In the first, the dust particles are forced to impinge on wetted surfaces; in the second, the particles are caught by a fine spray of water droplets. Wet collector types include:

- ▶ spray chamber scrubbers
- ▶ cyclonic scrubbers
- ▶ venturi scrubbers.
- ▶ *Spray chamber scrubbers*. In the simple spray tower, water sprays are injected into a suitable chamber counter current to the flow of emission stream. This system is essentially the same design as that used in the acid gas scrubbers described above and illustrated in Figure 7.4. The collection efficiency is quite acceptable for particle size over 10mm. The effectiveness of a conventional spray tower ranges from 94 per cent for 5mm particles to 99 per cent for 25mm particles.
- ▶ *Cyclonic scrubbers*. In a cyclonic spray tower, the dirty air is introduced into the lower portion of a vertical cylinder (see Figure 7.7). Water is introduced through an axially located multiple nozzle, which throws the water radially outward across the spiralling gas flow. In general, cyclonic scrubbers have a collection of between 90 and 100 per cent depending on particle size.
- ▶ *Venturi scrubbers*. These are a variation of the wet cyclone design in which the gas

together with the scrubbing liquid (generally water) is injected radially upstream of the throat section of the scrubber. This results in fine atomisation and good contacting with the scrubber chamber walls. As a result, the recovery efficiency may reach 99 per cent in the sub-micron range, and 99.5 per cent for 5um particles.

7.12.1.11 Electrostatic precipitators

In the simplest terms, a precipitator is a large box. The dust-laden gases are drawn into one side of the box. Inside, high voltage electrodes impart a negative charge to the particles entrained in the gas. These negatively charged particles are then attracted to a grounded collecting surface, which is positively charged. The gas then leaves the box up to 99.9 per cent cleaner than when it entered.

As illustrated in Figure 7.8, the particles from the continuing flow of dust build up on the collecting plates. At periodic intervals, the plates are rapped or the negative charge momentarily removed, causing the particles to migrate down the collection plates and ultimately fall into hoppers. The particles are then removed from the hoppers, usually by a rotary screw arrangement.

The prerequisite for the use of such precipitators is that the particles should easily hold an electrical charge (coal dust and metals are ideal, for example). They range in size from small-scale units to the building-scale equipment used in power stations and steel works for smoke, ore and coke dust recovery.

7.12.1.12 Fabric filters

Perhaps the most commonly encountered way to collect dust from a gas stream is to pass it through a porous filter medium in which the dust is trapped. In conventional bag filters, the dust-laden gas passes through filter bags made of woven or felted cloth. As the layer of collected material builds, the pressure difference required for continued gas flow increases, i.e. the system begins to clog up. Consequently, the accumulated dust must be removed at frequent intervals, either by mechanical shaking or by means of a reverse jet of air. A typical arrangement employing reverse

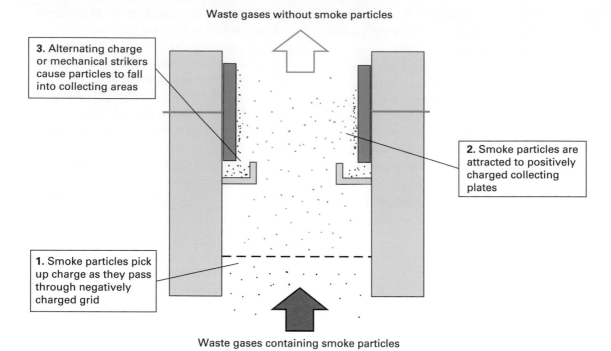

Figure 7.8 How an electrostatic precipitator works

or as it is sometimes known pulse jet cleaning is shown in Figure 7.9.

The fabric is selected having regard to collection efficiency, pressure drop, the required strength (determined by the type and frequency of cleaning), the temperature and the chemical characteristics of the particles. Fabric filters have efficiencies of 99 per cent or better when collecting 0.5 micron particles and can remove substantial quantities of 0.01 micron particles.

7.12.1.13 Dispersion of air pollutants; the design of chimneys

For many of the abatement technologies described above, the final step in the emissions treatment process is the discharge to the atmosphere of the remnant gases. This step can be crucial in relation to local air quality impacts and is the subject of considerable planning and statutory guidance in its own right, regardless of the treatment technique with which it is associated.

Chimneys are often necessary to ensure safe dispersal of the emissions and where this is the case the height, efflux velocity and temperature

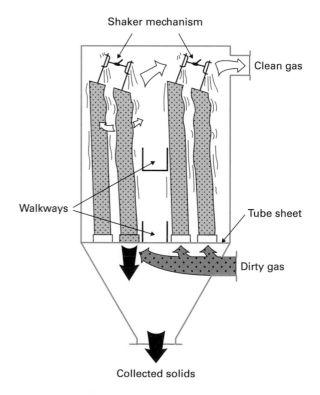

Figure 7.9 A bag filter

of emissions need to be set to ensure the optimum dispersion condition.

Guidance on chimney heights is given in the Environment Agency technical publication D1. As an approximate rule, chimneys should be 0.6 times higher than the buildings around them to avoid the effects of turbulence which might bring a plume down to ground level and a minimum efflux velocity of 15 m/s should be achieved. The air velocity in ducts containing water droplets should be no greater than 9 m/s. Doubling the effective height of a chimney will reduce the ground level concentration by a factor of 4.

The mixing and dilution of pollutants depend very much on the speed and direction of the prevailing wind. The concentration of a pollutant is inversely proportional to the wind speed, i.e. if the wind speed is doubled, the pollutant will be dispersed into approximately twice the volume of air, so the concentration at any point is halved.

Superimposed on the general wind velocity will be turbulence effects, which are random fluctuations in speed and direction. These turbulent 'eddy' currents are also formed when the temperature varies. If the air temperature rises, upward streams or 'thermals' are produced which add to the general level of turbulence. Turbulence is also formed due to ground roughness and buildings.

Atmospheric air temperature normally decreases with height. An airstream moving upwards from a chimney due to initial efflux velocity will typically decrease in temperature. Under ideal conditions, however, as the emissions stream rises, it cools but stays relatively warm compared with the surrounding air (which is also getting cooler with altitude). This temperature difference accelerates the rise (and hence dispersal) of the emission plume. So ideally at point of emission the airstream should be warmer than the surrounding air to facilitate upward movement and dispersal.

At times, however, the ambient air temperature variation with altitude is small. This can occur at night. The upward movement (and cooling) of an airstream under these conditions mean that the temperature difference between the airstream and the ambient air reduces with height. The result is a reduced rate of upward rise and dispersal (sometimes even a fall back to ground level). This is clearly undesirable in terms of emissions dispersion.

There are even occasions when, at increasing height, the air temperature actually increases. Because this is the reverse of the expected relationship. it is called 'an inversion'. Low-level inversions tend to form overnight when winds are light and skies are clear. On a frosty night the ground cools down to below air temperature. The air immediately above the cold ground will cool to temperatures lower than that of high air further away from the cold surface. On a very cold night the depth of the inversion layer will gradually extend upwards to greater heights. When the sun reappears, the air furthest from the cold ground surface warms more quickly, further exaggerating the temperature inversion. It may take some time for the ground to warm sufficiently for the inversion to disperse.

Inversions are of great importance to the study of pollutant dispersion. This is because the intermixing of air layers above and below an inversion is virtually excluded. The stronger the inversion, the greater the barrier to intermixing. If such a barrier forms above a chimney, the emissions stream will have insufficient buoyancy to penetrate the inversion and disperse upwards and instead will spread horizontally. Consideration of climatic conditions and the likelihood of inversions is therefore a crucial part in air quality modelling associated with both chimney design and conversely impact assessment associated with a proposed development.

7.12.2 Substitution

All the techniques described above are end-of-pipe abatement methods. However, as previously mentioned, process-based approaches to emissions control are becoming increasingly popular. The preferred option in such a strategy is to eliminate the source of polluting emissions either through elimination of the need for the process or through substitution of the material giving rise to the emissions.

Substitution solutions can often be challenging in the short term in terms of process change and investment in research and development or new equipment. However, in the longer term, the cost savings related to abatement controls can provide a rapid payback and significant on-going savings. One emerging issue is that care must be taken to avoid introducing new or different environmental impacts associated with the substitute material. In recent years this has pointed towards a life cycle assessment approach during the exploration of alternative designs and materials.

There have been two high profile examples of 'substitution' solutions, which have resulted in significant improvements in terms of local and national air quality in the past decade.

7.12.2.1 Low sulphur coal

The substitution of traditional coal supplies with low sulphur content coal in UK coal fired power stations has significantly contributed to the 65 per cent drop in SO_2 emissions since 1980. This is one example, however, where a shift in environmental impact has resulted. This is because the move to low sulphur coal changed the primary sources of supply from the UK to Argentina and Australia. The huge increase in transport requirements associated with this bulk commodity has presumably increased the carbon footprint of the supply chain significantly with adverse consequences in terms of contribution to climate change. So the strategy has reduced acid precipitation associated with pollution arising from UK power stations but has led to an increase in climate change impacts.

7.12.2.2 Solvent substitution

There have been numerous examples of process and product changes leading to the elimination of VOC emissions in a variety of applications. These include:

▶ the substitution of solvent-based inks with water-based inks in many printing applications;
▶ the use of propellant-free systems in personal deodorant products;
▶ the use of powder coatings and hot melt adhesives to replace solvent-based products.

As well as reducing the quantity of solvents used, there can be benefits gained if the properties of the remaining VOCs can be optimised by the following:

▶ replacing traditional solvents with others which have less environmental impact by virtue of lower toxicity, persistence or photochemical reactivity;
▶ moving to simpler mixtures, or even single components to facilitate recycling.

7.12.3 Emissions reduction at source

In some cases, elimination of emissions is not technically or economically feasible, however, there are numerous examples, particularly in relation to VOC emission sources, where emissions minimisation and recycling have proved effective.

7.12.3.1 Waste minimisation

A lot of work has been undertaken by the Environmental Technology Best Practice Programme (now incorporated under the WRAP programme) on solvent management techniques that aim to reduce fugitive losses (i.e. unplanned consumption). Case studies have shown losses from mixing, cleaning, leaks and untreated spillages amounting to as much as 25 per cent of total solvent purchased. These losses not only represent an environmental impact but often a significant, unnecessary expense to the organisation. Even simple housekeeping measures such as the use of dispensers rather than open containers for cleaning solvents can produce significant savings. The ETBPP recommends as a first step in any solvents management programme, the calculation of uncaptured emissions. With solvents, this can be a relatively straightforward and reasonably accurate calculation as it is often safe to assume that the majority of solvents purchased are released as emissions. Some are captured in extraction systems and some are not. The captured element can be calculated through monitoring stack or vent emissions. The uncaptured element is taken by default to be the total quantity of solvents purchased minus this captured quantity. Although not completely accurate, this provides an indication of the system losses that represent opportunities for emission reduction – often through relatively simple 'housekeeping' style initiatives.

7.12.3.2 VOC recycling

VOC recycling is often an attractive control technique given the high value of industrial solvents. Recovery methods range from adsorption, through to condensation and in many cases are supplemented by some kind of solvent cleaning stage to enable reuse of the material in the same process. This is particularly appropriate where low levels of contamination are experienced and high value solvents are used.

7.13 Water pollution control/ effluent treatment techniques

For an overview of key water pollutants, their primary sources and impacts see Chapter 1. Section 7.8 also provides an overview of the control techniques which are dealt with in a little more detail here.

As with the control of air emissions, substitution and reduction at source are appropriate techniques to reduce the volume and/or pollution potential of an effluent stream. As the principles are essentially the same whether we are considering air, land or water pathways we will focus in the following sections on abatement techniques, i.e. end-of-pipe controls.

Section 7.13.1 describing sewage treatment processes is based on the techniques used by municipal sewage treatment works. It is worth noting that the same basic processes are also applied to the treatment of organic industrial effluent.

Section 7.13.3 provides an overview of the treatment processes employed at industrial effluent treatment plants either to pre-treat effluents prior to discharge to the public sewer or as a complete treatment process prior to discharge to controlled waters.

7.13.1 Sewage treatment

Water service companies in England and Wales, and sewerage undertakers in Scotland and Northern Ireland, are responsible for the sewerage collection network and for the operation of treatment works and discharges. There are 200,000 miles of sewers in the UK, which between them collect over 2,400 million gallons (11,000 million litres) of urban waste water *every day*. There are around 9,000 sewage treatment works.

Sewage treatment is normally carried out in several stages. After passing through a screening process, the sewage flows into a sedimentation tank. From this point the sludge goes to a treatment process, such as anaerobic digestion while the liquid effluent goes to a biological oxidation process where the BOD is considerably reduced and nitrogen and phosphorus are removed. After further sedimentation the surplus sludge is again sent for sludge treatment and the liquid passes on for further tertiary treatment often through an ultraviolet filter. Figure 7.10 shows a typical sewage treatment process.

The following stages in sewage treatment will be considered:

▶ biological oxidation
▶ nitrogen removal
▶ phosphorus removal
▶ sludge treatment and disposal (including anaerobic digestion).

Essentially in a sewage treatment works the process is designed and managed to create the optimum conditions for each of these stages to take place.

7.13.2 Biological oxidation (aerobic treatment)

This biological oxidation process aims to optimise breakdown of the organic waste by bacteria in the effluent. In the process, oxygen is consumed by the bacteria for three reasons:

▶ to oxidise the organic matter;
▶ to grow the bacteria;
▶ to remove nitrogen.

Biological oxidation is the most commonly used form of secondary treatment for organic waste waters of BOD's 50 to 1,000mg/1 and greater. The three main processes employed are activated sludge, trickling filter and rotating biological

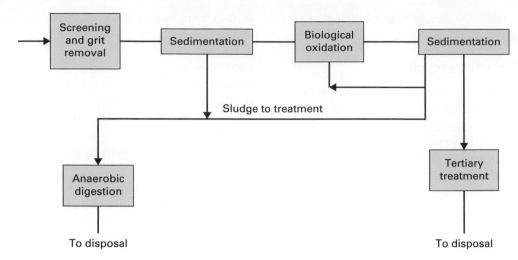

Figure 7.10 Flow diagram of a typical sewerage treatment process

contactor. Essentially the same breakdown of organic wastes takes place in each case, as shown in Figure 7.11.

The difference between the processes is simply in how they enhance oxygen content and circulate bacterial populations to accelerate organic breakdown.

▶ The activated sludge process involves getting oxygen into the wastewater either through a diffuser or a mechanical agitator and recycling about 10 per cent of the active biomass back to the process.
▶ Trickling filters involve trickling the wastewater over a bed of stone or plastic packing material (also called percolating filters).
▶ Rotating biological contactors involve a rotating disc partially immersed in the wastewater. In the latter two processes the micro-organisms grow attached to the surfaces.

The pros and cons of the different systems are summarised in Table 7.17.

7.13.3 Nitrogen removal

Nitrogen removal is carried out in two successive stages: nitrification and de-nitrification.

7.13.3.1 Nitrification

Nitrification is a two-step process involving the biological oxidation of ammonia to nitrate. First, one group of bacteria converts ammonia (NH_3) into nitrite (NO_2) in the presence of oxygen (aerobic conditions) and, second, a different group of bacteria converts the nitrite into nitrate (NO_3).

The micro-organisms involved are the autotrophic species *Nitrosomonas* and *Nitrobacter*, which carry out the reaction in the following steps. These are organisms that do not require organic material for food but can manufacture food from inorganic chemicals.

In the conversion of ammonium ions to nitrite:

$2NH_4 + 3O_2$ >–Nitrosomonas—> $2NO_2 + 2H_2O$ + 4H + new cells

and the conversion of nitrite to nitrate:

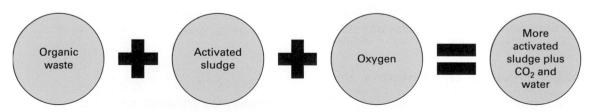

Figure 7.11 The biological breakdown of sewage

Table 7.17 Advantages and disadvantages of the different aerobic treatment methods

Treatment method	Advantages	Disadvantages
Activated sludge	Efficient (up to 95% reduction in BOD)	Can be complex and costly to build and run
Rotating biological contactor	Small, cheap to run	Only suitable for dilute waste waters
Trickling filter	Cheaper and simpler	Normally only 85% BOD activated sludge removal

$$2NO_2 + O_2 > \text{—Nitrobacter—} > 2NO_3 + \text{new cells}$$

They are slow-growing organisms and are sensitive to a number of factors: substrate concentration, pH, temperature, oxygen concentration and especially the presence of inhibitors. The extent of nitrification that occurs during treatment depends on the numbers of nitrifying bacteria present.

7.13.3.2 De-nitrification

In the second stage of nitrogen removal – de-nitrification – various bacteria convert nitrates into atmospheric nitrogen. De-nitrification occurs only in conditions where the dissolved oxygen is low (anoxic conditions) and when nitrate is present. The bacteria involved normally use the carbon in the wastewater as a carbon source. There are a number of different micro-organisms involved. Conversion to and the release of the inert atmospheric form of nitrogen are the end of the treatment process.

7.13.4 Phosphorus removal

Phosphorus is an essential element for all living organisms and a component, for example, of DNA. However, if too much nitrogen and phosphorus are present in water courses, algae blooms can develop, cutting down the light transmission in the water and so it is necessary to remove phosphorus from the waste water. Furthermore, unlike nitrogen, the key source of increased phosphorus build-up in receiving waters is not due to excessive fertiliser in land run-off, *but to sewage disposal and soluble polyphosphates used in detergents*.

Phosphorus can be removed either chemically by precipitation or biologically by concentration in bacterial cells followed by a settling out process.

7.13.4.1 Biological phosphorus removal

The biological phosphorus removal mechanism is reliant on the fact that certain bacteria are capable of removing simple fermentation substrates produced in the anaerobic stage of wastewater treatment and assimilating them into storage products within their own cells. These bacteria derive energy for their own growth from stored polyphosphates. Essentially in the aerobic zone of a sewage treatment process, these micro-organisms consume and store phosphates in their cell bodies.

Once the phosphorus is stored in the micro-organisms, it is important that the sludge is not inadvertently subjected to anaerobic conditions in subsequent treatment steps or the phosphorus may be released again as the bacterial cells die.

7.13.4.2 Chemical phosphate removal

This involves the use of chemical additives such as calcium, iron or aluminium that reacts with the phosphates to form a precipitant (which essentially becomes a suspended solid) which can then be allowed to settle out of the effluent stream. Chemical precipitation is almost always used in conjunction with a biological process when the latter alone cannot achieve a low enough phosphorus value.

The final treatment of the effluent prior to discharge to controlled waters is often to pass the effluent through an ultra violet source to kill off remnant bacteria.

7.13.5 Treatment and disposal of sewage sludge

Sewage treatment works are always less than 100 per cent efficient and as a result produce a sludge waste, which has to be disposed of. Beyond a certain point, sludge minimisation is impractical as the residence times required in the sewage treatment works would be increased excessively. A certain amount of re-cycling can be practised, for example, in activated sludge transfers, but treatment and disposal are inevitably required.

7.13.5.1 Disposal options

The disposal of sewage sludge presents problems since the phase-out of sea dumping. EC regulations also control the spreading of sewage sludge on agricultural land, where, though it has value as a fertiliser, it can also contaminate the land because of the presence of heavy metals and other pollutants (arising typically from industrial effluent discharges into the sewage system). Even so, significant quantities of sewage sludge are recycled by using it as a fertiliser in agriculture and forestry. Other disposal options include:

▶ use as a peat replacement;
▶ mixing with gravel to form an artificial topsoil for landscaping use;
▶ as a fuel in power generation plants;
▶ anaerobic digestion (see Section 7.13.5.2).

7.13.5.2 Anaerobic digestion

In the absence of oxygen, a number of groups of different anaerobic bacteria break down the substrate, in this case, sewage sludge, producing intermediate products consumed in turn by other bacteria. The by-products of this process are methane and carbon dioxide.

Anaerobic digestion is practised by all of the UK Water Authorities, to a greater or lesser extent and there is growing interest in the techniques as a source of renewable energy (the methane can be used as a fuel to generate electricity). Collaborative projects are well under way between the Water Authorities and the energy generator companies to increase the use of what could be a valuable energy source as part of the UK climate change programme.

7.13.6 Pre-treatment of industrial effluents

The treatment methods described above relate to the processes carried out by the sewerage undertakers to treat the mixed domestic/ trade effluent stream arriving at the sewage treatment works. For the more serious industrial pollutants or heavily contaminated effluents, there is normally a pre-treatment phase that takes place prior to discharge to the public sewer.

A wide range of treatment processes may be employed, depending on the nature of the industrial effluent and the discharge standards demanded by the trade effluent discharge consent. Typically, treatments fall under one of the following headings (though effluent treatment plants will often employ combinations of treatments to deal with pollutant mixtures).

7.13.6.1 Physical treatments

▶ Simple settlement or flotation tanks to remove organic or inorganic solids (often used in conjunction with chemical flocculants or precipitants which aid in separation of the solids).
▶ Adsorption using carbon filters to concentrate and remove impurities.
▶ Centrifuge and cyclones to separate out solid contaminants.
▶ Oil/water separators (flotation or tilted plate types) to remove oil and grease.
▶ Reverse osmosis (also known as ultra-filtration) – a membrane technology used to separate out dissolved solids such as salts (this method may also be used to regenerate process water for reuse).
▶ Solvent extraction – evaporation tanks with vapour recovery units.

7.13.6.2 Chemical treatments

▶ Ion exchange or manipulation to remove metal impurities or convert them into less harmful ions (e.g. conversion of hexavalent chrome to trivalent chrome reduces its toxicity significantly).

▶ pH regulation via the addition of acid (dilute hydrochloric acid is commonly used) or alkali (dilute sodium hydroxide is often used).

7.13.6.3 Biological treatments

Biological treatment is perhaps best established in the form of reed beds – artificial wetlands which often have additional wildlife benefits. Wastewater is introduced into the reed bed at the inlet and flows slowly through it via a horizontal path until it reaches the outlet zone. During its passage the wastewater will come into contact with a network of aerobic, anoxic and anaerobic zones. Reed rhizomes grow vertically and horizontally throughout the media, opening up the bed to provide new hydraulic pathways. The most common plant used is the common reed (*Phragmites australis*). This has the ability to transfer oxygen down through its leaf and stem structure, into the rhizomes and out through the roots. Aerobic bacterial populations proliferate in the area around the rhizomes. The reed beds thus facilitate the biological breakdown of organic waste. If well designed, they can be highly effective, low cost and aesthetically pleasing.

7.13.6.4 Sludge disposal

In most of the treatment methods outlined above, a solid waste is produced, often in the form of a filter cake with low moisture content (following the passing of the final sludge through a filter press to remove as much liquid as possible). This must be disposed of as a solid waste and, as it is essentially a means of concentrating pollutants from the effluent stream, it will often be classified as hazardous waste.

A new innovation in the treatment of heavy metals has recently been developed under a EU-funded research programme. The technology uses tiny magnetic particles known as super- paramagnetic composite particles (SPMC) that are added to the wastewater supply. They attract heavy metals such as arsenic, mercury and lead which are highly toxic to humans and the aquatic environment. The particles and the attached heavy metals are then separated by a second process, enabling the metals to be disposed of as a concentrated sludge, or even recovered and reused in other applications.

If commercialised, the method could be a viable low cost treatment with application in a variety of industries.

7.14 Land pollution/solid waste control

There are two distinct areas discussed in this section. The first (sections 7.14.1–7.14.6) covers remediation techniques used for the clean-up of contaminated land. The second (sections 7.14.7–7.14.12) covers the options available for treating waste that provide alternatives to landfill.

7.14.1 Contaminated land: remediation categories

There are a number of alternatives available in relation to the remediation of soil and groundwater contamination, falling into three broad categories:

▶ *Removal* – excavation of contaminated material and disposal at a licensed waste facility.
▶ *Containment* – involving on-site engineering to contain the parcel of contamination using methods such as surface capping, vertical and/or horizontal barriers to prevent migration of contaminants.
▶ *Treatment* – this may occur on or off site and may include physical, biological or chemical methods to reduce or eliminate the hazards associated with contamination. Although some overlap exists in the types of treatments used, for the sake of simplicity, the methods described below are separated into groundwater and soil techniques

7.14.2 Removal

Sometimes referred to as 'dig and dump', this was at one time a very common remediation strategy. However, in more recent years, the reduction in hazardous waste landfill space and the associated escalation in disposal costs, typically have made this a very expensive option. In addition, it is increasingly discouraged by regulators as in some ways it does not remediate contamination, merely moves it from one location to another, albeit to one where the risk of harm is controlled. It is still used on occasion on contaminated sites but typically only where relatively small quantities of material are involved or where there is an immediate pollution or health risk posed. For example, if contamination had already spread to an adjacent watercourse, it may be appropriate to dig out and remove contaminated soil immediately adjacent to the stream in order to prevent further pollution of controlled waters and to 'buy time' to remediate the rest of the contamination using the treatment techniques discussed below. It may even be appropriate to use some kind of temporary containment method such as ditches or vertical barriers to isolate the contaminated area from the watercourse until the remediation process is complete.

7.14.3 Containment

In some cases the simplest option for making contaminated land safe is simply to contain the contaminant in situ. This can only be the case where there is no risk of the contaminant spreading from site and that its location is such that the possibility of human contact is negligible. In most cases where containment is the chosen 'remediation option', removing the contaminant for treatment or disposal would increase the risk of pollution or human health impacts or the cost of remediation would be extremely high.

Containment measures involve on-site engineering to contain the parcel of contamination using methods such as surface capping (using concrete, imported soil or liners), vertical and/ or horizontal barriers to prevent migration of contaminants (using piling, jet grouting, installation of permeable membranes, etc.). Containment may only need to be in one plane, e.g. surface capping for a non-mobile contaminant such as asbestos, or might need to be three-dimensional, i.e. creating a containment box for highly mobile pollutants such as hydrocarbons or organochlorines. In all cases, clear identification and record keeping are required to ensure that no development is subsequently undertaken without the risk of contamination exposure being clearly understood.

7.14.4 Treatment techniques: groundwater

There are a wide variety of techniques available for treating contaminated groundwater both in situ and by pumping out for on-site or off-site treatment. Some common examples include the following.

7.14.4.1 Removal of floating product

Organic contaminants are frequently insoluble or only partially soluble in water and hence form a 'floating' layer on the top of the groundwater table. This feature means they lend themselves to recovery using skimmer pumps or passive separators placed in wells, pits or trenches. As the groundwater drains into the well or trench, the floating material stays on the surface and can be separated from the water. Water is also removed from the trench to continue the process of 'draw down' of groundwater (and hence contaminants) into the trench. Multiple such wells or trenches may be required to draw out the contaminants from a larger area.

7.14.4.2 Groundwater abstraction and on-site treatment

This involves pumping groundwater from wells/ boreholes at a site and treating it using a range of methods including:

▶ oil/water separation
▶ air stripping/aeration
▶ ultraviolet oxidation
▶ carbon adsorption filtration
▶ bioremediation.

Some of these techniques are discussed in the effluent treatment Section 7.13.6 above. Treated groundwater may then be either re-injected or discharged to sewer or controlled waters as appropriate.

7.14.4.3 Air sparging (in-situ air stripping)

This process forces air below the water table and uses the bubbles generated to mobilise dissolved contaminants upwards. The technique is effective at reducing volatile components in appropriate geological conditions. It has the additional benefit of increasing biodegradation rates in both groundwater and soil through an increase in oxygen content.

7.14.4.4 In-situ bioremediation

Bioremediation techniques for groundwater are very similar to those described for soil (see below). The only significant difference is that the oxygen and nutrients are injected below the water table rather than above it.

7.14.5 Treatment techniques: soil

7.14.5.1 Vapour extraction

Soil gas vapour extraction is a low cost but effective cleaning method for soil contamination by volatile organic substances, e.g. aromatic and halogenated hydrocarbons. A pump device sucks the contaminated soil gas out of the ground through a specially designed well. Typically, the soil gas is drawn through activated charcoal filters, which absorb the contaminants. The charcoal can then be recycled. The major advantage of this technique is that it works in situ. It can be used in conjunction with the air sparging techniques described in Section 7.14.4.3.

7.14.5.2 In-situ bioremediation

Bioremediation is based on the provision of an environment that enables indigenous or introduced microbes to metabolise the contaminants in the soil. Optimisation of the soil environment is achieved through the addition of nutrients, oxygen and water. Various methods are used to provide these additions, including horizontal or vertical infiltration points, surface applications and in-situ mixing (sometimes referred to as land farming).

Bioremediation is capable of degrading organic compounds into more basic degradation products such as carbon dioxide and water, which are generally less harmful or harmless to the environment. It has been used widely for hydrocarbon contamination in particular.

7.14.5.3 Soil flushing and soil washing

These are related techniques which physically 'wash' contaminants from the soil using water or a liquid solvent. The principal difference between the two techniques is that flushing involves in-situ applications and washing is essentially an 'excavate and replace' option employing dedicated plant. The latter technique had been proven to be highly successful on several plants in Germany at acceptable costs with plant capacities greater than 1,000 tonnes per day.

7.14.5.4 Thermal and chemical treatments

These are an extension to the soil washing techniques discussed above and involve removal of contaminated soil for on-site treatment using mobile plant. Thermal treatment, e.g. burning off hydrocarbons, may be appropriate for certain organic pollutants, while various chemical treatments may be used to mobilise and recover pollutants or to convert in situ to less harmful by-products, e.g. pH regulation of acid or alkaline contaminants.

7.14.6 Contaminated land remediation guidance

Project CL:AIRE is a public/private partnership involving the following stakeholders:

▶ government policy-makers
▶ regulators
▶ industry
▶ research organisations
▶ technology developers.

CL:AIRE provides a link between the main players in contaminated land remediation in the UK, to catalyse the development of cost-effective methods of investigating and remediating contaminated land in a sustainable way. It provides valuable independent assessment of treatment technologies in a variety of settings to both those responsible for contaminated land and clean-up contractors. The projects under the CL:AIRE banner include the following.

7.14.6.1 Technology Demonstration Projects (TDPs)

Each project has a fact sheet and project report produced to describe methods used and results obtained. The technology demonstration projects are for readily available commercial techniques.

7.14.6.2 Research projects

In addition to the TDPs there is a rolling programme of research projects under way which again cut across a range of analysis and remediation techniques but, in this case, aim to develop new knowledge and processes in the treatment of contaminated land. Further information on Project CL:AIRE is available from www.claire.co.uk.

7.14.6.3 The definition of waste: Development of Waste Industry Code of Practice (DoWCoP)

According to the CI:AIRE website, the DoWCoP is an initiative to improve the sustainable and cost effective development of land. The DoWCoP (published in 2008 and updated in 2011) provides a clear, consistent and streamlined process which enables the legitimate reuse of excavated materials on-site or their movement between sites with a significantly reduced regulatory burden. In many instances, the DoWCoP can provide an alternative to Environmental Permits or Waste Exemptions when seeking to reuse excavated materials.

The DoWCoP enables the direct transfer and reuse of clean naturally occurring soil materials between sites. It creates the conditions to support the establishment and operation of fixed soil treatment facilities, which have a key role to play in the future of sustainable materials management. It allows the reuse of both contaminated and uncontaminated materials on the site of production and between sites within defined Cluster projects.

7.14.7 Waste disposal options

As discussed in Chapter 1, waste production in the UK has been gradually declining in recent years. UK government data is several years out of date but Environment Agency data shows the amount of waste disposed to landfill reducing by 48 per cent between 2000 and 2010. UK government data published in 2011 shows total waste figures at 288.6 million tonnes in 2008, down from some 335 million tonnes of waste in 2004. Of the 2008 total, 45 per cent was recovered while 48 per cent was deposited onto or into the land.

The continued role of landfill in current waste management practices reflects the fact that landfill is the most adaptable and least expensive waste management option in most areas of England and Wales, although landfill tax charges have altered and continue to influence this choice. We will now briefly consider the principal disposal options for controlled wastes and the advantages and disadvantages with each.

7.14.8 Landfill

As well as being a non-productive use of land and a source of noise, odour and vermin nuisance, there are two potential pollution hazards associated with landfill:

▶ leachate contaminating the groundwater;
▶ production of methane waste gas.

The principal aim of lining (and capping) a site is to contain the leachate, thus preventing pollution of surrounding land and waters. Lining may also assist in leachate control by reducing groundwater infiltration into the landfill. Lined landfill sites can usually accept a wider range of wastes than would otherwise be possible. Artificial liners are constructed of impermeable materials. Natural lining materials, such as heavy clay soils, exhibit low permeability and may be acceptable for certain waste types.

Where the use of a liner is envisaged, the suitability of a site for lining will have been evaluated in the site investigation stage and, if found to be suitable, plans for its lining will have formed part of the working plan for the landfill operation.

The use of liners and improved leachate collection and treatment systems is demanded under the more stringent protection standards enforced via the EC Landfill Directive. Other implications of the directive include:

▶ the banning of clinical wastes from landfill;
▶ phased banning of disposal of tyres and liquid wastes to landfill;
▶ the requirement for leachate and groundwater monitoring at least twice a year for 30 years after cessation of disposal activities;
▶ the required installation of leachate collection and drainage systems;
▶ the collection and treatment of landfill gas where appropriate;
▶ the progressive reduction (through a series of targets over the period 2010–2020) in amount of biodegradable municipal waste disposed of via landfill.

Factors such as these have increased and will continue to increase the cost of landfill disposal, which has traditionally been the cheapest disposal option. This, in combination with the landfill tax, will reduce or eliminate the differential with other disposal options and may open up previously uneconomic recycling and recovery options.

7.14.9 Incineration

There are a number of advantages presented in relation to the incineration of waste:

▶ The reduction in the final disposal space required (down to the remnant ash volume); savings of 90 per cent volume are possible in relation to uncompacted waste, 40–50 per cent in relation to compacted waste. It should, however, be noted that incineration ash is often a concentrated source of heavy metals and other pollutants and must be disposed of as hazardous waste.
▶ The elimination hazardous components – PCBs, clinical waste, etc. – can only safely be disposed of through high temperature incineration.
▶ The generation of energy associated with 'waste to energy' plants.

In some incinerators all three advantages are achieved. However, not all waste incineration generates energy and not all incineration is at high enough temperature to destroy high hazard components.

Particularly in the context of the climate change challenge, there is considerable interest in power generation from waste (either directly from incineration or using landfill gas). The technology to use waste as a fuel for power generation is well proven and estimates of untapped potential are as high as the energy equivalent of 26 million tonnes of coal per year! Although incineration is now widely used as a method of waste disposal and is increasing in importance as a combined waste disposal/renewable energy solution, there remain serious differences of opinion regarding its credibility as a replacement energy source for fossil fuel power stations. There are several important objections:

▶ the emissions from the incineration process may contain highly toxic substances such as dioxins and furans which are produced when chlorine-containing compounds such as plastics and bleached paper are burned;
▶ incinerating organic wastes generates nitrogen oxides which lead to acid rain;
▶ incinerating organic wastes releases carbon dioxide, a greenhouse gas;
▶ incineration wastes valuable resources, which, though they may currently go to landfill, could be diverted for recovery/recycling;
▶ once a capital investment has been made in incineration facilities, there is less incentive for waste producers to pursue waste minimisation opportunities.

Increasingly stringent EU standards covering emissions standards from incinerators should mitigate the health hazards but some lobby groups remain firmly opposed on principle.

7.14.10 Recycling

Recycling is an important part of any waste management strategy as it keeps material resources in use and avoids or radically reduces disposal quantities. It is worth remembering, however that, as the waste hierarchy indicates, it is less desirable than either reuse or waste minimisation. Generally, recycling is effective when the following conditions apply:

▶ There exists a reliable supply of suitable waste material.

▶ It is feasible and economically viable to collect the material and transport it to a reprocessing site.

▶ It is technologically possible and economically viable to reprocess the waste into raw materials and products.

▶ Markets exist for the raw materials and products produced by the recycling process.

It should be noted, however, that initially some or all of these criteria may not exist and it may take investment, legislation, fiscal measures and education to put all the pieces in place. Furthermore, recycling can be expensive, with significant costs associated with materials collection as well as reprocessing. Recyclers will often pay for materials of significant intrinsic value (e.g. scrap metals) but waste producers may still have to pay for the collection of bulky low value items such as cardboard. In many cases, however, the collection costs will remain lower than disposal to landfill (and with increases in landfill costs, the case for paid recycling services strengthens).

In terms of the waste management services there are some other obstacles to recycling which include the following:

▶ Bulk waste disposal contracts can discourage recycling, particularly when penalties are imposed for reduced volumes of waste.

▶ Industry standards and legal requirements can limit the use of low grade recycled materials, limiting local markets for the recycling products.

▶ Space and other practical limitations (like contamination) make it difficult to separate and store materials that could be recycled.

▶ In many cases the market for recycled materials cannot exist until quality is proven and quantities pass threshold levels – this is the recycling Catch-22.

Difficulties aside, however, recycling is a crucial part of the waste management strategies of many countries with significant opportunities in multiple areas. Pre-consumer recycling (of factory waste, for example) tends to be particularly viable, because the waste material is often available in quantity, is readily separated from other waste and there is often a market close at hand for the end product.

Post-consumer waste is more difficult to manage but some Local Authorities in the UK have already demonstrated considerable success in recycling household waste. Driven by EU targets, in 2013 the UK national average reached 43 per cent recycling of municipal waste (in Germany, this figure is nearer 70 per cent). Table 7.18 lists some of the waste streams that have been subject to recycling activities.

Table 7.18 Examples of waste streams subject to recycling

Aerosols	Furniture	Plastic cups
Agricultural waste	Garage waste	Refrigerators
Aluminium	Gases	Rubber
Ash	Glass	Sewage sludge
Batteries	Metals	Silver
Building aggregates	Nuclear fuel	Solvents
Building materials	Oil	Textiles
Cardboard	Packaging materials	Toner cartridges
CFCs and HCFCs	Pallets	Tools
Computer disks	Paper	Tyres
Electrical appliances	Plastics	Vehicles
Fluorescent tubes		

Several government-supported organisations have been set up to promote and advise the business and public sectors on waste minimisation and recycling or recovery alternatives to landfill. WRAP – www.wrap.org.uk – the Waste and Resources Action Programme – is probably the best-known and established advisory body promoting waste reduction and recovery techniques in both the public and private sector.

7.14.11 Composting

Composting is essentially the recycling of organic waste, most of which can be broken down and used as fertiliser, on either a domestic or a commercial scale. It requires the separation of biodegradable waste and the control of moisture and temperature levels for maximum efficiency. As well as producing a valuable and marketable end product, it has the additional benefit of reducing the generation of landfill gas by removing the source material from the landfill stream.

Composting processes primarily fall into two categories: windrow composting, for garden-derived wastes (often referred to as 'green wastes'), and 'in-vessel' composting, which is required to process material containing food waste, which has either been collected separately, with garden waste or has been mechanically recovered from mixed municipal waste streams.

Windrow composting is an established technique for dealing with collected garden wastes in the UK. Wastes are stored in rows that are turned and watered to maintain optimum aeration, temperature and moisture content for rapid composting. In-vessel systems involve the initial composting of wastes in an enclosed vessel or tunnel, typically followed by a period of further composting outdoors. The advantage of these processes is that the vessel is designed to achieve and maintain specified temperatures to facilitate pathogen destruction in accordance with the requirements of the Animal By-Products Regulations. This legislation governs the management of wastes arising from animal sources, including food and catering wastes, to prevent animal by-products from presenting a risk to animal or public health through the transmission of disease. As Local Authorities strive to divert increased quantities of biodegradable waste from landfill, it is anticipated that in-vessel composting will become increasingly important.

The principal disadvantages associated with composting are the impacts associated with collection, the nuisance potential (particularly odour) associated with the process and the availability of markets for the final product.

7.14.12 Anaerobic digestion

Anaerobic digestion is a biological process where biodegradable wastes, such as food waste, are converted into a 'digestate' and biogas. The wastes are decomposed by microbes in the absence of oxygen. This is different to composting which is an aerobic process, taking place in the presence of oxygen.

Anaerobic digestion systems are enclosed, engineered vertical or horizontal vessels. The biodegradable material is macerated and water is often added to provide suitable moisture and flow properties. Waste remains in the vessels for 2–3 weeks and reaches temperatures of up to 60°C. As the waste degrades, biogas is produced, comprising mainly methane and some carbon dioxide. This is collected in tanks and used on- or off-site. Biogas can be used in a number of ways but is usually burned to produce electricity; the heat from the process may be used on site or by neighbouring users. Excess electricity not required by the plant can be exported for distribution in the grid and excess heat can be used for district or industrial heating.

Anaerobic digestion is used in the UK for treating agricultural manures and slurries, as well as sewage sludge, however, there is limited experience on its application to municipal biowastes in the UK. As with composting, however, it is a technique set to become much more widely used as waste and renewable energy targets begin to put pressure on municipal waste authorities.

7.15 Nuisance control techniques: noise

A wide range of measures can be introduced to control the source of, or limit exposure to, noise and may include one or more of the following categories:

▶ *Engineering*: reduction of noise at point of generation (e.g. by using quiet machines and/or quiet methods of working); containment of noise generated (e.g. by insulating buildings which house machinery or providing purpose-built barriers around the site); and protection of surrounding noise-sensitive buildings (e.g. by improving sound insulation in these receptor buildings and/or screening them by purpose-built barriers).

▶ *Lay-out*: adequate distance between source and noise-sensitive buildings or area; screening by natural barriers, other buildings, or non-critical rooms in a building.

▶ *Administrative*: limiting operating time of source; restricting activities allowed on the site; specifying an acceptable noise limit.

The Environmental Permitting Horizontal Guidance for Noise Part 2 deals with noise assessment and control. It suggests that once all administrative/ maintenance issues have been implemented, the minimisation of noise levels at sensitive receptors can be achieved in three ways:

▶ reduction at source;
▶ ensuring adequate distance between the source and receiver;
▶ the use of barriers between the source and the receiver.

It considers noise (and vibration) transfer from a source-pathway–receptor perspective as

illustrated in the diagram taken from the guidance note shown in Figure 7.12.

The BAT approach to managing noise that it presents is based on the noise control hierarchy shown in Figure 7.13.

7.15.1 Examples of noise management techniques

Noise control is a specialist subject, especially when considering the design of acoustic enclosures, screens, etc. but some commonly employed techniques are described below:

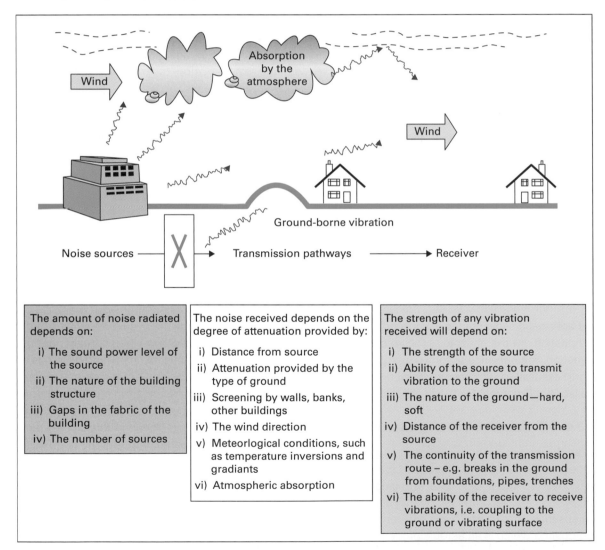

The amount of noise radiated depends on:

i) The sound power level of the source
ii) The nature of the building structure
iii) Gaps in the fabric of the building
iv) The number of sources

The noise received depends on the degree of attenuation provided by:

i) Distance from source
ii) Attenuation provided by the type of ground
iii) Screening by walls, banks, other buildings
iv) The wind direction
v) Meteorlogical conditions, such as temperature inversions and gradiants
vi) Atmospheric absorption

The strength of any vibration received will depend on:

i) The strength of the source
ii) Ability of the source to transmit vibration to the ground
iii) The nature of the ground—hard, soft
iv) Distance of the receiver from the source
v) The continuity of the transmission route – e.g. breaks in the ground from foundations, pipes, trenches
vi) The ability of the receiver to receive vibrations, i.e. coupling to the ground or vibrating surface

Figure 7.12 A source-pathway-receptor model of noise nuisance

Source: From the EA Horizontal guidance note for noise Part 2. Contains Environment Agency information © Environment Agency and database right.

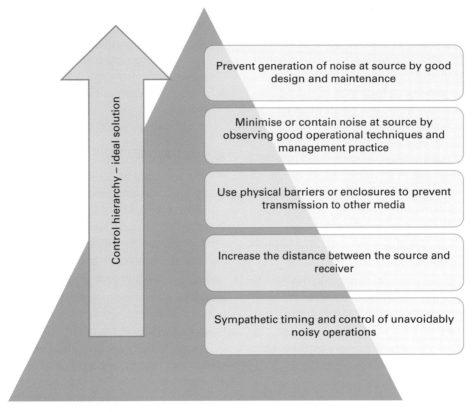

Figure 7.13 The noise control hierarchy

▶ Change the working method to use equipment or modes of operation that produce less noise. For example, in demolition works, hydraulic shears may be used in place of hydraulic impact breakers. In driving sheet steel piles, where ground conditions allow, the jacking method may be used which produces a fraction of the noise produced by traditional hammer-driven piling. When breaking out pavements or hard standing, alternatives to pneumatic breakers and drills can be used. The use of chemical splitters or falling-weight breakers can be considerably quieter.

▶ Reduce the need for noisy assembly practices, e.g. fabricate off-site at less noise-sensitive locations.

▶ Locate noisy plant as far away as possible from noise receptors.

▶ Adopt working hours to restrict noisy activities to certain periods of the day. More stringent standards may be applied for the evening (generally taken as 1900 to 2300 hours) and night (2300 to 0700 hours) than for the daytime.

▶ Arrange delivery times to suit the area – daytime for residential areas, perhaps night-time for inner city locations.

▶ Route construction vehicles to minimise the likelihood of nuisance.

▶ Keep access roads well maintained to avoid the additional traffic noise generated through jolting over rough surfaces.

▶ Use mufflers or silencers to reduce noise transmitted along pipes and ducts.

▶ Minimise drop heights into hoppers, lorries and other plant.

▶ Consider using rubber linings on tipper vehicles in very noise-sensitive locations.

▶ Ensure acoustic enclosures (engine cowls, etc.) are in place and securely fixed (to prevent rattling noise).

▶ Shut off equipment that is only periodically in use.

▶ Ensure plant is adequately maintained and lubricated to avoid frictional noise, etc.

295

▶ Consider the use of screening using specially created screens or capitalising on the location of buildings, material storage areas, etc.

▶ Tree planting may provide effective mitigation of visual impacts, and may psychologically reduce the subjective response to the noise. Acoustically, however, this usually has an almost negligible effect.

▶ Use of 'smart' reversing alarms, which produce sound at a volume relative to the background level, for example, 5 or 10 dB above, rather than at a fixed volume; or using other safe systems of work which obviate the need for reversing alarms, e.g. camera systems.

▶ Careful siting, use and volume control of public address systems.

In addition to these practical examples, the Environmental Permitting guidance note identifies ten generic types of noise-control strategies which account for the majority of the equipment used by the noise control industry. These are listed below:

▶ acoustic enclosures
▶ acoustic louvres
▶ noise barriers
▶ acoustic panelling
▶ acoustic lagging
▶ vibration damping
▶ impact deadening attenuators
▶ steam and air diffusers
▶ vibration isolation mounts.

While vibration damping and isolation mounts control vibration, they can often make a key contribution to noise control. It is worth noting that for many noise sources there will be more than one appropriate control option. The guidance provides simple summary sheets and indicative costs for each of the equipment headings. For full details, refer to the H3 Horizontal Guidance Note for noise assessment and control (available for download from the EA website).

7.16 Nuisance control techniques: odour

The Environmental Permitting Horizontal Guidance for Odour Part 2 – Assessment and Control identifies the following categories of techniques used for odour control:

▶ adsorption using activated carbon, zeolite, alumina;

▶ dry chemical scrubbing using solid phase systems impregnated with chlorine dioxide or potassium permanganate;

▶ biological treatment including soil bed bio-filters, non-soil bio-filters – such as peat/heather, wood-bark, compost – bio-scrubbers;

▶ absorption (scrubbing) using spray and packed towers, plate absorbers;

▶ incineration using existing or dedicated boiler plant, thermal or catalytic systems;

▶ other techniques including odour-modifying agents, condensation, plasma technology, catalytic iron filters and ozone and UV.

Defining a field of application for different abatement technologies is difficult, given the number of variables that must be considered. However, Figure 7.14, taken from the H4 Guidance Note, gives an indication of where abatement techniques are typically used as a function of gas flow rate and odorant concentration. Obviously the demarcations between technologies is not set, and other factors such as detailed gas composition and level of control required may play a very important role in the decision-making process.

Adsorption, absorption and incineration have already been covered in Section 7.12 on air pollution control above, so this section will focus on the other groups mentioned by the guidance note.

7.16.1 Dry chemical scrubbing

Dry chemical scrubbers are a relatively recent addition to the odour abatement market and are essentially a refinement of the adsorption techniques described earlier in relation to other air pollutants. Systems typically consist of an oxidising chamber and a polishing stage.

▶ *The oxidising chamber* contains a support material which is impregnated with oxidising material (e.g. chlorine dioxide, potassium permanganate, etc.). The odorous gas passes up through the oxidising chamber where it is adsorbed and then oxidised to non-odorous by-products.

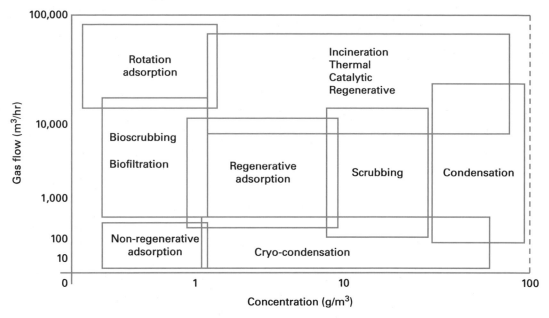

Figure 7.14 Application of odour abatement techniques to different levels of contaminant and rate of air flow
Source: from EA H4 Horizontal Guidance Note). Contains Environment Agency information © Environment Agency and database right.

▶ *The polishing stage* comprises activated carbon which is used to remove any unoxidised odorous compounds.

Dry chemical scrubbers are ideal for extremely low flow, relatively high concentration odorous gas streams. They can be purchased as stand-alone systems so that one unit can be installed next to one source. This is advantageous because there is no need for a complicated ducting system which keeps costs down.

7.16. Biological treatment

The phenomenon whereby organic compounds can be metabolised and degraded by naturally-occurring micro-organisms into non-odorous reaction products is widely used as the basis for odour control devices. As with water and land bio-remediation, the micro-organisms involved are reasonably robust, provided that there is a constant supply of carbon and oxygen (and that temperature and humidity are maintained within tolerance limits).

Biological treatment falls into two basic categories:

▶ *Bio-filtration*: a bio-filter typically consists of a large bed of soil (earth), compost or fibrous peat through which the malodorous air is passed.
▶ Bio-absorption: a typical example of bio-absorption is a packed tower system (see Figure 7.15) in which the packing material supports a microbial film (this design is sometimes referred to as a bio-scrubber).

A high efficiency (>99 per cent) of odour reduction with minimal secondary pollution (wastes generated) can be achieved with a soil bio-filter provided a suitable colony is established and maintained, though manufacturers are unlikely to provide guarantees in excess of 95 per cent. However, the footprint area of the filter is often large and a bio-scrubber may have to be considered where space is at a premium. The adaptation period of micro-organisms is slow in comparison to the fluctuations in many industrial processes and hence bio-filters are best suited to processes with an output stream in which changes of concentration take place sufficiently slowly to allow the microbial population to adapt. Designs clearly have to be carefully matched

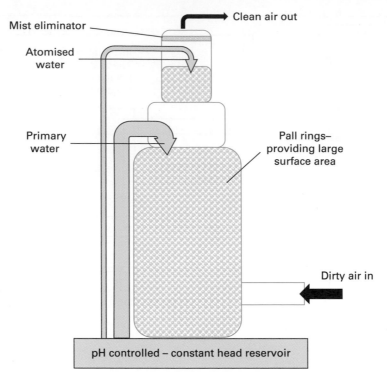

Figure 7.15 A sketch of a bio-scrubber unit

to the throughput rate and concentration of malodorous air.

Problems encountered with bio-filters are often related to the bed becoming blocked by compaction. This can be caused by driving heavy machinery over the bed, which should be avoided. The bed may also crack due to insufficient irrigation during dry weather. The distribution pipes may flood due to inadequate drainage or raising of the water table. Care must be taken to ensure that contaminated liquid effluents do not leak out of the bed.

Bio-scrubber problems include the build-up of biomass within the scrubber, resulting in blockage of the circulating water. Means of easy access to the interior of the scrubber for removal of this is desirable. There can sometimes be an odour from the liquor circulation holding tanks, and if chemicals or disinfectant get into the scrubbing liquor, they may be lethal to the 'scrubbing microorganisms'.

7.16.3 **Other techniques**

This category includes two further examples of odour control techniques.

7.16.3.1 **Odour modification**

This normally involves the addition of another chemical to mask, counteract or neutralise the original odour. Chemicals may be released at the boundary of the site creating the odour nuisance or even released directly into the emissions stream if there is a point source release. The approach is more often seen as a temporary control measure than a permanent solution as there are a number of disadvantages to the approach, summarised in the Environmental Permitting guidance as follows:

▶ The use of odour masking compounds can be problematic, as the malodorous emission may vary in concentration or nature with time. These variables make it difficult to ensure that unpleasant odours are 'blotted out' at all emission levels.

▶ The odour of the modifying agent can itself become a source of annoyance.

▶ In terms of the relationship between perceived intensity and concentration (of malodorant and odour modifier), the concentration will decrease downwind from the source, but the effect on their perceived intensities will not be uniform with the decreasing concentration. This is because, for every substance the perceived intensity decreases to different extents for the same decrease in concentration. Other factors such as differing diffusion characteristics of the modifier and the odour itself may cause the odour to separate from the modifying agent at a distance, thus producing two distinctly different odours at different points. In other words the two odours may fade/disperse at different rates so what might be effective masking at the point of generation may be ineffective at a distance from the source.

- ▶ The on-going cost of the modifying agent can be very expensive and maintenance costs can be high as fine spray nozzles can be prone to blockage.
- ▶ While no direct evidence has been found which links the administration of odour-modifying agents to ill-health, there is evidence that some of those exposed perceive that there could be an issue.
- ▶ The approach should not be considered, even as a temporary measure, where the odorous emission carries a risk to health or the odour itself serves as a safety warning, e.g. hydrogen sulphide emissions from sewage treatment works.

7.16.3.2 Condensation

Where an odorous gas stream contains high concentrations of vapour, condensation (as previously discussed in Section 7.12.1.4 on air pollution control) might be considered for odour reduction. There are various methods for bringing about condensation and they fall broadly into the following categories:

- ▶ *Direct contact condensers* – a cooling liquid (at ambient temperature or slightly below) is sprayed into the vapour-laden gas flow. This is the cheapest and simplest method.
- ▶ *Indirect contact (surface) condensers* – these are cooled with water or a cooling liquid. Coolant usually circulates in the tubes and vapour condenses on the outer surfaces and drips into a storage tank.

- ▶ *Air-cooled surface condensers* – air-cooled finned tubes are used to cool down the vapour sufficiently for condensation to occur.
- ▶ *Pressurised condensers* – gas is compressed before cooling. This is the most efficient condensation method but also the most expensive.

The main problem associated with condensation is the high energy requirement, ensuring that the cost, both capital and operating, tends to be high. Effluent disposal typically also needs to be considered. The process of condensation is mainly used as a pre-treatment to other odour abatement technologies.

7.17 Nuisance control techniques: dust

Dust may be generated from a variety of industrial, construction and agricultural sources. Where it is generated at a point source as part of an air emissions stream the particulate control measures described in Section 7.12 on air pollution control may be applicable. However, often dust as a nuisance arises from non-point sources such as construction sites. In such instances, different control techniques must be applied. In 2005, the Greater London Authority GLA produced a document entitled *London Code of Practice: The Control of Dust and Emissions from Construction and Demolition*, which provides guidance and standards related to dust control on construction sites. The principles are applicable to all construction sites and other locations and activities where similar sources of dust nuisance occur. The following areas of best practice guidance are selected from the guide.

7.17.1 Haul routes and site entrances/exits

Unpaved haul routes can account for a significant proportion of fugitive dust emissions, especially in dry or windy conditions, when the generation of dust through the movement of vehicles is exacerbated. Control options include temporary surfacing or, perhaps more commonly, damping down, i.e. spraying water onto the haul roads to suppress dust. Other measures to reduce dust generation include the setting of site speed limits, covering dusty loads and regular vehicle and/or wheel washing (especially when leaving site).

7.17.2 Excavations and earthworks

In addition to damping down, earthworks may be temporarily covered when no active work is underway or under adverse (hot and windy) conditions. Sites should be re-vegetated as soon as possible to stabilise exposed soils or covered with hessian, mulch, etc. to prevent dust generation.

7.17.3 Stockpiles and storage mounds

Wind-blown material from stockpiles is a common source of nuisance dust. A number of measures may be taken to minimise dust generation including:

- Make sure that stockpiles are maintained for the shortest possible time.
- Minimise drop heights to control the fall of materials.
- Do not build steep-sided stockpiles or mounds or those that have sharp changes in shape.
- Keep stockpiles or mounds away from the site boundary, sensitive receptors, watercourses and surface drains.
- Wherever possible, enclose stockpiles or keep them securely sheeted.
- Take into account the predominant wind direction when siting stockpiles to reduce the likelihood of affecting sensitive receptors.
- Seed, re-vegetate or turf long-term stockpiles to stabilise surfaces or use surface binding agents that have been approved by the Environment Agency.
- Re-use hard core material where possible to avoid unnecessary vehicle trips.
- Erect fences or use windbreaks (e.g. trees, hedges and earth-banks) of similar height and size to the stockpile to act as wind barriers and keep these clean using wet methods.
- Store fine or powdery material (under 3mm in size) inside buildings or enclosures.

7.17.4 Waste management

In addition to the legal requirement to 'store waste appropriately' under the 'Duty of Care' there are a range of actions that can help reduce dust generation, for example:

- minimising storage time and the movement of bulk wastes;
- prohibiting on-site burning of waste;
- ensuring that all wastes are removed from site with appropriate covers in place.

The full code of practice is available for download from the GLA website (www.london.gov.uk).

7.18 Sustainable procurement strategies

In Section IX of Chapter 6 we considered supplier company and product prioritisation methodologies. These methodologies essentially form the basis of a sustainable procurement strategy in that they enable an organisation to systematically address supply chain impacts (and risks) through a number of approaches that were covered in Chapter 4 under the three basic categories of:

- supplier specification, i.e. dictating standards relating to company operations or product design and/or materials content;
- partnership approach, i.e. working with suppliers to collectively find solutions to unsustainable or high impact products or processes;
- selection approach, i.e. using available information to choose between suppliers and/ or products such that the sustainability impact of our choice is reduced.

The choice of which strategy or strategies an organisation employs will depend on its level of commitment, its purchasing power, its relationship with its first tier suppliers and the amount of information available in relation to the environmental impact of products and services purchased. In practice, most sustainable procurement strategies will use elements from all three of the above categories and will be undertaken under the ethos 'as far as is reasonably practicable' with the constraining factors typically being:

▶ the cost of goods and services;

▶ the ease of access to the types of information required to make product or supplier selection;

▶ the availability of alternative companies, products or materials.

An example evolution of a procurement strategy evolving towards increased sustainability might be as depicted in Figure 7.16.

7.19 Biodiversity protection and enhancement strategies

In Chapter 1 we discussed the principles of biodiversity and ecosystem services and their relevance to organisations. In Chapter 6 we reviewed the Corporate Ecosystems Review process developed by the World Resources Institute, which highlights that organisations represent a risk *to* biodiversity but are also themselves at risk *from* biodiversity loss.

At first glance it may be difficult to see what an organisation can do to protect and enhance biodiversity beyond the more general heading of environmental management, i.e. implement an EMS that ensures consistent standards of pollution control and resource efficiency are applied. It is certainly true (and sometimes forgotten) that a standard environmental management approach may help protect biodiversity by:

▶ reducing the scale of waste, pollution and nuisance generated by an organisation which may help reduce direct pressure on individuals, species or habitats;

▶ reducing the quantities of raw materials and utilities consumed by the organisation which may have implications in terms of pressure on biodiversity throughout the supply chain.

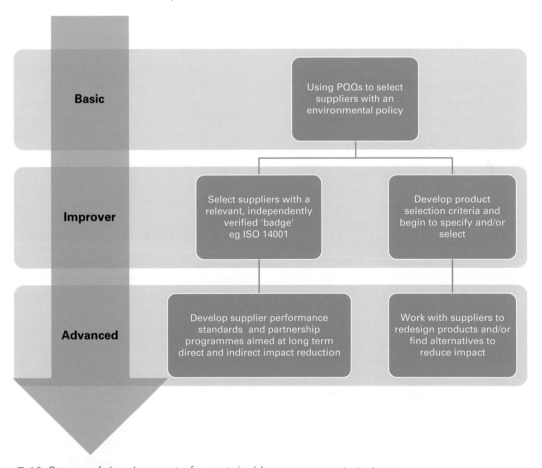

Figure 7.16 Stages of development of a sustainable procurement strategy

However, there are further actions that may be appropriate (depending on the nature and circumstances of the organisation) which go beyond the normal scope of a pollution control and resource-efficiency focused environmental programme. These actions may be grouped under two main headings – direct and indirect – which refer to the source of the 'pressure' on biodiversity resources. There has also been a growing interest in recent years in the use of biodiversity offsetting schemes.

7.19.1 Direct initiatives by organisations

For those organisations with land holdings of their own or that are involved in the provision of land development and/or management services such as construction or grounds maintenance, there are a number of initiatives that can be undertaken to enhance or maintain biodiversity.

A UK-wide programme of *local biodiversity action plans* has been established with 'lead partners' identified to coordinate and encourage conservation action for priority species and habitats. For example, the RSPB is the lead partner for 36 species (25 birds and 11 other species where the organisation's reserves or expertise are important components of management plan delivery).

Organisations with land holdings that contain areas of conservation interest can get involved with local biodiversity action plans and actively participate in land management practices that enhance or protect priority species or habitats. Even where land holdings currently contain no such conservation value, organisations may choose to manage areas in such a way as to enhance habitat value. Examples are even available of organisations without any 'green areas' managing such things as window boxes, roof gardens and building-mounted bird and bat boxes to enhance biodiversity locally. Local conservation bodies such as the Wildlife Trust are often involved as advisory partners in such projects. The Wildlife Trust has even developed a 'biodiversity benchmarking standard' which allows organisations to seek recognition through an audit process of the organisation's site-based

conservation efforts. If successful, the Wildlife Trust Biodiversity Benchmark logo can be used as a public statement of biodiversity commitment by the organisation.

7.19.2 Indirect initiatives by organisations

In addition to direct on-site activities, organisations have other options to make a positive contribution to biodiversity protection and enhancement. Two key possibilities present themselves:

▶ *voluntary support of local biodiversity initiatives* – through the provision of funds, personnel or other support in kind, e.g. hosting of fund raisers, participation in invasive species clearance projects, etc. This approach has the additional benefit of positive public relations with local communities.

▶ *Supply chain assessment initiatives* – through the expansion of selection criteria for goods and services to incorporate consideration of the associated biodiversity impacts. Simple selection criteria may be appropriate, e.g. an insistence that all timber and related products (including paper and board) are sourced from suppliers accredited by the Forestry Stewardship Council. Alternatively more complex guidance or decision-making profiles can be compiled and provided to purchasing departments that increase the probability that selections will be those least likely to have a negative biodiversity impact.

7.19.3 Biodiversity offsetting

Biodiversity offsets are conservation activities designed to deliver biodiversity benefits in compensation for losses, in a measurable way. DEFRA has produced a series of guidance documents aimed at supporting both developers and planners in considering and implementing biodiversity offsetting projects. The approach was promoted in the DEFRA (2010) report, *Making Space for Nature*, which looked at the health and resiliency of ecosystems in England and made a number of improvement proposals aimed at protection and enhancement of natural systems. Biodiversity offsetting, if conducted within the principles considered below, is considered a

Box 7.2 Biodiversity units

Imagine the construction of a new housing estate on a greenfield site. The initial assessment looks at the biodiversity units lost as a result of the development. This clearly requires habitat survey work as part of the planning process. A total number of biodiversity units lost will be calculated as shown in Table 7.19.

Table 7.19 Biodiversity loss calculation using the DEFRA guidance

Habitat type	Distinc- tiveness	Con- dition	Hectares	Number of units
Lowland meadow	6	2	6	(6x2x6) 72 bio- diversity units lost

Source: Adapted from DEFRA (2010b).

The developer would then need to find a way of creating 72 biodiversity units either as part of the development or elsewhere to offset the habitat loss incurred. While there is some flexibility in the nature of the offset, there are also some basic requirements relating to matching the biodiversity loss. The more distinctive the habitat loss, the more prescriptive the offset requirement as shown in Table 7.20.

Table 7.20 Offset requirements for different classifications of habitat distinctiveness

Distinctiveness of habitat lost	Distinctiveness of habitat provided by an offset
High	High – and usually the same habitat type
Medium	Medium or high
Low	Medium or high

Source: Adapted from DEFRA (2010b).

For hedgerows there is also a multiplier for good quality hedgerow lost, i.e. for every 100 metres of good quality hedgerow, 300 metres of newly planted hedgerow must be provided as an offset.

The offsetting approach allows planners and developers alike to provide and 'quantify' good quality 'compensation' that may contribute to a national biodiversity protection strategy rather than treating individual developments in isolation.

mechanism for improving the condition and connectivity of natural spaces and ecological communities. The principles set by DEFRA are that offsets should not be an 'excuse' to cause biodiversity harm and should deliver real benefits for biodiversity by doing the following:

▶ seeking to improve the effectiveness of managing compensation for biodiversity loss;
▶ expanding and restoring habitats, not merely protecting the extent and condition of what is already there;
▶ using offsets to contribute to enhancing England's ecological network by creating more, bigger, better and connected areas for biodiversity (as discussed in *Making Space for Nature*).

In essence, biodiversity offsetting should only be used where biodiversity loss is absolutely unavoidable and all other measures have been explored. There should be a genuine 'net increase' in habitat as a result of the offsetting activities and, wherever possible, improvements should be in line with national and regional strategies.

The approach is being piloted (2012–2014) in a number of English planning authority areas where developers are being offered biodiversity offsetting as part of the planning consent agreement. Developers can provide an offset themselves if they are able to do so, or they can commission someone else (an offset provider) to do it for them. In either case, a quantifiable amount of biodiversity benefit must be generated to offset the loss of biodiversity resulting from the development. The losses and gains are measured in the same way, even if the habitats concerned are different. In the biodiversity offsetting pilot, the measurement is done in 'biodiversity units', which are the product of the size of an area, and the distinctiveness and condition of the habitat it comprises. The assessment of biodiversity units lost and gained is calculated using the approach set out in the DEFRA guidance.

The approach is illustrated by the example shown in Box 7.2.

IV FURTHER RESOURCES

Table 7.21 presents some further resources.

Table 7.21 Further resources.

Topic area	Further information sources	Web links (if relevant)
Assessment techniques	DEFRA publication, 2012. Environmental Protection Act 1990: Part 2A – Contaminated land statutory guidance. DEFRA. BS 4142:1997 Method for rating industrial noise affecting mixed residential and industrial areas National Planning Practice Guidance – Noise	www.gov.uk/defra www.bsigroup.com planningguidance.planningportal.gov.uk/
Carbon management strategies	BREEAM guidance on sustainable building design	www.breeam.org
Contaminated land remediation techniques	Project CL:AIRE guidance and case studies	www.claire.co.uk/
Noise control techniques	Environment Agency Horizontal Guidance note H3, Part 2 Noise assessment and control	www.gov.uk/environment-agency
Odour control techniques	Environment Agency Horizontal Guidance note H4, Odour Management – how to comply with your environmental permit	www.gov.uk/environment-agency
Dust control techniques	Mayor of London best practice guide – the control of dust and emissions from construction and demolition	
Biodiversity strategies	World Business Council for Sustainable Development – business ecosystems training tool Wildlife Trust biodiversity benchmark DEFRA biodiversity offsetting guidance for offset providers and developers	www.wbcsd.org www.wildlifetrusts.org/biodiversitybenchmark www.gov.uk/biodiversity-offsetting

Developing and implementing programmes to deliver environmental performance improvement

Chapter summary

In earlier chapters we have considered a range of techniques to assess environmental priorities, manage environmental risk and identify opportunities for improvement in design and operation of an organisation's products and services. In Chapters 8, 9 and 10 we will consider the general principles involved in driving the implementation of long-term change strategies within organisations. The same basic principles apply to fairly minor, short-term changes in an organisation, but in the case of sustainability programmes, it is even more important to invest in communication, awareness and consultation as part of an overall approach intended to gather momentum as it progresses. We will consider this overall approach under the following four distinct areas:

- Implementing an environmental improvement programme

- Creating a business case for environmental change

- Communicating with stakeholders (in Chapter 9)

- Change management principles (in Chapter 10).

I IMPLEMENTING AN ENVIRONMENTAL IMPROVEMENT PROGRAMME

Environmental improvement programmes appear at a variety of scopes, timescales and administrative coverage. However, whether we are considering the implementation of an environmental management system, the introduction of a new waste management initiative or a product design improvement, we might consider the steps shown in Figure 8.1 as generally relevant within most organisations.

8.1 Identify ways to improve environmental performance

There are many tools and techniques used to identify opportunities for improvements. Many are considered in other chapters, e.g. auditing, life cycle analysis, carbon management programmes, etc. The point here in terms of generic steps to implementing an improvement programme is that some kind of review is conducted that highlights risks or opportunities for improvements

that generate clear actions to achieve those improvements.

8.2 Develop a programme to achieve improvement

For anything but the simplest initiatives, this stage is likely to involve the establishment of a leadership team with clear roles and responsibilities. There will then typically be some kind of project programme established, with key tasks and timescales identified and with critical paths highlighted. One or more persons will be appointed to track progress against the plan and highlight deficiencies or slippages as they arise. Polices, objectives and targets may be set and communication and review methods defined. Whether or not an organisation has a formal environmental management system, this stage will contain elements familiar to us from an ISO 14001 approach. In addition to these planning elements (to use the ISO 14001 terminology), we may set out procedures, train relevant members of staff or at least engage in awareness raising communications to let people know what is going on and why.

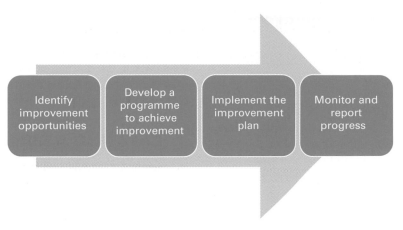

Figure 8.1 Steps in the implementation of an environmental improvement programme

The role of objectives and targets is important in this respect. They provide a clear sense of direction and motivation for actions to be completed. There are, however, some general principles around the effective use of targets. Objectives can, and perhaps should, be quite generic and open-ended to encourage innovation and long-term commitment. For example, a general commitment to eliminating the disposal of waste to landfill is a commonly encountered environmental objective. However, an objective like this alone, while clearly setting out our ambitions, gives little indication of the urgency and commitment required by staff today. Supporting targets can provide this, especially if they fulfil the general principle of being 'SMART' targets:

▶ *Specific* – ideally clearly linked to an environmental aspect within defined parameters, e.g. waste arising from a particular activity on a site.
▶ *Measurable* – ideally quantifiable using readily available data and presented in a format easily understood by all those who might be expected to help achieve them, e.g. percentage of total waste sent to landfill.
▶ *Achievable* – challenging or 'stretch' targets can be useful in some organisations to motivate innovation and a 'big push' in terms of improvements. However, overly ambitious or seemingly arbitrary targets can on occasion be demotivating leaving people with a sense of resignation to failure. It is perhaps a good idea to start with targets that

have been arrived at through some kind of calculation of opportunity for improvement so that participants have a fair sense of 'if we do this we should reach the target'. With a track record of success and a greater general awareness of the issue being dealt with stretch targets can then be used to launch a more concerted effort requiring a leap into innovation and the search for solutions.
▶ *Responsibilities allocated* – this links back to the 'specific' statement above and is also a more general point for action planning. Individuals or business units responsible for achieving the targets (or completing the actions) should be clearly stated.
▶ *Time-bound* – a clear point of achievement should be set and if long term, it is normally wise to set milestones over much shorter intervals to ensure that the programme keeps 'on track'.

The other key principle related to the use of targets in helping drive an improvement programme is that progress against them is regularly communicated and easily visible to all those involved. The presentation of key performance indicators linked to the current targets in easily accessible formats, close to the places where actions may need to be taken would be the ideal. An example linked to the objective given above might be a regularly updated performance graph with the target point clearly visible in the area where waste segregation takes place.

307

8.3 Implement the improvement plan

Again to use the ISO 14001 terminology, this is the implementation and operation phase, where tasks are completed, procedures followed and progress tracked. There may be an audit programme or less formal checks to ensure that things are proceeding according to plan. Slippage in progress against planned tasks and key milestones should be highlighted through review meetings or other communications.

8.4 Develop methods to monitor and communicate progress

Task-based monitoring of progress will normally occur in relation to the completion of actions set out in the improvement plan, however, in addition to management activities, there will normally be some tracking of performance changes linked to the goals of the improvement programme. For example, if we are implementing a waste minimisation programme we will not only track the completion of tasks such as installing a compactor, we will also be tracking the quantities of waste produced (probably normalised in some way to account for variable activity levels).

This area has been touched upon in Section 8.2 and is considered in some detail in Chapter 5.

8.5 Linking to a formal EMS and the continual improvement loop

Clearly, from the references made to ISO 14001 in the sections above, it is very common for organisations to link improvement programmes together in a structured manner under the auspices of an environmental management system. Although many organisations begin by using an EMS to reduce risk and enhance control of environmental aspects, the ethos of both ISO 14001 and EMAS is that the organisation should be continually striving to reduce their impact on the environment. This continual improvement loop, as represented in Figures 8.2 and 8.3, constitutes the final element of a long-term programme aimed at the delivery of environmental performance improvement. Essentially as each programme of improvement actions is completed, there is a review phase that links back to the starting point of 'identification of environmental improvements' and the cycle begins again.

II CREATING A BUSINESS CASE

In Chapter 4 we reviewed the wide range of business drivers creating both pressures and opportunities in relation to organisations, that are influencing the way that they manage not only pollution control and compliance but also environmental performance improvements. We grouped these drivers under the key headings of:

▶ legal
▶ financial
▶ market
▶ social.

For individuals involved in promoting environmental or sustainability initiatives in organisations, often a critical step is to establish a 'business case' for environmental improvements based on a combination of issues related to these 'driver' categories. How this is done will vary widely depending on the current goals and priorities of an organisation, the nature of change being considered and the current levels of awareness and support for environmental initiatives. In Section IV of Chapter 4, we considered the development of the business case in relation to:

▶ cost implications
▶ sales implications
▶ indirect benefits.

It is possible to identify some further common elements that may prove useful to consider for inclusion in a business case justifying support for any initiative which has as one of its primary goals the reduction of environmental impact.

8.6 Present financial information in clear cost-benefit analysis terms

Business decisions have to consider financial implications. This is not simply common sense in

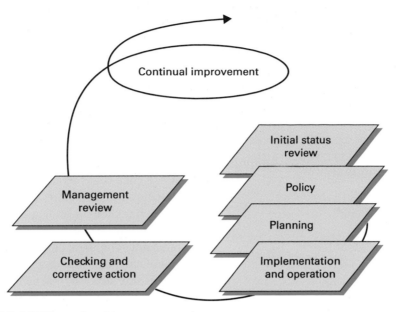

Figure 8.2 The ISO 14001 continual improvement loop

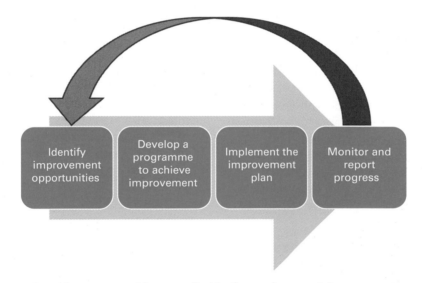

Figure 8.3 The continual improvement loop applied to the environmental programme

terms of the profitability of the organisation, it is a legal requirement for company directors. Any business case for an environmental initiative, no matter how large or small, should aim to provide a quantitative cost-benefit analysis of some kind.

This is sometimes easy to do. For example, consider the numbers in Table 8.1 which evaluate the cost of the energy savings associated with an investment in a new piece of equipment that is more efficient that its predecessor.

Table 8.1 A simple example of a cost benefit analysis

Cost of new equipment installation	£2,000
Cost of disposal of old equipment	£500
Savings in running costs per year associated with new equipment	£500
Return on investment timescale	5 years
Estimated total net project lifetime savings (based on the predicted life of the new equipment – in this case 20 years)	£7,500

Such calculations are often done in a more accurate format with adjustments made for inflation and predicted costs of key services, in this case, rising energy costs perhaps.

Where such cost-benefit analyses are more difficult is when initiatives are aimed at reducing risk rather than changing resource consumption or influencing sales. For example, the cost of installing spill response equipment in an oil storage area:

Cost of new spill response equipment	£500
Annual cost of replacement/ restocking	£100
Annual disposal cost of used/ contaminated spill response equipment	£100

These costs would need to be compared with the potential savings associated with a spill that was inadequately contained due to a lack of appropriate equipment. An estimate could be presented in the form of clean-up costs associated with a previous incident or with maintenance works required to remove contaminated hard standing generated by repeated minor spillages. Obviously, the cost-benefit analysis is less clear-cut in this instance and for that reason the latter elements of the business case become important.

8.7 Present ethical benefits in relation to key stakeholders

As with financial benefits, ethical or public relations benefits should be presented in a clearly justified manner. Where there are stakeholder groups with a direct interest in the proposed environmental improvement, clarity regarding the expected benefits can help guide how to present the initiative, not only within the business but to those stakeholders outside the organisation. Stakeholder groups may be very wide-ranging from local residents to suppliers, customers, regulators and many others. For further details on stakeholders, see Chapter 9.

8.8 Consider non-financial drivers (including clarity regarding environmental benefit)

Non-financial drivers may cover a range of things from legislative compliance through to demonstration of environmental benefits. For example, if government policy and consultation on legislation suggest that a particular course of action may be required in the mid to long term, a case could be made that improvements made now will be easier to manage and cause less disruption than waiting until mandatory requirements appear. Conversely, clarity regarding actual environmental benefits, e.g. the creation of new habitat as part of a wildlife focused landscaping scheme, makes the non-financial benefits clear as opposed to the rather woolly statements about 'environmentally friendly' developments or investments.

8.9 Benchmark performance standards and consider trends

The benchmarking of performance standards with competitors and the consideration of trends to demonstrate where on the 'response curve' the organisation is positioning itself are often a critical stage in securing support from senior figures in an organisation. Any investments or changes to the way an organisation operates should, where possible, be put into the wider business context of general and sector-specific pressures and opportunities and the current response status of competitors. As benchmarking is often a difficult process due to a lack of available information, the third party schemes assessment schemes, discussed in Section VI of Chapter 5, are increasingly useful.

Business strategies may focus on leading the sector and typically selling goods and services at a premium as 'best in class'. Alternatively, organisations may have focused at the middle or lower end of a market and therefore be unwilling to invest in any initiatives that are not mandatory. The latter is a rather short-sighted strategy as even those organisations producing lower-cost, tighter-margin products and services need to be considering any initiatives that generate resource efficiency improvements or reductions in other business costs. If selling into an informed customer market, it may be that even those companies selling at the budget end of the spectrum may win market share or at least enhance profitability through environmental performance improvements. Benchmarking our organisation against familiar competitors can provide decision-makers with a clear sense of

how the sector is responding and where they wish to position the organisation in relation to trends affecting the sector as a whole. As always in presenting a business case, it is important to be as specific as possible and to quantify any benefit claims made.

III FURTHER RESOURCES

Table 8.2 presents some further resources.

8.2 Further resources

Topic area	Further information sources	Web links (if relevant)
Improvement plans and environmental management systems	ISO 14004:2004 EMS General guidelines on principles, systems and supporting techniques	
Creating a business case	The Times 100 business case studies – especially the business and the environment section	businesscasestudies.co.uk/case-studies
	WRAP resource efficiency case studies	www.wrap.org.uk

CHAPTER 9

Communicating effectively with internal and external stakeholders

Chapter summary

This chapter considers the central issue of communication in the context of environmental management. We begin by considering the varying interests and communication needs of environmental stakeholders and the role of non-governmental organisations as champions of environmental causes. We then consider the basic principles of good communication techniques with a clear understanding that the role of an environmental practitioner is, in large part, that of a communicator. We evaluate some of the internal and external communication methods used by practitioners within organisations in the environmental management context. Finally, we conclude with a section on corporate reporting and consider some of the main sources of guidance in relation to environmental and sustainability reporting.

I COMMUNICATION AS PART OF ENVIRONMENTAL MANAGEMENT

In Chapter 8 we introduced the following four areas for consideration in relation to any long-term programme aimed at improving environmental performance and moving an organisation towards the goal of sustainability:

Implementing an environmental improvement programme
Creating a business case for environmental change
Communicating with stakeholders
Change management principles.

It is the third of these areas that will be considered here, exploring:

▶ the range of stakeholders who hold a wide variety interests in the environmental performance and credibility of an organisation;
▶ a selection of methods to ensure effective consultation and communication with them.

II IDENTIFYING ENVIRONMENTAL STAKEHOLDERS AND THEIR INTERESTS

As we have seen, the pressure to manage environmental concerns within an organisation and to be able to demonstrate such management is increasing. The term 'stakeholders' is used to describe anyone who has an interest in the organisation in question. Almost by definition, all stakeholders will have slightly different interests but here we are concerned broadly with those individuals or groups who may be interested in the environmental performance/credibility of the organisation. In a typical organisation, such stakeholders might include employees, customers, regulators, neighbours, shareholders, non-governmental organisations (NGOs) and, for high profile organisations at least, the wider public. The specific interest groups will vary from one organisation to another but typically arises from one or more of the stakeholder groups shown in Table 9.1.

9.1 The role of non-governmental organisations (NGOs)

Non-governmental organisations are independent, often charity-funded bodies with a range of interests and agendas. In general terms, their role is to promote one or many issues and frequently initiate a change in behaviour/policy. The change sought by an NGO may be in relation to the following:

▶ government, e.g. *Friends of the Earth*'s campaign to tighten the commitments made within the Climate Change Bill; Greenpeace's campaign against nuclear power;
▶ business sectors, e.g. *Greenpeace*'s campaigns against the Japanese whaling industry and the Amazonian soya industry;
▶ a single business, e.g. *Friends of the Earth*'s campaign against Shell with regards to its operations in Nigeria;
▶ a group of consumers, e.g. *People for the Ethical Treatment of Animals (PETA)* campaign against fur products;
▶ society as a whole, e.g. the *Fairtrade Foundation*'s trademark campaign to encourage all consumers to consider the trade arrangements in relation to key products and in addition to raise awareness in more general terms about the use of consumer purchasing decisions to drive change.

In some instances, the NGO will also play an active role in managing a particular resource, thereby combining their educational/campaigning role with a more direct management function.

The *National Trust* in the UK, for example, owns and opens to the public over 300 historic houses and gardens and 49 industrial monuments and mills. They also look after forests, woods, fens, beaches, farmland, downs, moorland, islands, archaeological remains, castles, nature reserves, villages – many on a free public access basis.

The *Royal Society for the Protection of Birds (RSPB)* manages 200 UK nature reserves covering 130,000 hectares which are home to 80 per cent of our rarest or most threatened bird species.

Table 9.1 Stakeholder groups and example areas of interest

Stakeholder	Example interests
Government	Wants industry to respond to government policy and measures, including the uptake of voluntary initiatives such as ISO 14001 or EMAS.
Regulators	Want companies with regulated activities to demonstrate compliance and the capability of continually delivering compliance.
Customers	Business customers may be exerting pressure for suppliers to demonstrate environmental responsibility and improve environmental performance. Some may even be taking life cycle environmental considerations into account with consequences for procurement decisions. Competitors may be developing and demonstrating effective environmental management thus increasing the pressure.
Consumers	The so-called 'green pound' is a notoriously difficult thing to quantify, however, some market sectors have emerged in recent years that depend on consumer interest in environmental issues. Furthermore, judging by purchasing patterns associated with high profile incidents, consumers increasingly will not buy from companies that have a poor environmental record or have products associated with high profile environmental problems.
Financial institutions	The developing environmental agenda and environmental concerns are presenting new risks for financial institutions as providers of capital or insurance. For example, investors may require more information on potential environmental liabilities, capital expenditure associated with environmental problems and the effect of environmental issues on profits. Public reporting of environmental performance in annual accounts is becoming an issue.
Public	How companies manage their environmental issues and demonstrate the effectiveness of this management influences the public image of the company – possibly affecting sales, complaints (especially from the local community), and in some cases even their licence to operate. Green groups and the media can have a key role in influencing this aspect of a company's reputation (positive or negative).
Parent companies or corporate functions	Subsidiaries or individual sites of multi-site organisations are often subject to expectations from parent companies or group functions to demonstrate compliance with a set of performance standards aimed at providing assurance in relation to corporate responsibility or reputation management/brand positioning.
Directors	Company directors not only have a business interest in pollution control and resource efficiency, but also a personal interest in legal compliance. If a company is shown to be poorly managed or negligent in its control of environmental issues, Directors can be held personally liable and prosecuted directly under the UK Environmental Protection Act, 1990.
Employees	Employees are unlikely to be keen to work for employers with poor environmental performance or a bad image. For some employees at least, good environmental practices and a sense of personal involvement in implementing improvements will act as a workplace motivator.
Non-governmental organisations (NGOs)	Lobby groups and local interest groups come in a wide variety of sizes, geographical spread and specific interests. At each scale, however, they may have a considerable influence on an organisation's operations and as such should certainly be considered key stakeholders.

9.1.1 NGOs as protectors of the environment: the advantages

There are a range of advantages in relation to the promotion of their particular issue, which may include:

▶ independence from government/business interests and therefore able to command a higher degree of credibility in terms of public opinion;

▶ autonomy from the electoral process means that calls for change may be made that would be considered politically sensitive;

▶ international interests which allow a wider perspective of both problems and potential responses;

▶ greater flexibility in confronting polluters or unacceptable behaviour – no requirement to work within defined remits or codes of practice (try imagining your local environment

315

agency officer in an inflatable boat trying to block harpooning of whales in the Southern Ocean);

▶ a strong 'grass-roots' membership which promotes direct action and personal engagement in environmental issues;

▶ in established NGOs a strong media profile ensures widespread coverage of actions or campaigns.

9.1.2 NGOs as protectors of the environment: the disadvantages

Although there are many advantages to NGO involvement in environmental protection, there are also difficulties faced by some or all of the parties involved:

▶ insecurity of funding and the consequent difficulties in maintaining long-term campaigns;

▶ limited access to information and/ or experience of the issues faced by organisations/government in addressing multiple priorities;

▶ charity-based funding requires the diversion of valuable resources into the crucial role of fund raising and/or a reliance on volunteer support (which may vary in reliability and effectiveness);

▶ in smaller organisations in particular there may be greater difficulty in cooperation/ coordination with other NGOs or government bodies;

▶ there can be a distrust/barrier to cooperation that develops between an NGO and the parties that they are scrutinising/ highlighting. This may undermine the potential value of the NGO in raising important issues to those involved in the problem and/or create a defensive 'digging in' on the part of the polluter or target organisation.

▶ In many cases NGOs are dependent on media coverage of their activities to generate public awareness – this is a notoriously fickle situation which can lead to the demand for high profile stunts that, while sometimes effective, may also serve to alienate parts of the target audience.

III EFFECTIVE STAKEHOLDER COMMUNICATIONS

All organisations communicate with a variety of stakeholders including employees, customers, regulators, neighbours and the wider public. Messages are continually being sent out whether intentionally or not – even an absence of communication can send a message! In this section, we will examine the issues surrounding environmental communication with particular emphasis on the provision of information to those outside the organisation.

This section covers the following key areas:

▶ Internal and external stakeholder needs
▶ Communication objectives
▶ What to communicate
▶ Internal communications – goals and tools
▶ External communications
▶ Green claims (including DEFRA's green claims guide), product labelling and company certifications
▶ Corporate reporting – the Global Reporting Initiative and CSR.

9.2 Identification of stakeholders' needs

From a corporate communications viewpoint, it is a useful exercise to compile a list of key stakeholders and their specific areas of interest/concern in relation to the organisation. Organisations that are particularly active in this area will often consult with prospective stakeholders to ensure that they understand clearly what the interests of each group are.

In the EMS section in Chapter 6, the ISO 14001 requirements relating to internal and external communications were described. These relate to stakeholders within the organisation and stakeholders external to it. Internal communications were described as those relating to employees, corporate bodies, etc. External communications were described as those relating to any third party with an interest in the environmental performance of the organisation but not directly involved with the organisation's activities. A summary of these requirements follows.

9.2.1 Internal communications

In order to meet the requirements of ISO 14001, an organisation must demonstrate that it does the following:

▶ communicates its environmental policy to all employees and makes it available to the public;
▶ communicates roles and responsibilities related to the EMS;
▶ reports EMS performance to senior management;
▶ has procedures in place to ensure that all employees understand the environmental issues associated with their work activities;
▶ has mechanisms for communication between different levels of the organisation;
▶ has mechanisms in place to communicate relevant procedures and requirements to suppliers and contractors.

In practice, many organisations go beyond these basic requirements in an attempt to engage employees in environmental improvement programmes. Two-way communication associated with feedback on performance of the EMS, progress against objectives and targets and improvement suggestions may all be important components of an effective internal communication programme.

9.2.2 External communications

In order to meet the requirements of ISO 14001, an organisation must demonstrate that it does the following:

▶ has procedures in place for receiving, recording and responding to relevant communications from external interested parties;
▶ has considered processes for external communication in relation to its significant environmental aspects and recorded its decision about whether or not to do so.

It is the second area of external communication that forms the basis of the additional requirements under EMAS for a public environmental statement. EMAS requires that a statement, designed for the public and written in a concise and accessible manner should be produced at the end of each audit cycle, i.e. a minimum of once every three years. A summary statement containing information on key performance indicators must normally be produced annually. The full statement should include:

▶ a description of the company's activities at the site;
▶ an assessment of all the relevant significant environmental issues;
▶ a summary of figures on emissions; waste arising; consumption of raw materials, energy and water; noise and other significant aspects;
▶ other factors regarding environmental performance and/or environmental policy for the site;
▶ the deadline for submission of the next statement;
▶ the name of the accredited environmental verifier who provides assurance that all the information presented in the report is accurate and representative;
▶ information on any significant changes since the previous statement.

Whether EMAS-accredited or not, there is growing interest in this type of communication through company environmental reports for reasons explained below.

9.3 Clarity regarding communication objectives

In order to communicate effectively we need to be clear about the goals we hope to achieve, the audience we hope to reach and the amount of time and money we can afford to spend.

There are a variety of reasons for communicating both internally and externally. Whatever the stakeholder concerns or the actual content of the communication, the goals may broadly be seen to include the following.

9.3.1 To ensure rules are followed

Employees, contractors and suppliers need to be aware of any procedures or restrictions in place to be able to follow them. It is also almost always the case that people are much more likely to do what is required if they understand the reasons for requests or rules.

9.3.2 To encourage involvement in environmental improvement initiatives

An extension of the above which may be especially important where the action required is voluntary, e.g. recycling of paper within an office area. Again, people need to understand the issue, be clear about what is expected from them and be motivated through a sense of the importance of the action required.

9.3.3 To minimise risk due to inappropriate action

Communication can help ensure that levels of awareness exist that reduce the likelihood of individuals taking intentional or unintentional action that could lead to unnecessary environmental impacts. It is almost impossible to develop procedures to cover every eventuality within an organisation, so such an approach to the management of environmental aspects is normally appropriate at some level.

9.3.4 To identify opportunities for improvement

As with risk control, individuals need to be aware of environmental issues and priorities in order to be able to identify improvement opportunities. However, in addition, it is also important that, having identified an opportunity for improvement, people have an easy mechanism to inform decision-makers or those who should take action, so that the potential benefits can be realised.

9.3.5 To improve community relations

For organisations with significant involvement with, reliance on or influence over, local communities, good communication is fundamental to good relations. In many cases, where poor relations exist, problems tend to be centred on a sense of threat on the part of communities that may be unfounded. Active communications on the part of organisations to keep people informed of issues that may affect them has been shown to be effective in allaying fears and preventing problems associated with incomplete or erroneous information.

9.3.6 To promote the company profile in the public arena

Companies may choose to promote their environmental performance in order to demonstrate that they are responsible, or that their products or services have an environmental benefit over competitors. This is the realm of marketing but, in relation to environmental performance, there must be a clarity and credibility to environmental claims and communications that provide a sense of reassurance and integrity rather than appearing to be motivated exclusively by sales.

9.3.7 To raise awareness and credibility with customers and/or consumers

This is linked to the objective described above but may require the provision of very specific information related to products or service standards. The aim is to address customer concerns and facilitate selection on the basis of relevant performance data.

9.3.8 To ensure good relations with regulators

Open and frequent communication with regulators, that includes voluntary consultation as well as obligatory reporting or formal applications, is likely to help develop relationships and a sense of confidence on the part of individual inspectors. Ultimately, the aim is to ensure that the regulator perceives the organisation to be working in a responsible and competent manner in relation to environmental issues.

9.3.9 To comply with statutory requirements

Many consents have routine reporting requirements as part of the conditions of the licence to operate. In addition, publicly listed companies in the UK must meet mandatory greenhouse gas emissions reporting standards from 2013. The UK government has indicated that this is just the beginning of mandatory reporting for organisations with a review and possible roll-out underway by 2016.

9.4 Generic motives for communication

The examples listed above can be grouped under four key headings describing generic motives for communication:

▶ informing
▶ instructing
▶ motivating
▶ consulting.

Understanding what we wish to achieve from our communications strategy is the first step to making it effective. Both the information delivered and the most effective method of delivery will vary significantly depending on the target audience and the goal of the communication. Clearly an organisation may wish to gain several or all of the outcomes listed above. If that is the case, however, though significant overlap may exist, it is likely that several different approaches and communication channels will be required to transfer appropriate information to and from each target group.

9.5 What to communicate?

This depends largely on the reasons for communicating and the target audience. It is generally true that any communication should be appropriate to the recipient in terms of relevance, detail and format. In most cases, there is little point in sending a lengthy technical report to a local community representative. The following examples of information could be relevant and form the basis of any communications made:

▶ company environmental policy;
▶ company environmental procedures;
▶ information relating to company environmental aspects;
▶ information relating to general environmental awareness;
▶ best practice information from within the organisation or from outside sources;
▶ potential benefits associated with environmental performance improvements – financial as well as environmental;
▶ performance data, e.g. emissions totals, complaints, audit results, etc. and also performance trends;
▶ objectives and targets related to operational performance;
▶ product-related information – materials composition, key performance indicators related to use, recommended disposal methods, etc.

Whatever the basis of the information, tailoring the message to the audience is perhaps the most important part of the process. There are no absolutes here but some principles that seem to hold true across the board are shown in Figure 9.1.

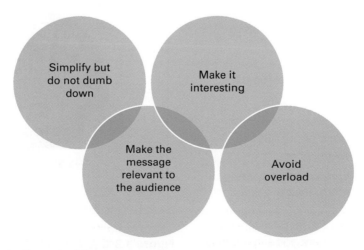

Figure 9.1 Communication principles

▶ Simplify but do not dumb down – almost everyone prefers concise and accessible information but that does not necessarily mean trivialising the subject.

▶ Make the message relevant to the audience – generally people are most receptive to information relating to their experience and to issues that they have the capability to influence. How it relates to them and what they can do about them may also be part of the message.

▶ Avoid overload – it is better to stick to fewer strong messages that get through than swamp the audience and risk none getting through.

▶ Make it interesting – it is sometimes necessary to be imaginative, innovative, artistic and even downright dramatic to achieve this.

9.6 Internal communication methods

As indicated above, there are a wide variety of reasons to communicate equally variable messages to stakeholders as part of an environmental management programme. It is an unfortunate truth that it is extremely common within organisations for communication to be a weak link. This is often caused by a lack of clarity over communication goals and identification of appropriate content. There is a third step involved, however, once we have clear objectives and an idea of content, namely choosing our communication method. Here we will focus on some of the methods commonly used to communicate within organisations.

9.6.1 Training and workshops

Training tends to be a crucial internal communication method in many organisations. There are, however, a range of possibilities available under the training banner.

Instructive sessions tend to be one-way and focused on transfer of information to assure competency. *Awareness sessions* tend to be more interactive and, although instructive, aim to provoke comment, involvement and suggestions from the audience. *Workshops* are the most open

forum and may be used to resolve conflicts of interest, identify opportunities, share examples of good practice and gain agreement for action plans. Workshops in particular are a useful mechanism for identifying concerns or issues from third parties and are increasingly used in the early stages of EIA to gauge stakeholder concerns.

9.6.2 Posters

Posters can be very effective communication tools if employed well. Unfortunately, they are all too frequently put up and abandoned on notice boards that are already rarely seen because of location. High impact images rather than detailed messages, that are changed regularly and in prominent and easy to read places can, however, have dramatic results (Figure 9.2). We need to take a leaf out of the advertiser's book when using posters – themes and humour can create situations where people are looking for the next update. Dramatic pictures that illustrate a theme rather than providing instruction or detailed

Figure 9.2 Greenpeace campaign poster
Source: Greenpeace UK.

information can be effective awareness raising tools. A modern equivalent of the humble poster is the PC screen saver.

9.6.3 Videos/e-learning, etc.

Videos and interactive e-learning packages can be effective self-learning tools or can complement other training methods as part of a course. It must be recognised, however, that simply making such tools available to people does not guarantee their use. But, with the right incentives, or if used as part of a training course, well-produced video material can be a highly effective communication tool.

9.6.4 Face-to-face communication

A greatly underestimated communication tool! Whether informally in discussions on a particular topic, or more formally in group meetings, the value of face-to-face communication should not be forgotten. It allows immediate questions, comments and clarifications and, perhaps most importantly, allows the communicator to gauge understanding, interest and likely response. All these are crucial to effective communication – as organisations that are becoming overly dependent on e-mail are finding to their cost.

9.6.5 Brochures and newsletters

Brochures and newsletters can be effective internal communication tools to provide a regular drip feed of information that is packaged in a way that encourages people to read them. A common approach is to combine staff news with environmental information.

9.7 External communication methods

As with internal communications, there are a wide variety of goals relating to highly variable stakeholder group interest. Here we will focus on some of the common methods/tools used to communicate with external stakeholder groups.

9.7.1 Brochures and newsletters

As external communication tools, brochures particularly come into their own, providing

accessible information in manageable quantities so that people or groups can request information on particular issues without the need to sift through additional unwanted information first.

9.7.2 Community liaison committees

These can be used to great effect in higher profile organisations where local communities might have an interest in the nuisance or pollution potential associated with operations. Generally, a sense of being consulted and the provision of a mechanism that allows concerns to be voiced improve relationships and the willingness to listen to proposed developments or operational change. The key to success seems to be to ensure that an adequate cross-section of interested parties is represented and that communication methods to the wider stakeholder group is, in some way, managed effectively (so that the message does not stop with the committee members).

9.7.3 Helplines/effective complaints procedures

As with the consulting and informing roles played by the committees mentioned above, an open and effective means for dealing with inquiries and complaints sends a significant message to interested parties – we care about our impact on you! The way that organisations manage this varies depending on the scale and nature of activities involved. A basic minimum might be an efficient complaints and inquiries procedure set out as part of an environmental management system. Or a dedicated helpline.

9.7.4 Regulator consultation and reporting

Particularly for permitted sites, a concerted attempt to develop both formal and informal links with regulators is often seen as a wise communications goal. Formal measures may be defined in the permit as part of scheduled or emergency reporting arrangements but even here, timely submission and clarity of provision can make a big difference in terms of regulator perception. Informal measures generally involve the development of professional relationships between key staff in the organisation and the regulatory body. While this is not always possible, it is nearly always true that knowing the person you are dealing with leads to improved communications.

9.7.5 Green claims

This heading covers a variety of communication tools including, among others, public relations materials, press coverage, product/service marketing, corporate credibility schemes, e.g. ISO 14001. Over the years the general public and interested parties in particular have become increasingly wary and even suspicious of 'greenwash', i.e. unsubstantiated claims of 'eco-performance', 'eco-friendly' or 'green' products, services or companies. This trend is, of course, itself an indicator of increased public awareness and interest in environmental issues which means that more companies are interested in making some kind of credible statement about themselves, their activities or their products.

In response to high levels of consumer and organisation confusion about what constituted good information in relation to environmental performance, in February 2011, DEFRA produced a publication entitled *Green Claims Guidance*, aimed at any organisation attempting to communicate environmental performance to third parties. It essentially suggests that any 'green claim' should bear scrutiny using the following steps:

> Ensure the content of the claim is relevant and reflects a genuine benefit to the environment. Present the claim clearly and accurately. Ensure the claim can be substantiated.

The guidance goes on to set out checklists and ways in which each of the steps could be achieved, regardless of the nature of the claim or whether it relates to a product, service or organisation. As part of the substantiation step, the following 'credibility schemes' are mentioned.

9.7.6 Corporate credibility schemes, e.g. ISO 14001

External assessment in the form of ISO 14001 or EMAS certification can be a very credible way of communicating environmental commitment and performance standards. Customers or consumers may be content to use certification as a single, simple gauge of environmental performance. Such responses make certification an extremely powerful and increasingly popular communication tool.

9.7.7 Product credibility assessment

Third party assessment schemes such as the EU Eco-label scheme are a prominent example of this approach, where organisations use third party benchmarking to rate the environmental performance of their product. Consumer guides that include ratings linked to environmental performance may also be important but are likely to be less useful in terms of company communication because of the lack of control over issue timetable, coverage, etc.

IV CORPORATE ENVIRONMENTAL REPORTING

Environmental reporting has traditionally been a voluntary method of communicating environmental performance to an organisation's stakeholders. Public reporting can be important, in terms of internal communication and engagement, in emphasising company commitment to environmental performance. As an external communication tool, it may be used to provide a comprehensive overview of company policies, impacts and performance.

It is increasingly difficult for larger organisations in particular, to justify not producing some kind of annual report. In recent years there has been debate over whether environmental reporting should be made mandatory in the UK. Denmark, New Zealand and the Netherlands have already started introducing legislation on environmental reporting. From April 2013 in the UK, all 1,800 FTSE listed companies are legally required to report on greenhouse gas emissions. Following a review in 2015, the UK government has confirmed its intention to consider extending this requirement to all 'large' companies (around 24,000 businesses with more than 250 employees). This mandatory greenhouse gas reporting is being introduced because of the clear belief that the requirement to report on performance provides a strong motivation to improve. The same principle could be applied across the whole of the environmental performance/sustainability spectrum.

For the moment at least, comprehensive voluntary reporting is likely to remain a minority activity without regulation demanding it. There are however, an increasing number of organisations reporting for a complex set of self-oriented objectives that might include the desire to do the following:

▶ Educate stakeholders.
▶ Explain/promote the organisation's achievements.
▶ Reassure regulators that further legislation is not required.
▶ Persuade stakeholders of the 'green credentials' of the organisation (whether valid or not!)
▶ Counter negative press or claims by environmental groups.
▶ Legitimate the industry or product sector.
▶ Express personal commitment by the board.
▶ Signal to financial stakeholders that environmental risk is managed sensibly.

(after Gray in Brady *et al.*, 2011)

9.8 Reporting standards and guidance

The format of environmental reports varies widely both in format and coverage. DEFRA, among others, has produced guidance and recommendations on the format and content of a credible corporate environmental report (see the 2103 guidance in the further resources below). DEFRA guidance suggests that reports should contain the following common elements:

▶ Introduction from the Chief Executive
▶ The organisation's Environmental Policy
▶ Background information about the organisation
▶ Description of management systems
▶ Key environmental impacts
▶ Environmental performance indicators
▶ Targets for improvement/progress against targets

▶ Compliance assessment (against regulatory requirements and/or voluntary standards such as ISO 14001 or EMAS).

Such standard headings suggest a consistency in report formatting but this is not necessarily the case. Companies may include the recommended content but present the information in any number of innovative ways to facilitate stakeholder access. The internet allows organisations to be highly flexible in the type and amount of information available, and they can utilise the interactive and multimedia capabilities to attract a wider range of stakeholders and potentially receive greater feedback from them. There is also growing interest in incorporating social issues into the reports in keeping with the social accountability ethos of sustainable development (see below).

Organisations that have made a commitment to produce annual environmental reports will require new information and data each year, which can be used to demonstrate progress. Inevitably perhaps the trend will be towards impact-related performance. This is an area of great importance, and one highlighted by the UK government as a priority. There are still only a minority of reporting companies which quantify their impact, for example, through the provision of data on waste, emissions to water and air, energy use or production of greenhouse gases.

The challenge for companies is to provide information which is consistent, reliable and benchmarked, to allow investors to assess policy and practice according to sector. Developments, such as the Global Reporting Initiative (see below and for details visit www.globalreporting.org), and other industry efforts, mean that such a reporting framework could well emerge in the near future. DEFRA has also published a number of guidance documents relating to environmental reporting both in general terms and in relation to specific performance issues.

The following examples (and others) are available from the DEFRA website:

▶ DEFRA Environmental Reporting Guidance Including Mandatory Greenhouse Gas Reporting Guidelines (June 2013);
▶ Environmental Key Performance Indicators – Reporting Guidelines For UK Business

▶ Guidelines for Company Reporting On Greenhouse Gas Emissions

▶ Green Claims Guidance (covered in Section 9.7.5).

The International Standards Organisation has also produced a guidance standard relating to corporate environmental communications. Although not designed for certification purposes 'ISO 14063:2006 Environmental management: Environmental communication – Guidelines and examples' may be a useful reference for organisations on general principles, policy, strategy and activities relating to both internal and external environmental communication.

9.8.1 The Global Reporting Initiative (GRI)

The first decade of the twenty-first century saw the emergence of corporate sustainability reports. In some cases they have comprised little more than a re-branding of the company's previous environmental reports. However, increasingly the trend is towards format and content more akin to the CSR reports mentioned below and covering the long-term social, ethical and environmental issues relevant to the organisation, presented within the context of sustainability. As with all the other report formats described above, policy statements should be matched with clear priorities, goals and progress reviews that are linked to both to the organisation's own view of what sustainability means and the concerns of identified stakeholder groups. The GRI guidelines cover the scope demanded of a high quality sustainability report and are a widely recognised benchmark for any organisation considering this kind of reporting.

The concept of the GRI is simple:

▶ define a standard sustainability reporting format;

▶ encourage voluntary third party assurance that the reporting standards have been achieved;

▶ collate reports in a searchable database to enable benchmarking of corporate reports and sustainability performance.

Detailed guidance, case studies and the searchable database of corporate reports are all available at www.globalreporting.org. The guidance provides a series of environmental issue and business sector-specific protocols that are aimed at standardising data collation and ensuring transparency regarding the definition of reporting boundaries.

The guidance then specifies a series of 'standard disclosures' organised under the following headings.

9.8.1.1 Strategy and profile

Disclosures that set the overall context for understanding organisational performance such as its strategy, company description, key impacts, risks and opportunities, key stakeholders and governance issues such as commitments to widely recognised standards such as ISO 14001 or the sustainability principles set out in Agenda 21 from the Earth Summit in Rio.

9.8.1.2 Management approach

Disclosures that cover how an organisation addresses the 'material aspects' (i.e. those relevant to them) outlined in the strategy and profile section. This is intended to provide context for understanding performance in specific areas.

9.8.1.3 Performance indicators

Indicators that elicit comparable information on the economic, environmental and social performance of the organisation. Core indicators are specified in each category but with guidance relating to the tailoring of the data to make it organisation specific. The main indicator groupings are shown in Table 9.2.

Reporting organisations are encouraged to follow this structure in compiling their reports. They are then required to specify the level to which they have complied with the GRI guidelines. To do this, they self-declare their compliance in relation to the number of criteria against which they report across the three standard disclosures categories. They then choose to either commission a third party verifier to confirm their assessment or request that the GRI do so. The system is not as rigorous as the verification process relating

Table 9.2 GRI performance indicator groupings

Economic indicator groupings	Environmental indicator groupings	Social indicator groupings
• Economic performance • Market presence • Indirect economic impacts	• Materials • Energy • Water • Biodiversity • Emissions, effluents and waste • Products and services • Legal compliance • Transport • Overall environmental protection expenditures	• Labour practices e.g. occupational health and safety injury and illness data • Human rights, e.g. risk assessments around child or compulsory labour in own and significant supplier operations • Society, e.g. percentage operations with implemented community engagement programmes • Product responsibility, e.g. type of product and service information provided to customers

Source: Adapted from Global Reporting Initiative Guidance.

to EMAS reports but does provide an easily accessed assurance of the standard of the sustainability report produced.

9.9 Corporate Social Responsibility (CSR) reporting

Corporate Social Responsibility (sometimes referred to as Corporate Governance) is even less well defined than sustainability but may be thought of as encompassing the responsibilities that an organisation has to all its stakeholders. For some organisations this may be interpreted exactly as sustainability, while for others with perhaps a greater focus on direct human issues, workforce and community issues may be emphasised more than the wider environmental impacts of company activities.

Amnesty International developed the concept of 'nested spheres of responsibility' as a way of expressing the variation in degree of responsibility that any organisation has. Figure 9.3 represents these nested 'spheres', with issues that are most easily controlled by the organisation being at the centre and those over which limited influence can be brought to bear at the periphery. Clearly, the organisation bears most responsibility for

those issues close to the core and least for those nearest the periphery. Without going into a detailed exploration of CSR it is perhaps simplest to consider short-term CSR strategies as being a 'moral business approach', while long-term CSR strategies are indistinguishable from a 'sustainable business approach'.

CSR reporting is becoming increasingly attractive as 'corporate governance' is being pushed or demanded by a range of stakeholders. In marketing parlance, it is not enough to behave in a socially responsible manner, you must be seen to be doing so! Credible reports must be 'spin-free', however, and tailored to the stakeholder group or groups at which they are aimed. Some companies with significant stakeholder concerns have chosen to have their reports independently verified in order to enhance credibility (in the same way as required by EMAS and recommended by the GRI). Assurance auditing of this type looks not just at the report but at all monitoring and data collection systems that feed into it in order to verify that the report is an accurate and representative picture of company performance. For further information on reporting assurance mechanisms, see Chapter 5, Section IV.

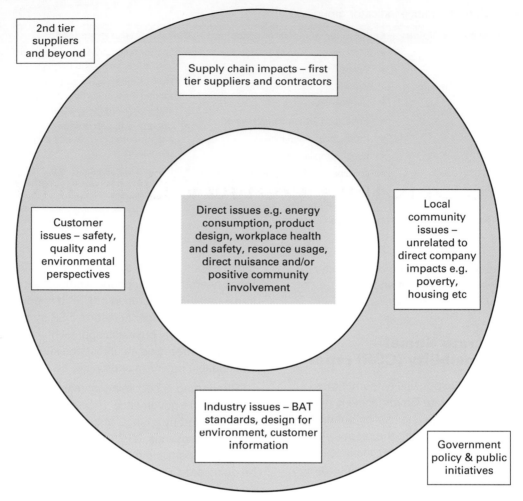

Figure 9.3 Nested spheres of corporate responsibility

V FURTHER RESOURCES

Table 9.3 presents further resources.

Table 9.3 Further resources

Topic area	Further information sources	Web links (if relevant)
Green claims	DEFRA, 2011. Green claims guidance	www.gov.uk/defra
Corporate reporting	DEFRA, 2013 Environmental reporting guidelines – including mandatory greenhouse gas reporting guidance	www.gov.uk/defra
	The Global Reporting Initiative, 2013. G4 – sustainability reporting guidelines. Reporting principles and standard disclosures	www.globalreporting.org
	Brady *et al.* (2011).	

CHAPTER 10

Influencing behaviour and implementing change to improve sustainability

Chapter summary

This chapter considers the obstacles that often arise in the course of implementing change in an organisation. Some general principles relating to change management are discussed in the context of an environmental improvement or sustainability programme. As an environmental practitioner we are championing a particular cause and course of change within our organisation. It is not unusual to occasionally feel like we are swimming against the tide. Understanding the resistance, objections and obstacles that present and recognising that they are likely to occur in relation to any change process can be helpful from both a personal perspective and, in ensuring that we are as effective as possible in overcoming them.

I CHANGE MANAGEMENT PRINCIPLES

In Chapter 8 we introduced the following areas for consideration in relation to any long-term programme aimed at improving environmental performance and moving an organisation towards the goal of sustainability:

Implementing an environmental improvement programme
Creating a business case for environmental change
Communicating with stakeholders
Change management principles.

It is the last of these areas that will be considered here, exploring what is sometimes referred to as 'culture change' which may be described as a fundamental shift in attitudes and behaviours in an organisation that eventually becomes self-sustaining and independent of management systems and/or individual champions.

Business management theory has produced a whole set of principles and approaches to managing change in organisations. We will attempt to present a summary of the key ideas here. For delegates seeking further information, the IEMA Practitioner Guide (2006) *Change Management for Sustainable Development*, written by Penny Walker, is an excellent resource written from the perspective of driving environmental change. It is available for purchase at www.iema.net. There are additional references in Section VI.

II UNDERSTANDING 'CULTURE CHANGE'

In the context of an organisation, 'culture' is a term used to express the shared values and attitudes of employees. It is important to note that while it may be useful to identify common trends and principles, any description of an organisation's culture is always a generalisation and will be an inexact representation of almost everyone as an individual! Nonetheless it is essential to have an understanding of the general trends if we are to engage in a process of change. The following characteristics might be important aspects of an organisation's 'culture':

▶ dominant styles of communication, e.g. consultative vs prescriptive;
▶ dominant thinking styles, e.g. creative and visual vs analytical and data-based;
▶ decision-making styles, e.g. inclusive and local vs top down/mandate-based;
▶ experience of change, e.g. successful and supported vs imposed and less than fully successful;
▶ company–employee relationship – this multi-faceted area encompasses the balance between employee and management perceptions. It might incorporate issues such as employees' sense of security and individual worth, as well as the company belief in a motivated and innovative workforce.
▶ scale and uniformity of the organisation – disparate and distinct areas of operation vs a common identity and shared goals.

The list above is just a few examples from what could be a very long set of 'culture criteria'.

'Culture change' as defined in Section I is the alteration of behaviour, attitudes and/or understanding by all or part of an organisation to the degree that it becomes the self-sustaining 'norm'. Examples at the scale of UK society might include radical changes such as the shift of values inherent in the abolition of slavery, through to the more minor, such as the adoption of household waste recycling practices. Within organisations, environmental examples can also be found across a spectrum that might be described as radical to incremental change. A radical change might include a major project to introduce a new manufacturing process and thereby eliminate a hazardous substance. An incremental change might be a change of procedure relating to the materials and waste management that aims to reduce resource consumption and minimise waste to landfill.

As we will see in the sections below on obstacles and planning for change in organisations, it is almost a prerequisite for successful 'culture change' to clearly understand 'what we want to achieve' and the 'defining characteristics' of the culture in which we are working.

III OBSTACLES TO CHANGE IN ORGANISATIONS

10.1 Subjective and objective obstacles

It is useful when planning a change initiative in an organisation to consider the obstacles or barriers that are likely to be faced. Penny Walker uses an approach developed by David Ballard to group barriers into individual or collective, and subjective or objective. Table 10.1 gives some examples in these categories.

Sometimes the same issues can appear in different categories, for example, some of the most commonly heard reasons for not adopting environmental improvement initiatives include:

▶ Not enough time.
▶ Too expensive.
▶ No benefit to the organisation.
▶ Customer constraints.
▶ Changes too risky – what if we get it wrong?
▶ Not enough expertise.
▶ People don't care.
▶ No commitment from senior management.

Less easily heard but perhaps often present in individual reactions to change programmes might be the following:

▶ This sounds like more work for me with no more reward.
▶ What's wrong with the way we currently do it?
▶ This won't work because . . . why won't they listen to me?
▶ This sounds like it will get in the way of me doing my job – my supervisor will not like it.
▶ I don't understand what benefit we'll get from doing it this way.

Depending on circumstances, any of these reasons could be objective or subjective and

Table 10.1 Examples of obstacles to organisational change

	Subjective	Objective
Individual	Based on one person's world view, assumptions or attitudes. Common examples: • It's too big an issue, no one will change including me • Climate change is a propaganda exercise used to raise taxes and fuel prices	Based on one person's role, authority, skills, resources, etc. Common examples: • I do not have any budget left to pay for the proposed changes • I do not understand the issue sufficiently to encourage others to change behaviours
Collective	Group culture, shared mind sets, e.g. • The company doesn't really care about the environment, it's just a public relations exercise • It's not my job, someone else should do it	Political, economic, social, technological, legal, environmental conditions, e.g. • Legal standards are too weak to prevent environmental damage • Economic health is gauged solely by gross domestic product or profit and loss sheets • Our print system is only designed to work with solvent-based inks

Source: Adapted from Walker (2006).

might be held individually or collectively. From a pragmatic change management perspective, it is the subjective (and especially unvoiced!) barriers that are most difficult to overcome. Nonetheless, both subjective and objective obstacles can be planned for and addressed over time.

Some key principles that may help in planning for and dealing constructively with these obstacles to change are considered below.

10.2 Constructively dealing with resistance

10.2.1 Resistance is natural

While again something of an over-generalisation, it might be wise to assume that the reaction to any change programme will be resistance on the part of many of those affected by the change (based on one or more of the obstacles discussed above). Those not averse to change will simply be better informed by strategies aimed at overcoming to resistance, but there may be a much greater chance of widespread acceptance if we take the approach championed by the Prosci Change Management Learning Centre which might be summarised as follows:

An appropriate stance might be to recognise that resistance is a natural response and instead of thinking about 'overcoming obstacles', think about 'supporting employees' through the change.
(Adapted from http://www.change-management.com/tutorial-7-principles-mod7.htm)

10.2.2 What does it mean for me?

Most resistance is, in one way or another, connected to questions about 'what it means for me' or slightly less directly 'how it benefits the business and therefore me'. The specifics will vary depending on the individual and the business context of course but if these questions are answered early, then the chances of people engaging in a change programme seem to increase significantly.

Until such questions are answered, research suggests that it does not really matter how aware people are of the arguments supporting a change, such awareness will not necessarily translate into a desire to act and resistance may well remain. So it is important when planning a change programme to consider how different employee groups will be affected and, in doing so, try to second-guess the nature of potential resistance. We can then prepare communications that address the reasons for resistance right at the beginning rather than waiting for obstacles to be presented.

10.2.3 Hearing it from the right person

It is particularly important for initial communications to be made by the 'right person'. Prosci's research has shown conclusively that people:

▶ need to understand the implications of change for them or they simply do not engage;

▶ respond much better when communications explaining the change objective and implications for them come from either a senior management figure or their direct supervisor (and ideally both). It appears that the likelihood of success reduces significantly if these initial 'buy-in' communications are made by the environmental manager or change champion.

10.2.4 Covert resistance

It is worth highlighting that the communication of resistance will vary widely between organisations with some workforces very quick to speak up with objections and others unwilling to say anything openly. Companies that have cultivated an overt 'can-do' attitude, especially if fairly autocratic in management style, may manifest the latter condition as individuals fear being seen as negative or obstructive. Even if everyone appears gung-ho, it is wise to consider that there is likely to be some 'closet resistance' which needs attention as much (if not more) than the more open form. Perhaps the best symptom of this more covert resistance is recurrent non-compliance or non-participation. Again it is wise to look carefully at why such behaviour is occurring rather than simply assuming more training, tighter supervision, etc. is required.

IV PLANNING FOR CHANGE

As with the principles of implementing an environmental improvement programme outlined in Chapter 8 and the wider principles of environmental management considered in the EMS section of Chapter 6, planned organisational change can be seen as a process of planning–doing–reviewing. Remember when looking at Figure 10.1 that this could apply at a range of scales from a minor modification to operational procedures, through to a long-term move from a fundamentally unsustainable business to a fully sustainable enterprise.

Box 10.1 An example: introducing a waste segregation system

For example, if we are trying to introduce a waste segregation system in a production area, we may, through discussion with people in the area, discover that there is considerable scepticism about municipal recycling schemes and concerns that the requirement to segregate waste may make already time-pressured tasks even more difficult to perform. We might therefore predict resistance around two key objections which could be summarised as 'there's no real benefit – it will all get landfilled anyway' and 'I'm not going to have enough time to do this.'

The first 'subjective' obstacle will need to be addressed both initially and in terms of on-going communication, with transparency and explanation (especially if arrangements are such that occasional loads do have to be landfilled). The second objection (which may or may not be objective) will need to be carefully considered in putting in place the arrangements for segregation. The best way to address this area of resistance would probably be to involve the operators themselves in deciding how the segregation process could work best.

Communication relating to the justification of the segregation process and assurances regarding clarity of disposal arrangements may best come from senior management or team leaders. The environmental team may then get involved alongside team leaders to help agree the segregation arrangements with the teams.

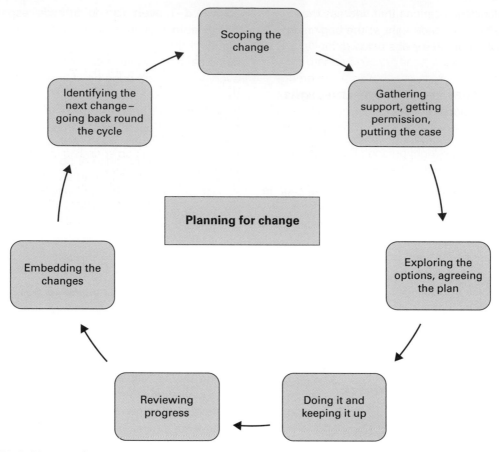

Figure 10.1 Planning for change

10.3 The change process

The sorts of things that might be considered in each stage of this cycle are as follows.

10.3.1 Step 1: Scoping the change

Defining the change goal – being clear about what we want to achieve, the timescale over which we want to achieve it and how we will know we have succeeded. This can be at any scale but, for example:

▶ *Overall change goal* – eliminate waste sent to landfill.
▶ *Timescale* – within 3 years.
▶ *Success marker* – all waste generated from the business will be catalogued by type, weight and disposal method. We will report monthly on percentage of total waste sent to landfill. When that percentage reaches

zero for three consecutive months, we will consider the change as implemented though we will continue to use the measure as a key performance indicator to ensure that the change is permanent.

10.3.2 Step 2: Gathering support, getting permission, putting the case

Identify whom you need to convince and then think carefully about how to present the case for change. This may be necessary with widely different groups of people and you will need to consider that each group (and perhaps even each individual) will have their own objections, interests and priorities which will all need to be taken into account to achieve real 'buy-in' to your change proposal.

You will therefore need to plan the best way to present the case for change to each group or individual. In some cases, group meetings or impersonal communications may be appropriate, while with key stakeholders or decision-makers it may be appropriate to hold one-to-one discussions. For some groups the financial arguments will be central, while with others the moral issues will need to take centre-stage. Whichever end of the spectrum you are dealing with, compile an argument based on key evidence and consider different ways of getting the message across. It is worth remembering that a two-minute conversation in a corridor can be as effective as a meeting presentation.

Do your research, think things through and then give it a go. However, be prepared to change tack if your initial approach is not working. Finally as a general rule, people like to have a sense of being listened to, so consider the possibility of making communications about change consultative in nature – asking people what they think the obstacles might be and how they might be overcome, rather than simply presenting 'the plan' as a *fait accompli*.

10.3.3 Step 3: Exploring the options, agreeing the plan

Penny Walker in the (2006) IEMA workbook, *Change Management for Sustainable Development*, talks about two contrasting approaches to the development and implementation of improvement plans.

10.3.3.1 The Decide-Announce-Defend (DAD) model

In this approach, a plan is agreed by a small group of people who then try to sell it to a larger group, often generating resistance and obstacles to implementation. Even where change is pushed through, there often remains a level of resentment through the implementation phase. This approach may be applicable where urgent change is required and there is completely solid senior management support for the initiative. However, when trying to engender a longer-term change in behaviour, the IAI approach is normally preferable.

10.3.3.2 The Involve–Agree– Implement (IAI) model

This is essentially a more consultative approach where stakeholders are involved in the process of both goal-setting and agreeing the best way of achieving those goals. Although often considered to be a much more time-consuming approach, if the DAD model creates considerable resistance, you can easily end up spending long periods defending the strategy and convincing people that have already adopted an opposition stance. Incorporating people's views tends to lead to more robust plans with participants more like to get and stay involved.

How you go about undertaking a consultation exercise prior to agreeing a plan for change will vary between organisations, but workshops, individual and group discussions, suggestion schemes, etc. can all be used to good effect. Often resistance to change revolves around a sense of a lack of control – of things 'being imposed on me, regardless of what I think'. An approach that affords people the opportunity to comment or suggest ways forward can help increase their sense of control. There will still be objections but you will have removed some of the grounds for resistance by consultation alone.

10.3.4 Step 4: Doing it and keeping it up

There is an element of leading by example here – ensuring that your own initiatives are seen through. Also it is important that others within the organisation begin to see a track record of successful change. For that reason, starting with some easy initiatives can help generate success stories that point the way forward.

As more people begin to get involved, the provision of support and communication mechanisms may be critical in ensuring that early enthusiasm does not quickly peter out. Environmental champions groups can be particularly useful if people are spread out in separate work locations and need to be able to share experiences and mix with peers to avoid the sense of being a 'lone voice in the wilderness'.

As the momentum grows, such mechanisms may become less important but in the early stages they can be critical to ensuring continued commitment by key people to the change process or improvement plan.

Feedback is also particularly important in maintaining momentum. Finding mechanisms to acknowledge actions taken and, particularly, to tie actions to progress achieved, can help keep individuals motivated, as well as communicating good practice and ideas to others.

10.3.5 Step 5: Reviewing progress

This is such an important part of any change or improvement programme that it is explored in detail in Chapter 5. Refer to that chapter for details, but essentially progress tracking should be via transparent and easily understood indicators that allow comparison of performance over the longer term but that are clearly linked to the improvement goals of the organisation.

As always, how you communicate these progress reviews is almost as important as what you measure. Giving a balanced presentation of 'how far we've come' with 'how far we've still to go' is important. It should be clear where the emphasis lies in terms of achievement or action required. If we are trying to encourage continued motivation, we do not want to give a sense of a 'job already completed' but we do need to ensure that efforts already made are given due recognition.

Formal reviews and independent assessment of progress can be useful and increase profile and credibility for the programme as a whole. But equally important are the regular informal checks related to 'what's working and what's not', 'how do key participants feel about progress, etc?'

10.3.6 Step 6: Embedding the changes

'If it hasn't embedded, it's not real change.' The acid test is whether you as the change leader can walk away from the programme confident that the whole process will continue without

the status of 'improvement plan' or the on-going drive from audits or other 'policing'-style initiatives.

Penny Walker describes it like this in relation to sustainability within an organisation:

> Embedding is partly about getting the changes intimately bound up with the written policies, procedures, targets and strategies – the artefacts of the organisation. It is also about cementing it firmly within the culture of the organisation – so it becomes the 'way we do things round here'. And it's about the core purpose and mission of the organisation – 'what we're here for'.

Several writers on this subject talk about the need to remember that 'change is a process not an event'. In other words, it is highly unlikely to happen immediately and will need significant efforts initially before gathering its own momentum. A wise head once described it thus: it's like rolling a millstone. It takes a huge effort to get it up off the ground and balanced on it edge. It takes more effort to start it rolling and to pick up a little speed. But once it's on its way, it develops a huge momentum of its own. Then you only need to check its course and give it a bit of extra help when approaching an incline to help it clear the slope and continue on its way.

10.3.7 Step 7: Identifying the next change – going back round the cycle

Once you feel that there has been real progress made, it is time to go back to the beginning of the cycle. The scope of your change ambitions may now be much broader, having gathered a team of supporters and made some real changes that you can use as 'references', demonstrating business and environmental benefits.

V DRIVING CHANGE: THE NITTY GRITTY

It is possible to get embroiled in all sorts of undoubtedly valid issues and management language related to the process of change management. As with environmental management systems, however, it is

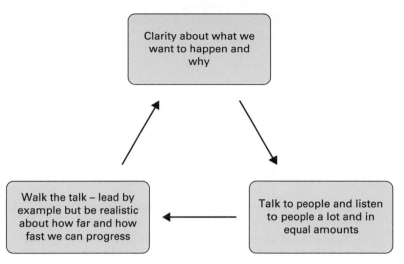

Figure 10.2 Driving change: the nitty gritty

worthwhile being clear about the basics of what is to be achieved and how best to make it happen. In essence, successful change management programmes might be considered to be built on three simple but critical actions/ attitudes by the change leader(s) as shown in Figure 10.2.

VI FURTHER RESOURCES

In addition to the (2006) IEMA Practitioner Guide, *Change Management for Sustainable Development*, written by Penny Walker (available for purchase at www.iema.net) and referred to several times through this chapter, the following websites in Table 10.2 may be of use to readers interested in learning more on this topic.

Table 10.2 Further resources

Topic area	Further information sources	Web links
Change management	The mindtools website has a host of free and subscription management information, a lot of which relates to change management.	www.mindtools.com
The psychology of culture change	There is a lot about change management on this website but of particular interest is the section on psychological contract theory.	www.businessballs.com/ changemanagement
Change management	An excellent tutorial series that covers much of the content of this chapter in much more depth and with links to further information and research.	www.change-management. com/tutorial-7-principles-mod1

Glossary

The following terms are used in the text and are commonly encountered in the environmental management context.

Abatement Control or reduction of pollution or removal of a nuisance by engineering (plant, equipment), operational (procedure) or regulatory (licence or order) means.

Acid rain The reaction of acid gases (especially sulphur dioxide and oxides of nitrogen) with atmospheric moisture thereby increasing the acidity of resulting precipitation. Such acidic precipitation over time results in increased acidity of receiving waters and soil, particularly in areas with naturally acidic soils and/or vegetation.

Ambient pollution Background levels of pollution that reflect the cumulative result of pollution from all relevant sources.

Aquifer Underground geological formation containing water that has typically accumulated over many years. Often used for potable water supply.

Best available techniques (BAT) Performance standard required under an Environmental Permit. 'Best' refers to the most effective techniques in achieving a high overall level of environmental protection. 'Available' means proven techniques allowing implementation in the relevant sector under economically and technically viable conditions. 'Techniques' refer both to technology used and the way in which the installation is designed, built, maintained, operated and decommissioned.

Best practicable environmental option (BPEO) The option which provides the most benefit or least damage to the environment as a whole, at acceptable cost, in the long term as well as the short term. The term recognises that in abating environmental impacts there are often trade-offs (e.g. reducing air pollution may increase production of solid waste or effluent) which may vary significantly depending on local receptors.

Bio-accumulation The build-up of persistent pollutants in the body tissues of some organisms, frequently increasing in concentration through the food chain.

Biochemical oxygen demand (BOD) A measure of the pollution in a body of water or effluent stream, based on the organic material it contains. Such organic material provides food for aerobic bacteria which require oxygen to break down their food source. The greater the volume of organic material, the higher the concentration of bacteria and the greater the oxygen demand. (BOD values give no indication of other pollutants such as suspended sediments or heavy metals.) If BOD exceeds available dissolved oxygen, oxygen depletion occurs and many aquatic organisms suffer – fish kills are not uncommon in such circumstances.

Biodiversity The variety of life on Earth, considered at the level of habitats, plant and animal species and in terms of genetic diversity within species.

Biomass and biomass energy Biomass is the total weight of living organic matter in a given area. Biomass energy is the energy available from organic material including wood, crops, crop residues, industrial and municipal organic waste, food processing waste and animal wastes. They may be used directly or converted before use (e.g. conversion of sugar beet waste into alcohol). Biomass energy is renewable if managed appropriately but as with fossil fuels produces smoke, soot and carbon dioxide when burnt.

Bund A type of 'secondary containment' in the form of an impervious wall around a tank or other primary container, designed to prevent pollution in the event of a leak or spill from the primary container.

Carbon footprint Defined by the Carbon Trust as 'the total set of greenhouse gas (GHG) emissions caused by an organisation, event, product or person'.

Chemical oxygen demand (COD)	A measure of the oxygen consumed in the chemical oxidation of organic and inorganic matter in water or effluent. It provides an indication of the impact of effluent on dissolved oxygen levels. A standard test uses potassium dichromate as an oxidising agent and measures oxygen consumed over a standard period of time.
Combined heat and power (CHP) (also known as cogeneration)	Energy technology which utilises the heat generated during electricity generation for space/water heating or steam generation. CHP may achieve energy generation efficiencies in excess of 85 per cent compared with typical coal-fired power generation efficiencies of 30–40 per cent (i.e. 30–40 per cent of the energy available in the fuel is converted into usable energy).
Corporate social responsibility (CSR)	A form of corporate self-regulation integrated into a business model whereby business monitors and ensures its active compliance with the spirit of the law, ethical standards and international norms. The goal of CSR is to embrace responsibility for the company's actions and encourage a positive impact through its activities on the environment, consumers, employees, communities and other stakeholders. CSR is often associated with 'triple bottom line' reporting whereby the business gauges its success in terms of positive results in relation to people, planet and profit.
Cumulative impacts	A combination of impacts arising from either: similar environmental aspects from different sources occurring simultaneously, or sequential aspects from the same source or different sources that do not allow adequate time for environmental recovery between occurrences. In either case, the sum of the impacts is greater than each individual environmental change.
Data	Information that might be used to assess priorities and monitor performance. The following key terms are often used: **Qualitative data** – information based on opinion, judgement or interpretation. **Quantitative data** – numerical measured criteria that may be subject to statistical analysis or benchmarking. **Absolute data** – the actual amount of something measured, e.g. tonnes of waste generated per year. **Normalised data** – based on absolute data but subject to some kind of referencing process to enable comparison between locations or over time.
Data verification and assurance	Assurance is defined as a formal guarantee – a positive declaration that a thing is true. Verification may be defined as the process of establishing the truth or accuracy of data. So verification processes provide assurance. Other relevant terms from accounting that are applied universally to any kind of data verification activities include: **Materiality** – a concept from accounting that relates to the importance, significance and relevance of information including the degree of accuracy of the data. **Responsiveness** – demonstrating reactivity to stakeholder concerns and meeting their information needs. **Completeness** – covering and including all relevant sources and activities relevant to an organisation in terms of environmental performance.
Eco-efficiency	Defined by the World Business Council for Sustainable Development as 'the delivery of competitively priced goods and services that satisfy human needs and bring quality of life, while progressively reducing ecological impacts and resource intensity throughout the life cycle to a level at least in line with the earth's estimated carrying capacity'.
Ecological footprint	A measure of human demand on the Earth's ecosystems. It represents the amount of biologically productive land and sea area necessary to supply the resources a human population consumes and to mitigate associated waste.
Eco-Management and Audit Scheme (EMAS)	The environmental management system model produced by the European Union as a voluntary standard that may be subject to third party verification. Since its revision in 2001 it has become closely aligned with ISO 14001.
Ecosystem	A community of interdependent organisms and the physical and chemical environment that they inhabit.
Ecosystem services	A term used to describe the value to humans of various aspects of natural systems. Normally expressed in relation to supporting, provisioning, regulating and cultural services.
Eco-toxicity	The potential for biological, chemical or physical stressors to affect ecosystems.
Effluent	Liquid waste stream typically released to drain or water body.

337

Glossary

Emission Term applied to the release of any waste substance or noise but most frequently used to describe releases to atmosphere.

Environment Surroundings in which an organisation operates, including air, water, land, natural resources, flora, fauna, humans and their interrelation. These can extend from within the organisation to the global system.

Environmental aspect Those elements of an organisation's activities, products and services which can interact with the environment.

Environmental audit Term used in a number of contexts to describe a variety of investigations into the environmental performance of a site or organisation. The scope of investigation and level of detail varies widely.

Environmental impact Any change in the environment, whether adverse or beneficial, resulting from an organisation's activities, products or services.

Environmental impact assessment The formal assessment and analysis of the potential impact of various forms of human activity on the environment. Specialist studies of existing environmental conditions, expected changes and proposed mitigation measures are legally required as part of the planning application process for defined development projects.

Environmental management system Those elements of an organisation's overall management system that monitor, control and improve performance against an organisation's policy or priorities. Internationally recognised models of management systems include ISO 14001 and EMAS.

Environmental quality standard A standard typically set by a regulatory body that specifies the quality of the ambient environment, e.g. the maximum concentration of a pollutant in the air or a body of water.

Eutrophication The process of algae proliferation in water due to a high concentration of nutrients. Often associated with run-off or effluent containing high concentrations of fertilisers or rapidly decomposable organic matter.

Fugitive emission Ad hoc releases of emissions (normally to air) e.g. from pipe/joint leaks, evaporative losses from storage tanks or during uncontained usage. Particularly relevant for volatile organic compounds.

Greenhouse effect The term used to describe the selective response of the atmosphere to different types of radiation. Incoming short wave radiation passes through unaltered, while returning long wave radiation is absorbed by the 'so-called' greenhouse gases.

Greenhouse gas (GHG) The group of about 20 gases responsible for the greenhouse effect through their ability to absorb long-wave radiation. Carbon dioxide (CO_2) is the most abundant but methane (CH_4), nitrous oxide (N_2O), the chlorofluorocarbons (CFCs) and tropospheric ozone also make significant contributions to the greenhouse effect.

Groundwater Water that accumulates in the pore spaces and rocks below the Earth's surface. It originates as precipitation and percolates down into aquifers or finds its way back to the surface via springs or connection to surface water courses. The upper limit of groundwater saturation is known as the water table.

Habitat The specific environment in which an organism lives. Although the term is often used in relation to a particular species, any given environment will be shared by a variety of interdependent organisms.

Hydrocarbon Organic compounds composed of hydrogen and carbon. They may be solid (e.g. coal), liquid (e.g. crude oil) or gaseous (e.g. natural gas) in form. They are used primarily as fuels but are also utilised as lubricants and as feed stocks for a variety of industrial materials (especially important to the plastics and fertiliser industries).

Indirect environmental impact Any change in the environment, whether adverse or beneficial, that is caused by third parties acting on behalf of, or in support of, an organisation's activities, products or services, e.g. power generation impacts associated with the supply of energy to an organisation for use in its manufacturing process.

ISO 14001 The environmental management system model produced by the International Standards Organisation against which organisations may choose to be assessed by third party certification bodies.

Key performance indicator (KPI) A parameter that measures the level of achievement in an area considered to be of particular importance to an organisation. Ideally KPIs will be normalised, i.e. referenced to some activity indicator to allow comparison over time, e.g. kilos of waste produced per product manufactured.

Life cycle analysis (LCA)	The compilation and evaluation of the inputs, outputs and the potential environmental impacts of a product system throughout its life cycle, i.e. through the consecutive and interlinked stages of a product system, from raw material acquisition or the generation of natural resources to the final disposal.
Nuisance	Interference with a person's use and enjoyment of the environment through something that annoys, bothers or causes damage to that person or their property. Includes noise, odour and visual intrusion.
Oxides of nitrogen	Term used to describe a group of gases (nitrous oxide N_2O, nitric oxide NO and nitrogen dioxide NO_2) formed by the combination of nitrogen and oxygen under high energy conditions (typically associated with high temperature incineration and internal combustion engines).
Ozone depletion	The breakdown of ozone in the upper atmosphere (stratosphere) by man-made chemicals containing bromine and chlorine, e.g. CFCs. This layer of ozone acts as a protective layer to life on Earth, filtering out harmful ultraviolet radiation from the sun.
PANs	Peroxyacetyl nitrates, or PANs, are powerful respiratory and eye irritants present in photochemical smog. PANs are secondary pollutants, which means they are not directly emitted as exhaust from power plants or internal combustion engines, but they are formed from other pollutants by chemical reactions in the atmosphere.
Photochemical smog	Smog produced by chemical reactions in the presence of sunlight on pollutants arising from combustion especially hydrocarbons and oxides of nitrogen (from vehicles). A variety of toxic chemicals are produced in the 'smog' including ozone (harmful to plants; irritant to eyes, nose and throat), aldehydes (smelly, poisonous and irritant to eyes, nose and throat) and peroxyacetyl nitrate (PAN – toxic, irritant to eyes, nose and throat).
Photosynthesis	The biochemical process by which plants use energy from sunlight to convert carbon dioxide and hydrogen (from water) into simple carbohydrates.
Policy	An agreed approach to a particular issue or group of issues that may be enacted or achieved through a number of policy instruments (see below). Policies may be set at international, national, corporate and even individual scales.
Policy instrument	Mechanisms used to achieve or implement a particular policy. Typical examples include fiscal (e.g. taxation), legislative (e.g. a new law), market (e.g. product design specifications) or voluntary instruments (e.g. ISO 14001).
Pollution	The introduction by man of substances or energy into the environment that are liable to cause hazards to human health, harm to ecological systems, damage to structures or amenity, or interference with legitimate uses of the environment.
Pollution linkages	The knock-on effects of pollution through physical or biological processes. Also described under the concept of pollution pathways, i.e. an initial source, pathway, receptor linkage then causing secondary impacts in other receptor groups.
POPs (persistent organic pollutants)	A group of chemicals that remain unchanged in the environment for many years and which may find their way into the food chain and human tissues via bioaccumulation. The so-called 'dirty dozen' POPs include the pesticides aldrin, chlordane, DDT, dieldrin, endrin, heptachlor, mirex and toxaphene, the industrial chemicals polychlorinated biphenyls (PCBs), hexachlorobenzene and the combustion by-products dioxins and furans.
Preliminary environmental review	The process of evaluation of legal requirements, management standards and environmental priorities for an organisation. Often carried out at the beginning of an environmental management system implementation programme and involving a 'gap analysis' with a relevant benchmark (typically ISO 14001).
Receptor	An entity that can be subject to an environmental impact caused by pollution or resource consumption. Includes water bodies, air, land, communities, ecosystems, individual organisms, human beings and property. Some receptors are particularly sensitive to certain impacts and are known as 'sensitive receptors'.
Risk	The potential for realisation of unwanted, adverse consequences to human life, health, property or the environment (a combination of the likelihood and consequences of a specific outcome, good or bad).
Secondary environmental impact	Any change in the environment arising as a direct consequence of an initial change caused by an organisation's activities, e.g. fish kill resulting from water pollution caused by a discharge by an organisation.

Glossary

Sensitive receptors Receptors that, for some reason, are particularly susceptible to impacts resulting from an organisation's activities. They may be human, e.g. children or the elderly, flora or fauna, e.g. rare species or habitats, or physical, e.g. aquifers used for potable water supply.

Significance assessment A systematic evaluation of the relative importance of an organisation's environmental aspects and impacts. It includes the assessment of each aspect in relation to a standard set of criteria related to regulatory requirements and stakeholder concerns. The process results in a logical and repeatable prioritisation of aspects which facilitates management decisions on controls and improvements.

Stakeholder Individuals, communities or organisations that have an interest in an organisation or who are affected by its policies, practices and performance.

Storm water Storm water is water that originates during precipitation events. It may also be used to apply to water that originates with snowmelt that enters the storm water system. Storm water that does not soak into the ground becomes surface run-off, which either flows directly into surface waterways or is channelled into storm sewers, which eventually discharge to surface waters.

Stratosphere The stratosphere is the second major layer of Earth's atmosphere, just above the troposphere, and below the mesosphere. The stratosphere is situated between about 10 km (6 miles) and 50 km (30 miles) altitude above the surface at moderate latitudes, while at the poles it starts at about 8 km (5 miles) altitude.

Sustainability Sustainability is a description of environmental balance where resource consumption and the absorption of wastes are balanced by its natural rate of replenishment and breakdown – the term 'environmental sustainability' is sometimes used to emphasise this meaning.

Sustainable development Sustainable development is the process of transforming human society into a form that allows 'sustainability' to exist. It involves consideration of the social and economic issues (as well as the environmental ones) that need to be addressed along the way.

Trade effluent Relates to effluent generated from a commercial or industrial process. Pollutant loads and discharge routes may vary widely. Typically, however, these discharges are subject to stricter legal controls than storm water run-off.

Troposphere The troposphere is the lowest portion of the Earth's atmosphere. It contains approximately 75 per cent of the atmosphere's mass and 99 per cent of its water vapour. The average depth of the troposphere is approximately 17 km (11 miles) in the middle latitudes. It is deeper in the tropical regions, up to 20 km (12 miles), and shallower near the poles.

Value chain A term coined by Michael Porter to describe the activities involved in transforming raw materials into usable and saleable products, including all raw material extraction/production processes, manufacturing and transport processes plus selling and distribution activities. These are the company operations that mirror the 'life cycle' of a product with each step 'adding value'.

Volatile organic compounds (VOCs) Organic compounds which evaporate readily, including acetone, ethylene, benzene, propylene and many solvent preparations. VOCs contribute to a number of air quality issues either directly (e.g. benzene is carcinogenic) or indirectly as components of photochemical smog.

Bibliography

Accountability (2008) AA1000 Assurance Standard, Accountability.

Beaman, A.L and Kingsbury, R.W.S.M (1981) 'Assessment of nuisance from deposited particulates using a simple and inexpensive measuring system', *Clean Air*, 11(2): 77–81.

Brady, J., Ebbage, A. and Lunn, R. (2011) *Environmental Management in Organizations: The IEMA Handbook*, 2nd edn. Available at: http://www.iema.net/suggested-reading#sthash.rqUp6uxx.dpuf.

British Standards Institute (1997) BS 4142:1997 Method for rating industrial noise affecting mixed residential and industrial areas.

British Standards Institute (2003) BS 7445–1:2003 Description and measurement of environmental noise: Guide to quantities and procedures.

British Standards Institute (2010) BS 8903:2010 Principles and framework for procuring sustainably.

Brundtland Commission (1987) *Our Common Future*. Oxford: Oxford University Press.

Carson, R. (1962) *The Silent Spring*, Boston: Houghton Mifflin.

Dale, R.A. (trans.) (2002) *Tao Te Ching*. London: Watkins.

Darnerud, P. O., Atuma, S., Aune, M., Cnattingius, S. and Wernroth, M.L. (1998) 'Polybrominated diphenyl ethers (PBDEs) in breast milk from primiparous women in Uppsala county, Sweden', *Organohalogen Compounds*, 35:411–414.

DEFRA (2006) *Procuring the Future*. London: DEFRA.

DEFRA (2009) *Guidance on How to Measure and Report Your Greenhouse Gas Emissions*. London: DEFRA.

DEFRA (2010a) *Odour Guidance for Local Authorities*. London: DEFRA.

DEFRA (2010b) *Making Space for Nature*. London: DEFRA.

DEFRA (2011) *Green Claims Guidance*. London: DEFRA.

DEFRA (2012a) *Environmental Protection Act 1990: Part 2A. Contaminated Land Statutory Guidance*. London: DEFRA.

DEFRA (2012b) *Biodiversity Offsetting: Guidance for Developers*. London: DEFRA.

DEFRA (2012c) *Groundwater Protection: Principles and Practice (GP3)*. London: DEFRA.

DEFRA (2013a) *Encouraging Businesses to Manage Their Impact on the Environment*. London: DEFRA.

DEFRA (2013b) *Environmental Reporting Guidelines – Including Mandatory Greenhouse Gas Reporting Guidance*. London: DEFRA.

DEFRA (2013c) *Making Sustainable Development a Part of All Government Policy and Operations*. London: DEFRA.

DEFRA (2014) *Reducing and Managing Waste*. London: DEFRA.

Eccles, R.G., Ioannou, I. and Serafeim, G. (2013) 'The impact of corporate sustainability on organizational processes and performance', *Harvard Business Review*, July 29.

Environment Agency (2002) *Horizontal Guidance note H3 Part 2 Noise Assessment and Control*. Bristol: Environment Agency.

Environment Agency (*2011a) Above Ground Oil Storage Tanks: Pollution Prevention Guidelines PPG2*. Bristol: Environment Agency.

Environment Agency (2011b) *Horizontal Guidance Note H4 Odour Management: How to Comply with Your Environmental Permit*. Bristol: London: Environment Agency.

Bibliography

Environment Agency (2013a) *Technical Guidance WM2.- Hazardous Waste: Interpretation of the Definition and Classification of Hazardous Waste*, 3rd edn. Bristol: Environment Agency.

Environment Agency (2013b) Environmental Permitting Regulations Operational Risk Appraisal (OPRA for EPR) version 3.8.

Environmental Protection UK (2010) *Pollution Control Handbook: The Essential Guide to UK & European Pollution Control Legislation.*

Global Reporting Initiative (2013) G4 – Sustainability Reporting Guidelines. Reporting Principles and Standard Disclosures. GRI.

Hanson, C. et al. (2012) *The Corporate Ecosystems Review: Guidelines for Identifying Business Risks and Opportunities Arising from Ecosystem Change.* Washington, DC: World Resources Institute.

Harrison, R. (2001) *Pollution: Causes, Effects and Control*, 4th edn. London: Royal Society of Chemistry.

Henriques, A. (2001) *Sustainability: A Manager's Guide.* London: British Standards Institute.

Humphrey, N. and Hadley, M. (2000) *Environmental Auditing.* Bembridge, Isle of Wight: Palladian Law Publishing Ltd.

Hyde, P. and Reeve, P. (2005) *Essentials of Environmental Management.* Wigston: Institute of Occupational Safety and Health.

IEMA (2001) *Managing Climate Change Emissions: A Business Guide.* Lincoln: IEMA.

IEMA (2002) *Environmental Purchasing in Practice: Guidance for Organisations.* Lincoln: London: IEMA.

IEMA (2005)*Managing Compliance with Environmental Law: A Good Practice Guide.* Lincoln: IEMA.

IEMA (2006) *Risk Management for the Environmental Practitioner.* Lincoln: IEMA.

ISO 14001:2004 Environmental management systems – requirements with guidance for use.

ISO 14004:2004 Environmental management systems – general guidelines on principles, systems and support techniques.

ISO 14005:2010 Environmental management systems – guidelines for the phased implementation of an environmental management system, including the use of environmental performance evaluation.

ISO 14006:2011 Environmental management systems – guidelines for incorporating ecodesign.

ISO 14031:2013 Environmental management – Environmental performance evaluation – guidelines

ISO 14040:2006 Environmental management – life cycle assessment – principles and framework

ISO 19011:2011 – Guidelines for auditing management systems.

Kemp, D. (1998) *The Environment Dictionary.* London: Routledge.

Owen, A. (2000) *Ecodesign: A Training Guide for Business.* Sheffield: Green Training Works Ltd.

Population Reference Bureau (2006) World Population data sheet. Available at: www.prb. org.

Porritt, J. (2007) *Capitalism as if the World Matters.* London: Routledge.

Reeve, R. (2002) *An Introduction to Environmental Analysis.* Chichester: John Wiley & Sons Ltd.

Shot in the Dark/Green Training Works (2000) 'Ecodesign Guide including the Design Abacus'. Available at: www. greentrainingworks.co.uk.

Stern Report (2006) *Stern Review on the Economics of Climate Change.*

Sustainable Development Commission (2009) Prosperity without Growth: The Transition to a Sustainable Economy. Available at: www.sd-commission.org.uk/publications.php.

UK Government (1990) *This Common Inheritance.*

UNEP (2010) *Global Diversity Outlook*, 3rd edn. Available at: www.cbd.int/gbo3/.

UNEP (2012) *UNEP Global Outlook on Sustainable Production and Consumption Policies.* New York: UN.

UNEP (n.d.) Global Biodiversity Assessment and World Conservation Union global survey.

Walker, P. (2006) *Change Management for Sustainable Development.* Lincoln: IEMA.

Waters, B. (2013) *Introduction to Environmental Management.* London: Routledge.

Welsh Government (2013) *One Wales: One Planet: The Sustainable Development Annual Report 2012–13*. Cardiff: Welsh Government.

Wilson, E.O. (1987) 'The little things that run the world', *Conservation Biology*, 1(4): 344–46.

World Business Council for Sustainable Development (1997) *Sustainable Production and Consumption: A Business Perspective*. Geneva: World Business Council for Sustainable Development.

World Business Council for Sustainable Development (2000) *Eco-Efficiency: Creating More Value*. Geneva: World Business Council for Sustainable Development.

World Business Council for Sustainable Development (2011) *Guide to Corporate Ecosystem Valuation*. Geneva: World Business Council for Sustainable Development.

Index

References in **bold** indicate a table, those in *italics* are for figures and glossary terms are shown in **bold, underlined.**

Index

Index

Water Industry Act (1991) 88
water pollution: abstraction and discharge to controlled
waters 89–90; Anti-Pollution Works Regulations 1999
92; control techniques 264–5; discharge consent
parameters 90–1; discharges of surface water run-off
90; environmental policies **49**; European legislation
88; example emission limits, water discharge
activity permit **92**; groundwater 28–9; international
agreements 87–8; legal requirements applicable
to effluent discharges in England **89**; monitoring
parameters 171–3; pollutants 26–8, **26**, 87; prescribed
substances 90; receiving water body monitoring 171;
source-based monitoring 171; UK legislation 88–92;
waste water treatment techniques 265; *see also*
effluents; groundwater pollution

water pumping station noise level (case study) 256
Water Resources Act (1991) **65**, 87, 88
WEE (Waste Electrical and Electronic Equipment)
see Waste Electrical and Electronic Equipment
(WEEE)
Wildlife and Countryside Act (1981) (amended 2000) 60,
118
Wilson, E.O. 33
World Business Council for Sustainable Development
146–8
World Resources Institute methodology: corporate
ecosystems review (ESR) 241, *242*; example risks
and opportunities **243**
World Wildlife Fund (WWF) 217
WRAP 48

354